Metallurgy for Physicists and Engineers

Metallurgy for Physicists and Engineers
Fundamentals, Applications, and Calculations

Authored by
Zainul Huda

CRC Press
Taylor & Francis Group
Boca Raton London New York

CRC Press is an imprint of the
Taylor & Francis Group, an **informa** business

CRC Press

Taylor & Francis Group

6000 Broken Sound Parkway NW, Suite 300

Boca Raton, FL 33487–2742

© 2020 by Taylor & Francis Group, LLC

CRC Press is an imprint of Taylor & Francis Group, an Informa business

No claim to original U.S. Government works

Printed on acid-free paper

International Standard Book Number-13 978-0-367-19838-1 (Hardback)

Visit the Taylor & Francis Web site at
www.taylorandfrancis.com

and the CRC Press Web site at
www.crcpress.com

Contents

Preface..xv
Acknowledgments...xvii
Author Biography..xix

PART I Fundamentals of Metallurgy

Chapter 1 Introduction 3

    1.1   Metallurgy, Materials, and Engineer..3
    1.2   Engineering Materials 4
            1.2.1   Classification of Materials...4
            1.2.2   Metals/Alloys 4
            1.2.3   Ceramics/Glasses 6
            1.2.4   Polymers/Plastics 6
            1.2.5   Composites 6
    1.3   Metallurgy—Classification and the Manufacturing
          Processes 7
            1.3.1   Classification of Metallurgy 7
            1.3.2   Extractive Metallurgy..7
            1.3.3   Physical Metallurgy..7
            1.3.4   Mechanical Metallurgy 8
             1.3.5   Manufacturing Processes in Metallurgy 8
    Questions 10
    References 11

Chapter 2 Crystalline Structure of Metals...13

    2.1   Amorphous and Crystalline Materials 13
            2.1.1   Amorphous Solids..13
            2.1.2   Crystalline Solids 13
    2.2   Crystal Systems—Structures and Properties 14
            2.2.1   Unit Cell, Crystal Lattice, and Crystal Systems 14
            2.2.2   The Simple Cubic Crystal Structure 15
            2.2.3   Crystal-Structure Properties 16
    2.3   Crystal Structures in Metals...16
            2.3.1   Body-Centered Cubic (BCC) Structure 17
            2.3.2   Face Centered Cubic (FCC) Structure 17
            2.3.3   Hexagonal Close-Packed (HCP) Crystal Structure.... 18
            2.3.4   Computing Theoretical Density of a Metal...............19
    2.4   Miller Indices 20
    2.5   Crystallographic Directions 20

| | 2.5.1 | Procedure to Find Miller Indices for a Crystallographic Direction | 20 |

2.5.1   Procedure to Find Miller Indices for a
Crystallographic Direction 20

2.5.2   Industrial Applications of Crystallographic
Directions 21

2.6   Crystallographic Planes 21

2.6.1   Procedure to Find Miller Indices for a
Crystallographic Plane 21

2.6.2   Industrial Applications of Crystallographic
Planes 22

2.7   Linear and Planar Atomic Densities 22

2.8   X-Ray Diffraction—Bragg's Law 23

2.9   Indexing a Diffraction Pattern for Cubic Crystals 25

2.10   Crystallites—Scherrer's Formula 26

2.11   Calculations—Examples of Crystalline Structures
of Solids 26

Questions and Problems 38

References 39

Chapter 3   Crystal Imperfections and Deformation 41

3.1   Real Crystals and Crystal Defects 41

3.2   Point Defects and Alloy Formation 41

3.2.1   What Is a Point Defect? 41

3.2.2   Vacancy 42

3.2.3   Interstitial and Substitutional Point Defects 43

3.2.4   Solid-Solution Alloys 43

3.2.5   Alloy Formation—Hume-Rothery Rules 43

3.3   Dislocations 44

3.3.1   What Is a Dislocation? 44

3.3.2   Edge Dislocation and Screw Dislocation 45

3.3.3   Dislocation Movement—Deformation by Slip 45

3.4   Slip Systems 46

3.5   Deformation in Single Crystals—Schmid's Law 48

3.6   Plastic Deformation—Cold Working/Rolling 50

3.7   Calculations—Examples on Crystal Imperfections
and Deformation 51

Questions and Problems 56

References 57

Chapter 4   Diffusion and Applications 59

4.1   Introduction to Diffusion and Its Applications 59

4.1.1   What Is Diffusion? 59

4.1.2   Industrial Applications of Diffusion 59

4.2   Factors Affecting Rate of Diffusion 60

4.3   Mechanisms and Types of Diffusion 61

4.3.1   Diffusion Mechanisms 61

4.3.2   Types of Diffusion 62

4.4   Steady-State Diffusion—Fick's First Law 62

4.5   Non-Steady State Diffusion—Fick's Second Law 63

4.6   Applications of Fick's Second Law of Diffusion 66

4.7   Thermally Activated Diffusion—Arrhenius Law 67

4.8   Calculations—Examples on Diffusion and Its
          Applications 69

Questions and Problems 74

References 76

PART II   *Physical Metallurgy—Microstructural Developments*

Chapter 5   Metallography and Material Characterization 79

5.1   Evolution of Grained Microstructure 79

5.2   Metallography—Metallographic Examination of
          Microstructure 80

5.2.1   Metallography and Its Importance 80

5.2.2   Metallographic Specimen Preparation 80

5.3   Microscopy 84

5.3.1   Optical Microscopy (*OM*) 85

5.3.1.1   The *OM* Equipment 85

5.3.1.2   Imaging in Optical Microscopy 85

5.3.1.3   Resolution of an Optical Microscope 86

5.3.2   Electron Microscopy 86

5.3.2.1   Scanning Electron Microscopy 86

5.3.2.2   Transmission Electron Microscopy 88

5.4   Material Analysis and Specification of Composition of an
          Alloy 89

5.4.1   Material Analysis/Characterization Tools 89

5.4.2   Specification of Composition in Atom Percent 90

5.4.3   Specification of Average Density of an Alloy 91

5.5   Quantitative Metallography—Grain Size Measurement 91

5.5.1   Manual Quantitative Metallography 92

5.5.1.1   The Chart Method of Manual
                   Quantitative Metallography 92

5.5.1.2   The Counting Method of Manual
                   Quantitative Metallography 92

5.5.2   Computer-Aided Quantitative Metallography—
               Computerized Image Analysis 93

5.6   Calculations—Examples on Metallography and
          Materials Characterization 94

Questions and Problems 102

References 104

Chapter 6   Phase Diagrams 105

    6.1   The Basis of Phase Diagrams 105
        6.1.1   Gibbs's Phase Rule 105
        6.1.2   What Is a Phase Diagram? 105
    6.2   Phase Transformation Reactions and Classification of
        Phase Diagrams 105
        6.2.1   Phase Transformation Reactions 105
        6.2.2   Classification of Phase Diagrams 107
        6.2.3   Unary Phase Diagrams 107
    6.3   Binary Phase Diagrams 107
        6.3.1   Basics of Binary Phase Diagrams 107
        6.3.2   Isomorphous Binary Phase Diagrams 108
        6.3.3   Mathematical Models for Binary Systems 109
    6.4   Binary Phase Diagrams Involving Eutectic Reactions 110
        6.4.1   Eutectic Binary Phase Diagrams With
                Complete Insolubility in Solid State 110
        6.4.2   Eutectic Binary Phase Diagrams With Partial
                Solid Solubility 111
    6.5   Binary Phase Diagrams Involving Peritectic/Eutectoid
        Reactions 113
    6.6   Binary Complex Phase Diagrams 114
    6.7   Ternary Phase Diagrams 114
    6.8   Calculations—Examples in Phase Diagrams 115
    Questions and Problems 129
    References 131

Chapter 7   Phase Transformations and Kinetics 133

    7.1   Phase Transformation and Its Types 133
    7.2   The Kinetics of Phase Transformations 133
        7.2.1   The Two Stages of Phase Transformation—
                Nucleation and Growth 133
        7.2.2   Nucleation 133
                7.2.2.1   Nucleation and Its Types 133
                7.2.2.2   Homogeneous Nucleation 134
                7.2.2.3   Heterogeneous Nucleation 138
    7.3   Growth and Kinetics 138
    7.4   Kinetics of Solid-State Phase Transformation 139
    7.5   Microstructural Changes in Fe-C Alloys—Eutectoid
        Reaction 140
        7.5.1   S-shape Curves for Eutectoid Reaction 140
        7.5.2   Pearlite: *A Quantitative Analysis* 141
        7.5.3   Time-Temperature-Transformation (TTT)
                Diagram 142
        7.5.4   Bainite and Martensite 144
    7.6   Continuous Cooling Transformation (CCT) Diagrams 145

7.7   Calculations—Examples on Phase Transformations
       and Kinetics 146
Questions and Problems 153
References 154

PART III   Engineering/Mechanical Metallurgy and Design

Chapter 8   Mechanical Properties of Metals 159

8.1   Material Processing and Mechanical Properties 159
8.2   Stress and Strain 159
8.3   Tensile Testing and Tensile Properties 162
       8.3.1   Tensile Testing 162
       8.3.2   Tensile Mechanical Properties 164
8.4   Elastic Properties: *Young's Modulus, Poisson's Ratio,
       and Resilience* 166
8.5   Hardness 167
       8.5.1   Hardness and Its Testing 167
       8.5.2   Brinell Hardness Test 167
       8.5.3   Rockwell Hardness Test 169
       8.5.4   Vickers Hardness Test 170
       8.5.5   Knoop Hardness Test 171
       8.5.6   Microhardness Test 171
       8.5.7   Hardness Conversion 172
8.6   Impact Toughness—*Impact Energy* 172
8.7   Fatigue and Creep Properties 173
8.8   Calculations—Examples on Mechanical Properties
       of Metals 173
Questions and Problems 181
References 182

Chapter 9   Strengthening Mechanisms in Metals 183

9.1   Dislocation Movement and Strengthening Mechanisms 183
9.2   Solid-Solution Strengthening 183
9.3   Grain-Boundary Strengthening—*Hall-Petch Relationship* 185
9.4   Strain Hardening 187
9.5   Precipitation Strengthening 189
9.6   Dispersion Strengthening—Mechanical Alloying 190
9.7   Calculations—Examples on Strengthening Mechanisms
       in Metals 191
Questions and Problems 198
References 199

Chapter 10 Failure and Design 201

    10.1 Metallurgical Failures—Classification and Disasters........... 201
        10.1.1 Fracture and Failure—Historical Disasters 201
        10.1.2 Classification of Metallurgical Failures 201
    10.2 Ductile and Brittle Fractures 202
    10.3 Ductile-Brittle Transition Failure 204
    10.4 Griffith's Crack Theory 205
    10.5 Stress Concentration Factor... 206
    10.6 Safety and Design.. 207
        10.6.1 The Factor of Safety in Design................................ 207
        10.6.2 Fracture Mechanics—*A Design Approach* 208
           10.6.2.1 K, K_c, and the *Plain Strain Fracture*
                    Toughness (K_{IC}) 208
           10.6.2.2 Design Philosophy of Fracture Mechanics 210
    10.7 Fatigue Failure.. 210
        10.7.1 What Is Fatigue?.. 210
        10.7.2 Stress Cycles.. 211
        10.7.3 Fatigue Testing—*Fatigue Strength and
           Fatigue Life* 212
        10.7.4 The Modified Goodman Law.................................... 213
        10.7.5 Avoiding Fatigue Failure in Component Design...... 214
    10.8 Creep Failure 214
        10.8.1 What Is Creep?... 214
        10.8.2 Creep Test and Creep Curve 214
        10.8.3 Effects of Stress and Temperature on Creep Rate.... 216
        10.8.4 Larson-Miller (LM) Parameter................................ 216
    10.9 Calculations—Examples on Failure and Design................... 217
    Questions and Problems 227
    References 228

Chapter 11 Corrosion and Protection.. 231

    11.1 Corrosion and Society 231
    11.2 Electrochemistry of Corrosion 231
        11.2.1 Oxidation-Reduction (REDOX) Reactions............. 231
        11.2.2 Electrochemical Corrosion and Galvanic Cell......... 232
        11.2.3 Standard Electrode Potential (E^o)............................ 233
    11.3 Galvanic Corrosion and Galvanic Series 234
        11.3.1 Galvanic Corrosion and Corrosion Potential 234
        11.3.2 Galvanic Series... 234
    11.4 Rate of Corrosion 235
        11.4.1 Rate of Uniform Corrosion 235
        11.4.2 Oxidation Kinetics 237
    11.5 Forms of Corrosion.. 238
        11.5.1 The Classification Chart of Corrosion 238

11.5.2   Uniform Corrosion and Galvanic Corrosion........... 238
11.5.3   Localized Corrosion... 239
11.5.4   Metallurgically Influenced Corrosion 241
     11.5.4.1   Intergranular Corrosion 241
     11.5.4.2   De-Alloying Corrosion 241
11.5.5   Mechanically Assisted Corrosion 241
     11.5.5.1   Erosion Corrosion..................................... 242
     11.5.5.2   Fretting Corrosion 242
     11.5.5.3   Corrosion Fatigue 242
11.5.6   Environmentally Assisted Corrosion 242
     11.5.6.1   Stress-Corrosion Cracking 243
     11.5.6.2   Hydrogen Embrittlement (*HE*).................. 243
     11.5.6.3   Liquid Metal Embrittlement...................... 244
11.6   Protection Against Corrosion 244
11.6.1   Approaches in Protection Against Corrosion........... 244
11.6.2   Design of Metal-Assembly...................................... 244
11.6.3   Cathodic Protection... 245
11.6.4   Surface Engineering by Coating 246
     11.6.4.1   Coating and Types of Coating Processes ... 246
     11.6.4.2   Coating by Electroplating.......................... 247
11.6.5   Corrosion Inhibitors 248
11.6.6   Material Selection 248
11.6.7   Heat Treatment/Composition Control 248
11.7   Calculations—Examples on Corrosion and Protection......... 249
Questions and Problems 256
References 258

Chapter 12   Ferrous Alloys 261

12.1   Ferrous Alloys—Classification and Designation 261
12.2   Plain Carbon Steels 261
12.2.1   Classification and Applications............................... 261
12.2.2   Microstructures of Carbon Steels............................ 262
12.2.3   The Effects of Carbon and Other Elements on
     the Mechanical Properties of Steel 264
12.3   Alloy Steels.. 265
12.3.1   Definition, AISI Designations, and Applications..... 265
12.3.2   Effects of Alloying Element on Steel....................... 266
12.3.3   Classification of Alloy Steels 267
12.3.4   High-Strength Low-Alloy (HSLA) Steels 267
12.3.5   Tool and Die Steels... 268
12.3.6   Hadfield Manganese Steels 268
12.3.7   Maraging Steels... 268
12.3.8   Stainless Steels 268
12.3.9   High-Silicon Electrical Steels 270
12.3.10   TRIP Steels 270

12.3.11 Mathematical Models for Alloy Steelmaking
(Secondary Steelmaking) 271
12.3.11.1 Effects of Different Processes on the
Steel-Holding Ladle Temperature 271
12.3.11.2 Mathematical Models for Liquidus
Temperature for Alloy Steels 271
12.3.11.3 Mathematical Models for Additions to
Achieve the Aim Composition 271
12.4 Cast Irons 272
12.4.1 Metallurgical Characteristics and Applications 272
12.4.2 Types of Cast Irons 273
12.4.3 White Cast Iron 273
12.4.4 Gray Cast Iron 273
12.4.5 Ductile or Nodular Cast Iron 274
12.4.6 Malleable Cast Iron 275
12.4.7 Alloyed Cast Iron 276
12.4.8 Austempered Ductile Iron 276
12.5 Calculations—Worked Examples on Ferrous Alloys 276
Questions and Problems 283
References 285

Chapter 13 Nonferrous Alloys 287

13.1 Nonferrous Alloys 287
13.2 Aluminum Alloys 287
13.2.1 Properties and Applications 287
13.2.2 Designations and Applications 287
13.2.3 Aluminum-Silicon Casting Alloys 288
13.2.4 Aluminum-Copper Alloys and Aerospace
Applications 290
13.3 Copper Alloys 291
13.3.1 The Applications of Copper Alloys 291
13.3.2 Copper-Zinc Alloys 291
13.3.3 Copper-Tin Alloys 292
13.4 Nickel and Its Alloys 293
13.4.1 General Properties and Applications 293
13.4.2 Ni-Base Superalloys and Creep Behavior 293
13.5 Titanium and Its Alloys 294
13.5.1 Properties and Applications 294
13.5.2 Microstructure and Properties of Titanium
Alloys 294
13.6 Precious Metals 296
13.7 Other Nonferrous Alloys 297
13.8 Calculations—Worked Examples on Nonferrous Alloys 298
Questions and Problems 304
References 306

PART IV Thermal Processing and Surface Engineering

Chapter 14 Recrystallization and Grain Growth 311

14.1 The Three Stages of Annealing 311
14.2 Recrystallization 311
14.3 Grain Growth 312
14.4 Calculations—Worked Examples on Recrystallization
and Grain Growth 314
Questions and Problems 319
References 319

Chapter 15 Heat Treatment of Metals 321

15.1 Heat Treatment of Metals—An Overview 321
15.2 Heat Treatment of Steels 321
15.2.1 The Processes in Heat Treatment of
Carbon Steels 321
15.2.2 Equilibrium Cooling of Carbon Steels 321
15.2.3 Heat-Treatment Temperatures Ranges and
Thermal Cycles for Carbon Steels 322
15.2.4 Full Annealing of Carbon Steel 323
15.2.5 Normalizing of Carbon Steel 324
15.2.6 Spheroidizing Annealing of Carbon Steel 325
15.2.7 Hardening of Steel 325
15.2.7.1 The Process of Hardening Steel 325
15.2.7.2 Hardenability and Jominy Test 326
15.2.7.3 Calculating Hardenability of Structural
Steels 327
15.2.8 Tempering of Steel 328
15.3 Heat Treatment of Aluminum Alloys 329
15.4 Heat Treatment of Copper Alloys 331
15.5 Calculations—Worked Examples on Heat Treatment of
Metals 332
Questions and Problems 338
References 339

Chapter 16 Surface Engineering of Metals 341

16.1 Surface-Hardening Processes 341
16.1.1 Basic Principle of Surface Hardening 341
16.1.2 Surface Hardening Keeping Composition
the Same 342
16.1.3 Surface Hardening Involving Change in
Composition 342
16.2 Surface Coating—An Overview 343

16.3 Electroplating and Its Mathematical Modeling.....................344
    16.3.1 Electroplating Principles..344
    16.3.2 Mathematical Modeling of Electroplating
           Process..344
16.4 Calculations—Examples in Finishing/Surface
    Engineering 345
Questions and Problems..348
References 349

Answers to MCQs and Selected Problems **351**
Index...**355**

Preface

This volume serves as a unique textbook on physical and mechanical metallurgy. I went through the academic literature on engineering metallurgy and found that the quantitative approach in the field is missing; this literature gap motivated me to write this book. The salient features of this text include principles, applications, and *200 worked examples/calculations* along with *70 MCQs* with answers. These attractive features render this volume suitable for recommendation as a textbook of physical metallurgy for undergraduate as well as master's level programs in metallurgy, physics, materials science, and mechanical engineering. This textbook is intended for engineering students at the sophomore level; however, all engineers, in all specialties, can benefit from the thorough knowledge and calculations in the field of engineering metallurgy.

Metallurgists/Engineers are frequently involved in the processing of alloys. In order to design and develop the alloys, a physicist must relate structures with properties; this relationship requires a strong knowledge of physical metallurgy. Keeping in view this requirement, this volume is designed to educate engineering students with up-to-date knowledge of physical metallurgy (focusing on its quantitative aspects) with particular reference to practical skills in metallography and material characterization.

The book is divided into four parts: *Part 1* introduces readers to metallurgy and materials science. Here, the crystalline structure of metals, crystal imperfections and deformation, and diffusion and its applications are covered in depth. The principles of crystallography are also applied to explain the deformation behavior of metals and metal processing. *Part 2* covers physical metallurgy with particular reference to microstructural developments. Here, material characterization and metallographic tools are first explained followed by in-depth discussions on phase diagrams, phase transformations, and kinetics.

Part 3 offers highly useful stuff for metallurgical and mechanical engineers by covering mechanical properties of metals; strengthening mechanisms in metals; failure and design; corrosion and protection; ferrous alloys and nonferrous alloys. In particular, engineering designers can significantly benefit from Chapter 10 (*Failure and Design*). Finally, thermal processing and surface engineering aspects of physical metallurgy are dealt with in *Part 4*. The topics covered include recrystallization and grain growth, heat treatment of metals, and surface engineering of metals.

There are a total of 16 chapters. Each chapter first introduces readers to the technological importance of the chapter-topic and definitions of terms and their explanation; and then its mathematical modeling is presented. The meanings of the terms along with their *SI* units in each mathematical model are clearly stated. There are *200 worked examples (calculations), 240* mathematical models (equations/formulae) and *over 160* diagrammatic illustrations in this book. Each *worked example* clearly mentions the specific formula applied to solve the problem followed by the steps of solution. Additionally, *exercise problems* are included for every chapter. The answers to all **70 MCQs** and selected problems are provided at the end of the book.

Zainul Huda
Department of Mechanical Engineering
King Abdulaziz University

Acknowledgments

I am grateful to God for providing me with the wisdom to complete the write-up of this book's manuscript and get it published.

Thanks to my wife for her continued patience during many nights while writing this book. I express my gratitude to Robert Bulpett, PhD, Professor Associate, Experimental Techniques Center, Brunel University, London, UK, for the valuable discussion with him on energy dispersive X-ray spectroscopy (EDS) during the write up of Chapter 5. I would like to acknowledge Mr. Shoaib Ahmed Khan, Financial Analyst, UniLever PLC, Jeddah, Saudi Arabia, for his assistance in developing high-quality graphical plots during the write up of Chapter 7.

Author Biography

Besides this text, **Zainul Huda** is the author/co-author/editor of five books, including the books entitled: *Manufacturing: Mathematical Models, Problems, and Solutions* (2018, CRC Press), and *Materials Science and Design for Engineers* (2012, Trans Tech Publications). Dr. Zainul Huda is Professor at the Department of Mechanical Engineering, King Abdulaziz University, Saudi Arabia. His teaching interests include materials science and engineering, manufacturing technology, plasticity and metal forming, metallurgy, and aerospace materials. He has been working as a full professor in reputed universities (including the University of Malaya, King Saud University, etc.) since February 2007. He possesses over 35 years' academic experience in materials/metallurgical/mechanical engineering at various universities in Malaysia, Pakistan, and Saudi Arabia.

Prof. Zainul Huda earned a PhD in Materials Technology (Metallurgy) from Brunel University, London, UK in 1991. He is also a postgraduate in Manufacturing Engineering. He obtained B. Eng. in Metallurgical Engineering from *University of Karachi*, Pakistan in 1976. He is the author/co-author/editor of 128 publications; which include six books and 34 peer-reviewed international SCOPUS-indexed journal articles (26 ISI-indexed papers) in the fields of materials/manufacturing/mechanical engineering published by reputed publishers from the USA, Canada, UK, Germany, France, Switzerland, Pakistan, Saudi Arabia, Malaysia, South Korea, and Singapore. He has been cited over 650 times in www.scholar.google.com. His author h-index is 12 (*i*10-index: 17). He, as *PI*, has attracted eight research grants, which amount to a total worth of USD 0.15 million. Prof. Huda is the recipient of *UK-Singapore Partners in Science Collaboration Award* for research collaboration with British universities; which was awarded to him by the British High Commission, Singapore (2006). He is also the recipient of VC's Appreciation Certificate issued by the Vice Chancellor, University of Malaya, Malaysia (2005); and a shield awarded by the Minister of Defense Production of Pakistan for presenting a paper on tank gun barrel material technology during an international conference in Taxila (1994).

Professor Huda has successfully completed 20+ industrial consultancy/R&D projects in the areas of failure analysis and manufacturing in Malaysia and Pakistan. He is the developer of Toyota Corolla cars' axle-hub's heat-treatment manufacturing process first ever implemented in Pakistan (through Indus Motor Company/Transmission Engineering Industries Ltd, Karachi) during 1997. Besides industrial consultancy, Dr. Huda has worked as plant manager, development engineer, metallurgist, and G. engineer (metallurgy) in various manufacturing companies, including Pakistan Steel Mills Ltd. Prof. Zainul Huda has delivered guest lectures/presentations in the areas of materials and manufacturing in the UK, South Africa, Saudi Arabia, Pakistan, and Malaysia. Prof. Zainul Huda holds memberships in prestigious professional societies, including: (a) Pakistan Engineering Council, (b) Institution of Mechanical Engineers (*IMechE*), London, UK, and (c) Canadian Institute of Mining, Metallurgy, and Petroleum (*CIM*), Canada. He has supervised/co-supervised to completion several PhD and master's research theses. His biography has been published in *Marquis Who's Who in the World, 2008–2016*. He is a member of the International Biographic Centre (IBC), UK's TOP 100 ENGINEERS—2015. He is the World's Top 20 (No. 14) research scholar in the field of materials and manufacturing (go to www.scholar.google.com and type Zainul Huda).

Part I

Fundamentals of Metallurgy

1 Introduction

1.1 METALLURGY, MATERIALS, AND ENGINEER

Metallurgy and materials are fascinating fields of science and engineering; they have epochs of history named after them i.e. Stone Age, Iron Age, Bronze Age, and now the Silicon Age. Advances in metallurgy and materials have preceded almost every major technological leap since the beginning of civilization. An engineer analyzes systems, designs them, and uses metals/materials to build a structure, or manufacture a machine (or its components). Metals and alloys provide the foundation for our modern way of life. This is why metallic materials play a pivotal role in almost all disciplines of engineering; which include: mechanical, civil, electrical, biomedical, chemical, aerospace, and the like.

Metallurgy involves the study, innovation, design, implementation, and improvement of processes that transform mineral resources and metals into useful products thereby improving the quality of our lives. *Materials technology* links scientific research with applied engineering to develop materials for real-world applications. This objective is achieved by the application of the principles of chemistry, physics, and mathematics along with computer technology to understand and relate structure to the properties of materials.

There exists a strong relationship between structure, properties, processing, and performance of a material (see Figure 1.1). For example, diamond and graphite are the two allotropic forms of carbon. Diamond has excellent performance as a very hard and wear-resistant material owing to its very strong carbon-carbon covalent bonds structure (Callister, 2007). On the other hand, graphite is a softer material due to the layered arrangement of atoms in its structure. A metallurgist/materials technologist thus designs processes to manipulate materials with required performance to meet the needs of modern technology. For instance, investment-cast superalloy gas-turbine (GT) blades have inferior performance (lower creep strength) owing to their equiaxed grain structure. Metallurgists developed the directional solidification (DS) technique for manufacturing columnar grained/single crystal GT blades assuring a much better creep strength suitable for application in modern gas turbine engines (Huda, 2017a).

The relationship presented in Figure 1.1 enables a materials technologist to design and develop materials for special applications; these include: low alloy high strength (LAHS) steels for stressed structural applications, soft magnet cores for electrical transformers, permanent magnets for motors, fiber optics for internet and communication technologies (ICT), ceramic tiles for resisting high temperatures (> 1600°C) in space shuttles, and the like (Schaffer *et al.*, 1999). These unique properties are derived by designing materials with specialization by controlling the structure of the material from the macroscopic level to the atomic level (in nanomaterials).

3

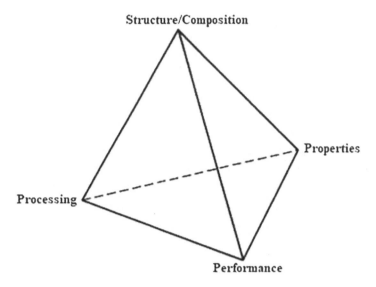

FIGURE 1.1   Relationship between processing-structure-properties-performance of materials.

1.2   ENGINEERING MATERIALS

1.2.1   CLASSIFICATION OF MATERIALS

All engineering materials can be classified into four major classes: (1) metals, (2) polymers, (3) ceramics, and (4) composites. Within each of these classes, materials are often further divided into groups based on their chemical composition or physical/mechanical properties (see Figure 1.2). The various classes and groups of materials are explained in the following subsections.

1.2.2   METALS/ALLOYS

Metals and alloys are inorganic materials having crystalline structures as well as strength and ductility, with mercury as an exception. In metallic crystalline structures, the valence electrons have considerable mobility and are able to conduct heat and electricity easily. Additionally, the delocalized nature of metallic bonds make it possible for the atoms to slide past each other when the metal is deformed indicating ductile behavior in metals and alloys. Strictly speaking, metals (elemental metals) are pure electropositive elements whereas alloys are composed of one or more metallic elements that may contain some non-metallic elements. Examples of metals include: iron, aluminum, zinc, copper, tin, and the like. Steels, cast irons, brasses, bronzes, and superalloys are some examples of alloys. Alloys (in formed shapes) find wide applications in building/automotive structures, power generation, aerospace, biomedical, electrical, and other sectors. Metals and alloys are broadly divided into two groups: *ferrous materials* and *nonferrous materials*. *Ferrous materials* contain a large percentage of iron; examples include: carbon steels, alloy steels, cast irons, and wrought iron. In particular, alloy steels are noted for their high strength and find applications in automobiles, aircrafts,

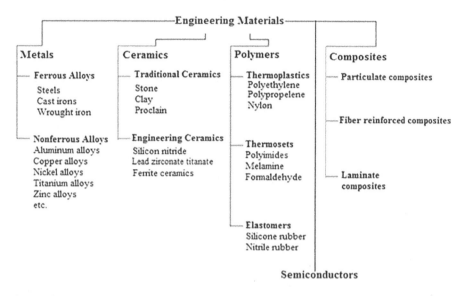

FIGURE 1.2 Classification chart of engineering materials.

FIGURE 1.3 Copper tubes as refrigerant line in a HVAC system.

pressure vessels, boilers, cutting tools, and the like. *Nonferrous materials* contain no or small amount of iron; examples of nonferrous materials include aluminum alloys, copper alloys, nickel alloys, zinc alloys, and the like. They generally have lower strength as compared to ferrous materials but they are better in corrosion resistance and higher in cost. Copper tubes (due to high thermal conductivity) are extensively used as refrigerant lines in heating ventilation air-conditioning (HVAC) systems (see Figure 1.3). Metals and alloys are discussed in detail in Chapters 12–14.

1.2.3 Ceramics/Glasses

Ceramics and glasses are inorganic materials having high melting temperatures. They have high compressive strength but their tensile strength is generally poor. There are two major categories of ceramics: traditional ceramics and engineering ceramics (Shackelford, 2008). Traditional ceramics include: stone (silica), clay, porcelain, fire-bricks, etc. They are widely used as construction products; which include tiles, wash-basins, and the like. They are also used as refractories; which are critical materials that resist aggressive conditions, including high temperature (up to 1750°C), chemical and acid attack, abrasion, mechanical impact, and the like. Engineering ceramics take advantage of specific thermal/mechanical/electrical/optical/biomedical properties of glass or ceramic materials. For example, engineering ceramics are applied as thermal barrier coatings (TBC) on gas-turbine blades. Ferrite ceramics are used in computers and microwave ovens.

1.2.4 Polymers/Plastics

Polymers are organic materials that are based on repeating molecules or macromolecules. A polymer contains many chemically bonded units that themselves are bonded together to form a solid. Plastics are *synthetic* polymers i.e. they are artificial, or manufactured polymers. The building blocks for making plastics are small *organic* molecules (generally hydrocarbons). Each of these small molecule is called a *monomer* (one part) because it is capable of joining with other monomers to form very long molecule chains—called *polymers* (many parts); this process is called *polymerization*.

Plastics are classified into two groups according to what happens to them when they are heated to high temperatures: (a) thermoplastics, and (b) thermosets (Painter and Coleman, 2008). *Thermoplastics* keep their plastic properties when heated i.e. they melt when heated and then harden again when cooled. *Thermosets*, on the other hand, are permanently "set" plastics; they cannot be melted once they are initially formed; they would crack or become charred when exposed to high temperatures. In addition to thermoplastics and thermosets, there is another class of polymers i.e. elastomers (e.g. rubbers).

1.2.5 Composites

A material that is composed of two or more distinct phases (matrix and dispersed) and has bulk properties significantly different from those of any of the constituents, is called a *composite*. Examples of composite components include: fiberglass tanks, helmets, sport shoes, sailboat masts, light-poles, compressed natural gas tanks, rescue air tanks, and the like. In particular, carbon-fiber reinforced polymer (CFRP) composites are increasingly being used in the wings and fuselage of aerospace structures (Huda and Edi, 2013). Based on reinforcement form, composites may be divided into three categories: (a) particle reinforced composites, (b) fiber reinforced (FR) composites (short fibers and continuous fibers), and (c) laminates (Callister, 2007).

1.3 METALLURGY—CLASSIFICATION AND THE MANUFACTURING PROCESSES

1.3.1 CLASSIFICATION OF METALLURGY

Metallurgy may be classified into the following four branches: (a) extractive metallurgy, (b) physical metallurgy, (c) mechanical metallurgy, and (d) manufacturing processes in metallurgy. Figure 1.4 illustrates the classification chart of various branches and sub-branches in metallurgy; which are discussed in the following subsections.

1.3.2 EXTRACTIVE METALLURGY

Extractive metallurgy, also called *chemical metallurgy* or *process metallurgy*, deals with the extraction of metals from their ores and refining them to obtain useful metals and alloys; the reactions of metals with slags and gases are also studied in this field. The first stage in extractive metallurgy generally involves crushing and grinding of ore lumps into small particles followed by mineral dressing operations. Based on the type of metals, extractive metallurgy may be either *ferrous metallurgy* or *nonferrous metallurgy*. Based on the type of processing, extractive metallurgy may be divided into three types: (a) hydro-metallurgy, (b) pyro-metallurgy, and (c) electro-metallurgy. Since the scope of this book does not permit a detailed discussion on extractive metallurgy, the reader is advised to refer to existing literature (Habashi, 1986; Bodsworth, 1994; Wakelin and Fruehan, 1999).

1.3.3 PHYSICAL METALLURGY

Physical metallurgy focuses on the physical properties and structure of metals and alloys. It involves the study of the effects of the chemical composition, heat treatment, and manufacturing process on the microstructure of the material so as to achieve components with optimal properties. This book covers almost all areas of physical

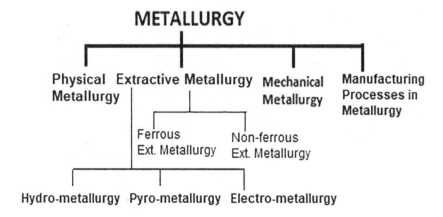

FIGURE 1.4 Classification of metallurgy into various branches.

metallurgy with particular reference to crystalline structure of metals; crystal imperfections and deformation; diffusion, microstructural and material characterization techniques; phase diagrams; phase transformations and kinetics; recrystallization and grain growth; and heat treatment of metals (see Chapter 2–9, Chapters 14–15).

1.3.4 MECHANICAL METALLURGY

Mechanical metallurgy involves the study of relationships between an alloy's mechanical behavior, the processing used to produce the alloy, and the underlying structure ranging from the atomic to macroscopic level. Because mechanical metallurgy is most strongly related to how metals break and deform, subfields (particularly, fracture mechanics, design against failure) have developed largely because of disastrous failures such as the *Versailles train* crash, the *Titanic* ship failure, and the like. Part 3 of this book (Chapters 8–13) focuses on mechanical metallurgy, including corrosion and the protective design.

1.3.5 MANUFACTURING PROCESSES IN METALLURGY

Manufacturing processes in metallurgy generally refer to the metal shaping and treating processes that are generally used for manufacturing tangible metallic products. There are nine manufacturing processes in metallurgy: (1) metal casting, (2) metal forming, (3) machining, (4) welding/joining, (5) heat treatment, (6) surface engineering, (7) powder metallurgy, (8) metal injection molding (MIM), and (9) metal additive manufacturing.

Metal Casting. Metal casting involves mold design and preparation, melting of metal in a furnace, and admitting molten metal into the mold cavity (where the metal solidifies at a controlled rate and takes the shape of the mold cavity) followed by the removal of the solidified metal (casting) from the mold. In most cases the casting is cleaned, heat treated, finished, and quality assured before shipping to market. There are a variety of different types of casting processes. The most common example of metal casting is the sand casting process (see Figure 1.5).

Metal Forming. The plastic deformation of a metal into a useful size and shape is called *metal forming*. Depending on the forming temperature, metal forming may be either hot working (HW) or cold working (CW). There is a wide variety of different types of metal forming processes; these include rolling, forging, extrusion, bar/wire drawing, shearing/blanking/piercing, bending, deep drawing, and the like. In *metal rolling* the metal stock is passed through one or more pairs of rolls to reduce the thickness or to obtain desired cross-section.

Machining. Machining involves the removal of metal/material from a work-piece by using a cutting tool. Machining can be considered as a system consisting of the work-piece, the cutting tool, and the equipment (machine tool). In conventional or traditional machining, material removal is accompanied by the formation of chips; which is accomplished by use of either a single-point or a multiple-point cutting tool. On the other hand, non-traditional or advanced machining are chip-less material-removal processes that involve use of energy (rather than a cutting tool) for material cutting (Huda, 2018).

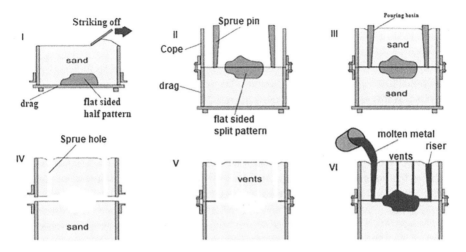

FIGURE 1.5 The steps in sand casting process in metal casting.

Welding/Joining. Welding ranks high among industrial manufacturing processes. Joining two or more pieces of metal to make them act as a single piece by *coalescence* is called *welding*. Metals can be permanently joined together by one of the following three technologies: (1) welding, (2) brazing, and (3) soldering. Among the three joining technologies, welding is the most important, the most economical, and the most efficient way to permanently join metals. Welding of two work-pieces usually requires them to be heated to their melting temperatures (Huda, 2017b). On the other hand, brazing and soldering involves joining of two work-pieces below their melting temperatures. Many different energy sources can be used for welding; these include electric arc, a gas flame, a laser, an electron beam, friction, ultrasound, and the like. Depending upon the type of energy source and techniques, a wide variety of (more than 50) welding processes have been developed.

Heat treatment involves the controlled heating, holding, and cooling of metals for altering their physical and mechanical properties. Depending on the mechanical properties to be achieved, heat-treatment processes for steels can be classified into: (a) full annealing, (b) normalizing, (c) hardening and tempering, (d) recrystallization or process annealing, and (e) spheroidizing annealing. These heat-treatment processes of steel involve application of iron-iron carbide phase equilibrium diagram and time-temperature-transformation (TTT) diagram. In *hardening* heat treatment, steel is slowly heated either at a temperature about 50°C above the line A_3 into the austenite region (in the case of hypo-eutectoid steels), or at a temperature about 50°C above the A_1 line into the austenite-cementite region (in the case of hypereutectoid steels). Then, the steel is held at this temperature for sufficiently long time to fully transform phases either into austenite or into austenite+cementite as the case may be. Finally, the hot steel is rapidly cooled (quenched) in water or oil to form martensite or bainite so as to ensure high hardness/toughness. In many automotive components, hardening is followed by tempering. This author has ensured the required high hardness in a forged hardened-tempered AISI-1050 steel for application in the axle-hubs of a motor-car by achieving tempered martensitic microstructure (Huda, 2012). Heat-treatment processes are discussed in detail in Chapter 15.

Surface Engineering. *Surface engineering,* also called *surface finishing/coating,* aims at achieving excellent surface finish and/or characteristics by use of either finish grinding operations or by plating and surface coatings to finish part surfaces. Applied as thin films, these surface coatings ensure corrosion resistance, wear resistance, durability, and/or decoration to part surfaces. The most common plating and surface coating technologies in use include: (a) painting, (b) electroplating, (c) electroless plating, (d) conversion coating, (e) hot dipping, and (f) porcelain enameling. Surface engineering is discussed in detail in Chapter 16.

Powder metallurgy (P/M) comprises a sequence of manufacturing operations involving metal powder production, powder characterization, powder mixing and blending, compaction, and sintering. In *P/M*, a feedstock in powder form is processed to manufacture components of various types; which include bearings, watch gears, automotive components, aerospace components, and other engineering components.

Metal Injection Molding (MIM). The metal injection molding (MIM) process starts with the combination of metal powders and a polymer binder, creating a feedstock for injection molding. Once introduced to the molding machine, the feedstock becomes viscous as the polymer binder is heated in the barrel of the molding machine. The feedstock is then injected into molds where it fills the mold cavity thereby creating a "green" part. The polymer binder is then removed from the metal part. After de-binding has been completed, the parts are sintered at very high temperatures in vacuum or continuous furnaces. During the sintering process, the high surface energy of the fine metal powders is released and the powders consolidate to form a solid metal part approaching 96–98% of full density. Secondary/machining operations can be applied to achieve a variety of final part specifications. Metal injection molded parts find applications in aerospace, automotive, commercial, dental, electronics, firearms, industrial, and sporting goods.

Metal Additive Manufacturing. *Additive manufacturing (AM)* covers a range of technologies that build three-dimensional (3D) objects by ***adding*** layer-upon-layer of material, whether the material is metallic, plastic, or ceramic. The ***metal AM*** process is an *AM* process that uses metal powders, including **stainless steels, aluminum, nickel, cobalt-chrome,** and **titanium alloys**, and the like. *AM* technologies include: 3D Printing, rapid prototyping (RP), direct digital manufacturing (DDM), layered manufacturing, and additive fabrication. *AM* essentially involves the use of a computer, 3D modeling software (CAD software), machine equipment, and layering material. Firstly, a CAD drawing is created. Then, the CAD data (CAD file) is read by the *AM* equipment; which lays downs or adds successive layers of liquid, powder, sheet material or other, in a layer-upon-layer fashion to fabricate a 3D object.

QUESTIONS

1.1. Encircle the best answer for the following multiple-choice questions (MCQs):

(a) Which one is a refractory metal? (i) aluminum (ii), tungsten, (iii) iron, (iv) zinc.

(b) Which natural material is a ceramic? (i) wood, (ii) stone, (iii) iron, (iv) lignin.

(c) Which material is a composite? (i) fiberglass, (ii) superalloy, (iii) PVC, (iv) glass.

(d) Which element is the most extensively used in computers? (i) Fe, (ii) *Al*, (iii) Si, (iv) W.

(e) Which field of metallurgy covers fracture mechanics? (i) extractive metallurgy, (ii) physical metallurgy, (iii) mechanical metallurgy.

(f) Which branch of metallurgy covers hydro-metallurgy? (i) extractive metallurgy, (ii) physical metallurgy, (iii) mechanical metallurgy.

(g) Which class of materials is PVC? (i) polymer, (ii) metal, (iii) ceramic, (iv) composite.

1.2. What are some of the important properties of metallic materials?

1.3. Compare hardness and thermal properties of ceramics and polymers.

1.4. What composite properties are attractive for aerospace and automotive applications?

1.5. Classify metallurgy into various branches and sub-branches. Briefly explain the four branches/fields of metallurgy.

1.6. List the manufacturing processes in metallurgy. Explain the following processes: (a) metal casting using sand casting method, and (b) metal additive manufacturing.

REFERENCES

Bodsworth, C. (1994) *The Extraction and Refining of Metals.* CRC Press, Boca Raton, FL.

Callister, Jr., W.D. (2007) *Materials Science and Engineering: An Introduction.* John Wiley & Sons Inc, New York.

Habashi, F. (1986) *Principles of Extractive Metallurgy.* CRC Press, Boca Raton, FL.

Huda, Z. (2017a) Energy-efficient gas-turbine blade-material technology—*A review. Materiali in Technologiji (Materials and Technology),* 51(3), 355–361.

Huda, Z. (2017b) *Materials Processing for Engineering Manufacture.* Trans Tech Publications, Pfaffikon, Zürich, Switzerland.

Huda, Z. (2018) *Manufacturing: Mathematical Models, Problems, and Solutions.* CRC Press, Boca Raton, FL.

Huda, Z. (2012) Reengineering of manufacturing process design for quality assurance in axle-hubs of a modern motor-car – a case study. *International Journal of Automotive Technology,* 13(7), 1113–1118.

Huda, Z. & Edi, P. (2013) Materials selection in design of structures and engines of supersonic aircrafts: A review. *Materials & Design,* 46, 552–560.

Painter, P.C. & Coleman, M.M. (2008) *Essentials of Polymer Science and Engineering.* DEStech Publications, Lancaster, PA.

Schaffer, J.P., Saxena, A., Antolovich, S.D., Sanders, Jr., T.H. & Warner, S.B. (1999) *The Science and Design of Engineering Materials.* WCB-McGraw Hill Inc, New York.

Shackelford, J.F. (2008) *Introduction to Materials Science for Engineers.* 7th ed. Prentice Hall Inc, Upper Saddle River, NJ.

Wakelin, D.H. & Fruehan, R.J. (1999) *Making, Shaping, and Treating of Steel (Iron-making).* 11th ed. Association of Iron & Steel Engineers (AISE), Warrendale.

2 Crystalline Structure of Metals

2.1 AMORPHOUS AND CRYSTALLINE MATERIALS

In a solid, atoms are strongly held together by forces depending on the type of bonding (ionic, covalent, or metallic bonding). The atoms can be arranged together as an aggregate either in a definite geometric pattern (to form crystalline solids) or somewhat irregular pattern (to form amorphous solids) (Rohrer, 2001). For example, when a pure metal is cooled from liquid state to solid state at room temperature, atoms are gathered together in a definite geometric pattern to form the crystalline solid. On the other hand, when molten sand (silica, SiO_2) is rapidly cooled, atoms are gathered together in somewhat irregular pattern to form an amorphous solid. A distinction between amorphous and crystalline solids is presented in Table 2.1. The difference between amorphous and crystalline solids is illustrated in Figure 2.1.

2.1.1 AMORPHOUS SOLIDS

In an amorphous solid, atoms are held apart at equilibrium spacing; however there is no long-range periodicity in atom location in the structure. An example of an amorphous solid is glass; additionally some types of plastic (e.g. wax) also have amorphous structure. An amorphous solid is sometimes referred to as a super-cooled liquid since its molecules are arranged in a random manner as in the liquid state. For example, glass is commonly made from silica sand. When sand is melted and the liquid is cooled rapidly, an amorphous solid (glass) is formed (see Figure 2.1a).

2.1.2 CRYSTALLINE SOLIDS

In crystalline solids, atoms are arranged in a definite three-dimensional (3D) geometric pattern with long-term periodicity. A crystal is a solid in which the atoms that make up the solid take up a highly ordered, definite, geometric arrangement that is repeated three-dimensionally within the crystal. The crystal structure of quartz (SiO_2) is shown in Figure 2.1b; which is similar to diamond crystal. Crystalline solids predominate (90%) of all naturally occurring and artificially prepared solids. For example, metals, minerals, limestone, diamond, graphite, salts (NaCl, KCl, etc.) all have crystalline structures. A detailed account of crystal structures is given in sections 2.2–2.4.

TABLE 2.1

Difference between Crystalline and Amorphous Solids

#	Crystalline solids	Amorphous solids
1	Atoms are arranged in a definite 3D geometric pattern with long-term periodicity.	No long-range periodicity in atom locations in the structure.
2	They have a sharp melting point.	They melt over a range of temperatures.
3	The crystalline structure is symmetrical.	The amorphous structure is unsymmetrical.
4	They fracture along specific cleavage planes.	They do not break along fixed cleavage planes.

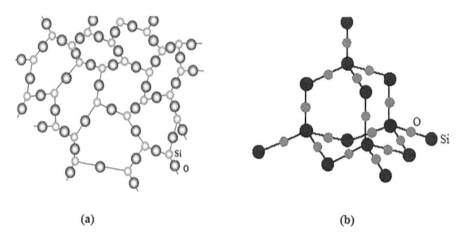

(a)   (b)

FIGURE 2.1   The structure of two solids: (a) amorphous solid (glassy silica), (b) crystalline solid (quartz).

2.2   CRYSTAL SYSTEMS—STRUCTURES AND PROPERTIES

2.2.1   Unit Cell, Crystal Lattice, and Crystal Systems

Ideally, a crystal is a repetition of identical structural units in a three dimensional space. A *unit cell* is the smallest repeating pattern of atomic arrangement in a crystal structure. A *lattice* refers to the periodicity in the crystal. The terms "*crystal lattice*" and "*crystal structure*" may be used interchangeably. A unit cell can be described in terms of the lengths of three adjacent edges (a, b, and c) and the angles between them (α, β, and γ) (see Figure 2.2).

Crystalline solids are usually classified as belonging to one of the following seven crystal systems: (1) cubic, (2) hexagonal, (3) tetragonal, (4) trigonal, (5) orthorhombic, (6) monoclinic, and (7) triclinic; these seven systems depend on the geometry of the unit cell, as shown in Figure 2.3. It is evident in Figure 2.3 that the simplest crystal structure is the cubic system, in which all edges of the unit cell are equal to each other, and all the angles are equal to 90°. The tetragonal and orthorhombic systems

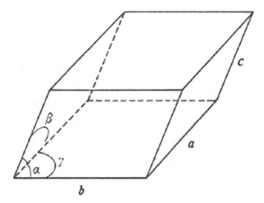

FIGURE 2.2 A unit cell with edges a, b, c and co-axial angles α, β, γ.

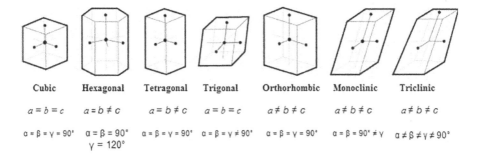

Cubic	Hexagonal	Tetragonal	Trigonal	Orthorhombic	Monoclinic	Triclinic
$a = b = c$	$a = b \neq c$	$a = b \neq c$	$a = b = c$	$a \neq b \neq c$	$a \neq b \neq c$	$a \neq b \neq c$
$\alpha = \beta = \gamma = 90°$	$\alpha = \beta = 90°$ $\gamma = 120°$	$\alpha = \beta = \gamma = 90°$	$\alpha = \beta = \gamma \neq 90°$	$\alpha = \beta = \gamma = 90°$	$\alpha = \beta = 90° \neq \gamma$	$\alpha \neq \beta \neq \gamma \neq 90°$

FIGURE 2.3 The seven crystal systems structures (see also Figure 2.2).

refer to rectangular unit cells, but the edges are not all equal. In the remaining systems, some or all of the angles are not equal to 90°. The least symmetrical structure is the triclinic system in which no edges are equal and no angles are equal to each other or to 90° (Barret and Massalski, 1980). The hexagonal system is worth noting; here two edges of the unit cell are equal and subtend an angle of 120°. Hexagonal crystals are quite common both in metals and ceramics; examples include crystal structures of zinc, graphite, and the like.

2.2.2 THE SIMPLE CUBIC CRYSTAL STRUCTURE

The simplest of all crystal structures is the *simple cubic (SC)* lattice. In the *SC* structure, all edges are of the same length and all planes are perpendicular (see Figure 2.4a). The *SC* unit cell has atoms at the eight corners of the cube; the corner of the unit cell is the center of the atom. The three orthogonal planes of the unit cell bisect each atom, so that one eighth of any atom is inside the unit cell (see Figure 2.4b). Notable examples of simple cubic system include: polonium (*Po*) and sodium chloride (*NaCl*) crystals.

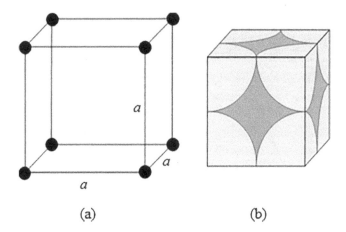

(a)   (b)

FIGURE 2.4 Simple cubic unit cell showing: (a) atoms at eight corners of the cubic cell, (b) one-eighth of each corner atoms is inside the unit cell.

It is evident in Figure 2.4 that there are eight corner atoms in the cube, and each corner atom contributes one eighth to the cell. Thus, the number of atoms per unit cell for the simple cubic (*SC*) system can be calculated as follows:

$$\text{No. of atoms per unit cell for SC} = 8 \times \frac{1}{8} = 1 \qquad (2.1)$$

Hence, there is one atom inside the simple cubic unit cell.

2.2.3 Crystal-Structure Properties

In general, there are three important crystal-structure properties: (a) number of atoms per unit cell, (b) coordination number (CN), and (c) atomic packing factor. The *CN* is defined as the number of atoms touching any given atom in the crystal lattice. The fraction of volume occupied by atoms (as solid spheres) in a unit cell is called the atomic packing factor (*APF*). Numerically,

$$\text{APF} = \frac{N_a V_a}{V_{uc}} \qquad (2.2)$$

where N_a is the number of atoms per unit cell, V_a is the volume of an atom, and V_{uc} is the volume of the unit cell.

2.3 CRYSTAL STRUCTURES IN METALS

All metals, except mercury, are crystalline solids at room temperature. In general, metals and many other solids have crystal structures described as body center cubic (*BCC*), face-centered cubic (*FCC*), or hexagonal close packed (*HCP*); the three crystal structures are discussed as follows.

2.3.1 BODY-CENTERED CUBIC (BCC) STRUCTURE

The body-centered cubic (*BCC*) unit cell has atoms at each of the eight corners of a cube plus one atom in the cube-center (see Figure 2.5a). Since each of the corner atoms is the corner of another cube, the corner atoms in each unit cell are shared among eight unit cells (see Figure 2.5b). The BCC crystal lattice/structure is illustrated in Figure 2.5c. Examples of metals crystallizing in BCC structure include: α-iron (α-Fe), chromium (Cr), vanadium (V), molybdenum (Mo), etc.

The number of atoms per unit cell for BCC crystal structure can be calculated by using Equation 2.1 and by referring to Figure 2.5b as follows:

$$\text{No. of atoms per unit cell for BCC} = (8 \times \frac{1}{8}) + 1 = 2 \qquad (2.3)$$

The coordination number for BCC lattice structure can be found as follows. Since the central atom in the BCC unit cell is in direct contact with eight corner atoms, the coordination number for BCC unit cell is: CN = 8 (see Figure 2.5c). It can be shown that the atomic packing factor for BCC crystal lattice is: APF = 0.68 (see Example 2.1).

2.3.2 FACE CENTERED CUBIC (FCC) STRUCTURE

Face-centered cubic (FCC) crystal structure has atoms at each corner of the cube and six atoms at each face of the cube (see Figure 2.6a). Each corner atoms contributes one-eighth to the unit cell whereas the atom at each face is shared with the adjacent cell (Figure 2.6b). The FCC crystal lattice is shown in Figure 2.6c. Examples of metals crystallizing in FCC structure include: γ-iron (γ-Fe), copper (Cu), nickel (Ni), aluminum (*Al*), platinum (*Pt*), and the like.

It is evident in Figure 2.6b that the face diagonal in an FCC unit cell is equal to $4R$. Thus we can develop a relationship between the atomic radius (R) and the lattice parameter (a) as follows:

$$(4R)^2 = a^2 + a^2 \qquad (\text{Pythagoras theorem})$$
$$\text{Face diagonal in an FCC unit cell} = 4R = \sqrt{2}\,a \qquad (2.4)$$

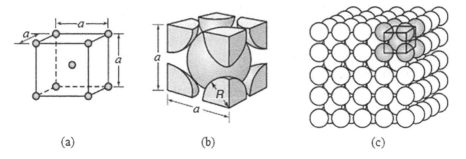

(a) (b) (c)

FIGURE 2.5 BCC crystal structure; (a) BCC unit cell, (b) BCC unit cell showing contribution of atoms to the unit cell, (c) BCC crystal lattice.

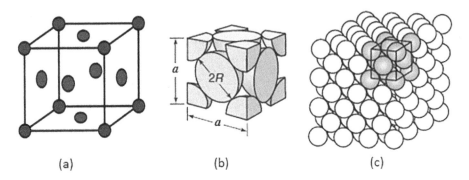

FIGURE 2.6   FCC crystal structure; (a) FCC unit cell, (b) FCC unit cell showing contribution of atoms to the unit cell, (c) FCC crystal lattice.

It can easily been shown that there are 4 atoms per unit cell in FCC crystal lattice (see Example 2.2). We can deduce coordination number of FCC lattice as follows. The face-centered atom is surrounded by four corner nearest-neighbor atoms, four face atoms in contact from behind, and the four (hidden) face atoms in the adjacent unit cell in front (see Figure 2.6c). Thus the coordination number of FCC is: CN = 4 + 4 + 4 = 12. It can be mathematically shown that the atomic packing factor for FCC lattice is: APF = 0.74 (see Example 2.3).

A comparison of the CN and APF values of the BCC with the FCC lattices indicates that the former are lower than the latter. It means that the BCC crystal lattice does not allow the atoms to pack together as closely as the FCC lattice does.

2.3.3   HEXAGONAL CLOSE-PACKED (HCP) CRYSTAL STRUCTURE

The unit cell of an HCP lattice is visualized as top and bottom planes; each plane has 7 atoms, forming a regular hexagon around a central atom. In between these two planes, there is a half hexagon of 3 atoms (see Figure 2.7a). In HCP unit cell, one lattice parameter (c) is longer than the other (a); this relationship can be expressed as:

$$c = 1.63a \qquad\qquad (2.5)$$

Each corner atoms in the HCP unit cell contributes one-sixth to the cell whereas the atom at each face is shared with the adjacent cell; the three middle-layer atoms contribute full to the cell (see Figure 2.7b). The HCP crystal lattice is shown in Figure 2.7c. Examples of metals crystallizing in HCP structure include: zinc (Zn), titanium (Ti), and the like.

It can been shown that there are 6 atoms per unit cell in the HCP lattice (see Example 2.4). The coordination number (CN) of HCP lattice can be deduced as follows. The face-centered atom has 6 nearest neighbors in the same close-packed layer plus 3 in the layer above and 3 in the layer below (see Figure 2.7c). This makes the total number of the nearest neighbors equal to 12. Hence, coordination number (*CN*) of HCP lattice = 12. It can be mathematically shown that the atomic packing factor for HCP lattice is: APF = 0.74 (see Examples 2.5–2.6). By reference to subsections 2.3.1 and 2.3.2, the crystal properties data for BCC, FCC, and HCP are summarized in Table 2.2.

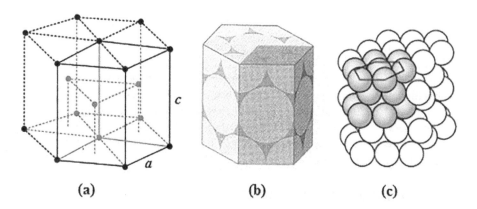

FIGURE 2.7 The HCP crystal structure; (a) HCP unit cell, (b) HCP unit cell showing contribution of atoms to the unit cell, (c) HCP crystal lattice.

TABLE 2.2
Crystal Structure Properties of Some Metals at Room Temperature

Crystal structure	No. of atoms/unit cell (n)	CN	APF	Examples
BCC	2	8	0.68	*Fe, Cr, V, Mo*
FCC	4	12	0.74	*Al, Ni, Cu, Pt*
HCP	6	12	0.74	Zinc, titanium

Table 2.2 shows that both FCC and HCP crystal structures have higher CN and APF values as compared to BCC lattice. The crystallographic data (in Table 2.2) indicate that FCC and HCP lattices are closed packed structures whereas BCC is not. In order to form the strongest metallic bonds, atoms in metals are packed together as closely as possible (Douglas, 2006).

2.3.4 Computing Theoretical Density of a Metal

The density of a metal is usually determined in a laboratory by dividing the mass of the metal by its volume. The experimental results so obtained may be verified by theoretical calculations based on the metallic crystal structure. Thus, the theoretical density can be calculated by:

$$\text{Density} = \rho = \frac{Mass\,of\,atoms\,per\,unit\,cell}{Volume\,of\,the\,unit\,cell} = \frac{n\,A}{V_{uc}N_A} \tag{2.6}$$

where n is the number of atoms per unit cell for the metallic lattice; A is the atomic weight, g/mol; N_A is the Avogadro's number (= 6.02 x 10^{23} atoms/mol). The computation of the density of a metal, based on its crystal structure, is illustrated in Example 2.7.

2.4   MILLER INDICES

The specification of directions and planes within crystalline solids is of great technological importance to a metallurgist since some materials' properties strongly depend on their crystallographic orientations. *Miller indices* are used as the notation system for defining planes and directions in crystal lattices. In order to define a crystallographic plane, three integers *l*, *m*, and *n*, known as *Miller indices*, are used. The three integers are written as (*hkl*) for the crystallographic plane whereas the crystallographic directions are represented by [*hkl*]. By convention, a negative integer is represented by a bar (e.g. negative $h = \bar{h}$).

2.5   CRYSTALLOGRAPHIC DIRECTIONS

2.5.1   Procedure to Find Miller Indices for a Crystallographic Direction

Any line (or vector direction) specified by two points in a crystal lattice is called a *crystallographic direction*. The step-by-step procedure to define a crystallographic direction by finding its Miller indices, is described as follows:

1.   Locate two points lying on the given crystal direction.
2.   Determine the coordinates of the two points using a right-hand coordinate system as defined by the crystallographic axes *a*, *b*, and *c*.
3.   Obtain the number of lattice parameters moved in the direction of a coordinate axis by subtracting the coordinates of the *tail point* from that of the *head point*.
4.   Clear fraction and/or reduce the results so obtained to the smallest integers.
5.   Represent the Miller indices as [*hkl*].

The previous set of procedures (1–5), used to find *Miller indices* for a crystallographic direction, are illustrated in Example 2.7. For cubic crystals, the *Miller indices* are the vector components of the direction resolved along each of the three coordinate axes and reduced to the smallest integers i.e. [100], [010], and [001], as shown in Figure 2.8. A family of crystallographic directions is usually denoted by < >. For example, the family of the crystallographic directions for the three orthogonal axes in cubic crystals can be denoted by <100>.

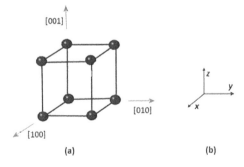

FIGURE 2.8   Crystallographic directions; (a) Miller indices for coordinate axes in a cubic crystal; (b) The coordinate axes: *x*, y, and z.

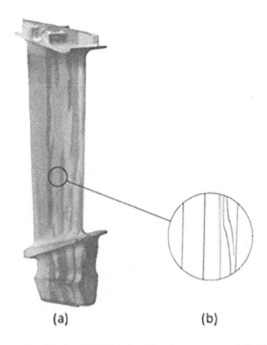

(a)   (b)

FIGURE 2.9   Gas-turbine blade; (a) DS blade, (b) columnar crystals in the DS blade.

2.5.2   INDUSTRIAL APPLICATIONS OF CRYSTALLOGRAPHIC DIRECTIONS

A notable example of crystallographic direction of industrial importance is the directionally solidified (DS) superalloy turbine-blade (Figure 2.9a). Here, the directional solidification (DS) manufacturing process is so designed as to grow the crystals in the [001] direction (Figure 2.9b). The columnar grained (CG) structure in the DS blade imparts superior creep strength for high-temperature applications in gas-turbine engines.

Another real-world example of crystallographic direction is in the processing of the grain-oriented electrical steel (GOES). The *GOES* contains around 3–4% silicon and is made by complex processing techniques (see Chapter 12, section 12.3.9). These processing techniques impart excellent magnetic properties in the rolling direction: [100] crystallographic direction. The *GOES* finds applications in transformers for power transmission as well as in generators for steam turbine.

2.6   CRYSTALLOGRAPHIC PLANES

2.6.1   PROCEDURE TO FIND MILLER INDICES FOR A CRYSTALLOGRAPHIC PLANE

A crystallographic plane refers to the orientation of atomic plane in a crystal lattice. It can be represented by using a unit cell as the basis (see Figure 2.2, Figure 2.8). In all crystal systems, except the hexagonal system, crystallographic planes are specified by three Miller indices as *(hkl)*. The *Miller indices (hkl)* are the reciprocals of the fractional intercepts (with fractions cleared) that the plane makes with the three orthogonal crystallographic axes *x*, *y*, and *z*.

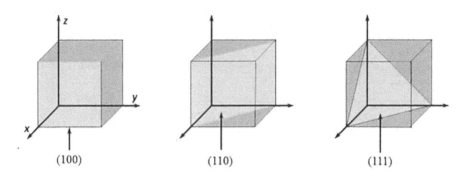

FIGURE 2.10   Miller indices for some important crystallographic planes.

In order to define the orientation of a crystal face or a plane by specifying its Miller indices, the following procedure should be followed:

1. Find the fractional intercepts that the plane makes with the crystallographic axes. It means find how far along the unit cell lengths does the plane intersect the axis.
2. Take the reciprocal of the fractional intercept of each unit length for each axis.
3. Clear the fractions.
4. Enclose these integer numbers in parenthesis (), and designate that specific crystallographic plane within the lattice.

The use of the previously listed steps (1–4), in the specification of crystallographic planes by determining their Miller indices, is explained in Examples 2.8–2.9. The crystallographic planes with Miller indices (100), (110), and (111) are illustrated in Figure 2.10.

2.6.2   INDUSTRIAL APPLICATIONS OF CRYSTALLOGRAPHIC PLANES

The specification of crystallographic planes in a crystalline solid is important for a metallurgist. In metals, plastic deformation occurs by slip along specified crystallographic planes—called *slip planes* (see Chapter 3). For example, in BCC metals, (110), (101), and (011) are the *slip planes*. Another industrial application of crystallographic planes is found in semiconductor fabrication. Here, it is easier to cleave the silicon wafer along (111) plane than along other crystallographic planes. The knowledge of crystallographic planes (diffraction planes) also greatly helps us to identify a metal by determining its crystal structure (see section 2.8).

2.7   LINEAR AND PLANAR ATOMIC DENSITIES

Linear and planar atomic densities are important consideration in *slip*—a mechanism by which metals plastically deform along *slip planes* (see Chapter 5). *Linear atomic density* is the ratio of the number of atoms contained in the selected length of line to the length of the selected section. Numerically (Tisza, 2001),

$$\rho_L = \frac{N_{linear}}{l} \qquad (2.7)$$

where ρ_L is the linear atomic density, atoms/mm; N_{linear} is the number of atoms centered in the examined crystallographic direction; and l is the length of the examined direction, mm.

The *planar atomic density* is the ratio of the number of atoms contained in the selected area to the area of the selected section. Mathematically,

$$\rho_P = \frac{N_{planar}}{A} \qquad (2.8)$$

where ρ_P is the planar atomic density, atoms/mm^2; N_{planar} is the number of atoms centered on a crystallographic plane, and A is the area of the plane, mm^2.

The significances of Equations 2.7–2.8 are illustrated in Examples 2.10–2.11.

2.8 X-RAY DIFFRACTION—BRAGG'S LAW

The X-ray diffraction technique is an important tool of modern crystallography; it allows us to determine the crystal structure of an unidentified crystalline material. The Bragg's law in this technique is used to relate the spacing between the crystallographic planes.

Consider a beam of X-rays that impinge on the surface of a crystal and interact with the atoms by interference phenomenon. Let us consider the beam of two X-rays entering the crystal with one of these planes of atoms oriented at an angle of θ to the incoming beam of monochromatic (single wavelength) X-rays (see Figure 2.11). The first ray (wave) is reflected from the top plane of atoms (plane 1), and the second wave is reflected from plane 2 (Figure 2.11). The inter-planar distance is d, and the wavelength is λ. The path difference between the two rays (difference in the distance travelled from the X-ray source to the detector) can be given by:

$$\text{Path difference} = d \sin\theta + d \sin\theta = 2d \sin\theta \qquad (2.9)$$

The constructive interference or *diffraction* occurs when the path difference is an integral multiple of the X-ray wavelength. Hence, Equation 2.9 can be rewritten as:

$$n\lambda = 2d \sin\theta \qquad (2.10)$$

where n is an integer, and θ is the Bragg angle. This equation is called *Bragg's law*. According to *Bragg's law*, the diffraction occurs at specific values of θ; which is determined by known values of n, λ, and d. For first order reflection, $n=1$; for second order reflection, $n=2$, and so on.

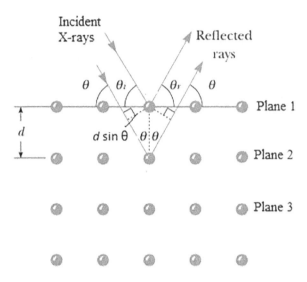

FIGURE 2.11   X-ray diffraction (note: $\theta_i = \theta_r = \theta$; diffraction angle = 2θ).

There are a number of experimental techniques used in X-ray diffraction (XRD); however, the commonly practiced technique is the *powder diffraction method*. In this technique, a sample powder is mixed with an amorphous binder. The powder-binder mixture (crystal) is subjected to XRD (Figure 2.11). The diffracted intensity is measured as a function of θ by using an XRD pattern (Figure 2.12) by employing an instrument: a *diffractometer*. The significances of Equation 2.10 and Figure 2.12 are illustrated in Example 2.12.

The inter-planar spacing (the distance between two adjacent and parallel planes of atoms) in a lattice depends on the Miller indices and the lattice parameters. For a cubic crystal system, the inter-planar spacing (d) is given by:

$$d = \frac{a}{\sqrt{h^2 + k^2 + l^2}}$$

(2.11)

where a is the lattice parameter (edge length of the unit cell), and h, k, and l are the Miller indices of the crystallographic plane. The value of d can be computed with the aid of XRD pattern by using Bragg's law. If the XRD pattern is indexed (h, k, and l are known), the lattice parameter for a metal (a) can be calculated by:

$$a = d\sqrt{h^2 + k^2 + l^2}$$

(2.12)

The significance of Equation 2.12 is illustrated in Example 2.13.

By combination of Equation 2.11 and Equation 2.12, we obtain:

$$a = \frac{n\lambda\sqrt{h^2 + k^2 + l^2}}{2\sin\theta}$$

(2.13)

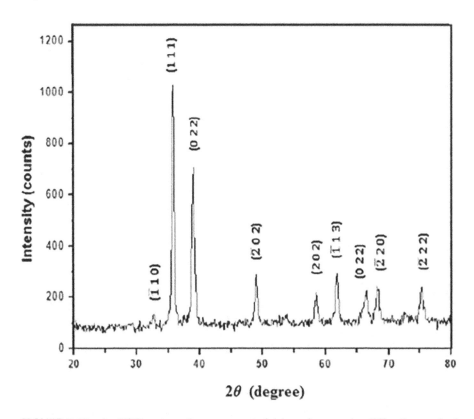

FIGURE 2.12 An XRD pattern for a nanomaterial by using powder diffraction method (note: intensity peaks occur at diffraction angles of 36°, 39°, 49°, 58°, etc.).

Equation 2.13 enables us to index a crystallographic plane and to determine crystal structure for a cubic system (see Examples 2.15–2.16).

2.9 INDEXING A DIFFRACTION PATTERN FOR CUBIC CRYSTALS

The process of determining the unit cell parameters from the peak positions in an XRD pattern is called *indexing*. It is worth noting that the XRD peaks are indexed according to certain reflection rules/condition. The conditions for allowed XRD reflections for cubic crystal structures are shown in Table 2.3.

The index number for a plane can be calculated by (Helliwell, 2015):

$$\text{Index number for plane } (hkl) = h^2 + k^2 + l^2 \tag{2.14}$$

In cubic crystal structure, the first peak in XRD pattern results from the planes with the lowest Miller indices. Thus, the index number of the first peak in the XRD pattern for three cubic structures can be computed by using Equation 2.14, as illustrated in Table 2.4.

TABLE 2.3

Conditions for Allowed Reflections for Cubic Crystal Structures

Crystal structure	Condition for allowed reflection
Simple cubic	All possible h, k, and l values
BCC	$(h + k + l)$ is even
FCC	h, k, and l are either all even or all odd

TABLE 2.4

Index Numbers for the First XRD Peak in Cubic Lattices

Crystal structure	The lowest Miller indices	Index number
Simple cubic	(100)	$h^2 + k^2 + l^2 = 1^2 + 0^2 + 0^2 = 1$
BCC	(110)	$h^2 + k^2 + l^2 = 1^2 + 1^2 + 0^2 = 2$
FCC	(111)	$h^2 + k^2 + l^2 = 1^2 + 1^2 + 1^2 = 3$

Examples 2.15–2.16 illustrate the significances of Equation 2.14 and Tables 2.3–2.4.

2.10 CRYSTALLITES—SCHERRER'S FORMULA

Most engineering materials are *polycrystalline* i.e. they are made of a large number of crystals held together through defective boundaries. These small crystals are called *crystallites* or *grains* (see Figure 2.13d). *Crystallite size* can be measured from an X-ray diffraction pattern by using *Scherrer's formula*, as follows (Langford and Wilson, 1978):

$$t = \frac{K\lambda}{B\cos\theta} \tag{2.15}$$

where t is the averaged dimension of crystallites, nm; K is the Scherrer's constant—an arbitrary value in the range of 0.87–1.0; λ is the wavelength of X-ray, nm; and B is the peak width ($B = 2\theta_{High} - 2\theta_{Low}$) of a reflection (in radians) located at 2θ (see Example 2.17).

2.11 CALCULATIONS—EXAMPLES OF CRYSTALLINE STRUCTURES OF SOLIDS

EXAMPLE 2.1 CALCULATING APF FOR BCC CRYSTAL LATTICE

Calculate the atomic packing factor (*APF*) for BCC crystal lattice.

SOLUTION

No. of atoms per unit cell for BCC = $N_a = 2$

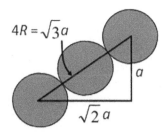

FIGURE E-2.1   Calculation for atomic radius, R, in terms of lattice parameter, a for BCC.

$$\text{Volume of an atom} = V_a = \frac{4}{3}\pi R^3 \qquad (E-2.1a)$$

By reference to Figure 2.5b and Figure E-2.1, $R = \frac{\sqrt{3}}{4}a$

By substituting the value of R in Equation E-2.1,

$$V_a = \frac{4}{3}\pi R^3 = \frac{4}{3}\pi\left(\frac{\sqrt{3}}{4}a\right)^3 = \frac{\sqrt{3}}{16}\pi a^3 \qquad (E-2.1b)$$

The volume of BCC unit cell = $V_{uc} = a^3$ (E-2.1c) (see Figure 2.5a).

By subtitling the values from Equations E-2.1a, E-2.1b, and E-2.1c into Equation 2.2, we get:

$$APF = \frac{N_a V_a}{V_{uc}} = \frac{2 \times \frac{\sqrt{3}}{16}\pi a^3}{a^3} = \frac{\sqrt{3}}{8}\pi = 0.68$$

Hence, the atomic packing factor for BCC crystal lattice = 0.68.

EXAMPLE 2.2 CALCULATING THE NUMBER OF ATOMS PER UNIT CELL FOR FCC

Calculate the number of atoms per unit cell for FCC lattice.

SOLUTION

In the FCC unit cell, there are eight one-eighths atoms at the corners and six halves at the faces.

The contribution of eight one-eighth corner atoms to the unit cell = $8 \times \frac{1}{8} = 1$ (see Figure 2.6b).

The contribution of six halves (atoms) at the faces of the unit cell = $6 \times \frac{1}{2} = 3$ (see Figure 2.6b).

The number of atoms per unit cell in FCC = 1 + 3 = 4.

EXAMPLE 2.3 COMPUTING FOR APF FOR FCC LATTICE

Calculate the atomic packing factor (*APF*) for FCC crystal lattice.

SOLUTION

No. of atoms per unit cell for FCC = N_a = 4 (see Example 2.2)
By using Equation 2.4,

$$R = \frac{\sqrt{2}a}{4}$$

$$R^3 = \frac{\sqrt{2}\,a^3}{32}$$

Volume of an atom = $V_a = \frac{4}{3}\pi R^3 = \frac{4}{3}\pi[\frac{\sqrt{2}\,a^3}{32}] = 0.185\,a^3$

Volume of FCC unit cell = $V_{uc} = a^3$

By using Equation 2.2,

$$APF = \frac{N_a V_a}{V_{uc}} = \frac{4 \times 0.185\,a^3}{a^3} = 0.74$$

Hence, **the atomic packing factor (*APF*) for FCC lattice = 0.74**.

EXAMPLE 2.4 COMPUTING NUMBER OF ATOMS PER UNIT CELL FOR HCP LATTICE

Calculate the number of atoms per unit cell for HCP lattice.

SOLUTION

In the HCP lattice, each corner atoms contributes one-sixth to the unit cell whereas the atom at each face is shared with the adjacent cell; the three middle-layer atoms contribute full to the cell (see Figure 2.7b):

Contribution of six corner atoms in the top layer $= 6 \times \frac{1}{6} = 1$

Contribution of two face atoms (one in the top layer and the other in the

bottom) $= 2 \times \frac{1}{2} = 1$

Contribution of six corner atoms in the bottom layer $= 6 \times \frac{1}{6} = 1$

Contribution of the three middle-layer atoms = 3

Number of atoms per unit cell in HCP lattice = Total Contribution = 1 + 1 +
1 + 3 = 6

FIGURE E-2.5 The six triangular divisions of the hexagonal face.

EXAMPLE 2.5 COMPUTING THE VOLUME OF HCP UNIT CELL

Calculate the volume of unit cell in HCP lattice in term of the lattice parameter.

SOLUTION

By reference to Figure 2.7a, the volume of HCP unit cell (V_{uc}) can be expressed as:

V_{uc} = (area of the hexagonal face) x (height of the hexagonal face)
(E-2.5a).
In order to determine the area of the hexagonal face, we can draw a new figure (Figure E-2.5).

$$\text{Area of hexagonal face} = 6 \times \left(\text{area of each triangle} \right)$$
$$= 6 \times (\frac{1}{2} \; x \, \text{base} \; x \, h) = 3\,a\,(h) = 3\,a\,(\frac{\sqrt{3}}{2}\,a)$$
$$\text{Area of hexagonal face} \; = \; 3\frac{\sqrt{3}}{2}\,a^2$$

By substituting the value of the area of hexagonal face in Equation E-2.5a,

$$V_{uc} = (3\,\frac{\sqrt{3}}{2}\,a^2) \times (c) = (3\,\frac{\sqrt{3}}{2}\,a^2) \times (1.63a) \; = \; 4.23a^3$$

EXAMPLE 2.6 COMPUTING THE ATOMIC PACKING FACTOR (APF) FOR HCP CRYSTAL

By using the data in Example 2.5, calculate the APF for HCP Crystal structure.

SOLUTION

No. of atoms per unit cell for HCP = N_a = 6; Volume of HCP unit cell = V_{uc} = 4.23 a^3
By reference to Figure 2.7(b), it can be shown that volume of an atoms (V_a) can be related to the lattice parameter for HCP by: $V_a = 0.522 \; a^3$.
By using Equation 2.2,

$$\text{APF} = \frac{N_a V_a}{V_{uc}} = \frac{6 \times 0.522\,a^3}{4.23\,a^3} = 0.74$$

EXAMPLE 2.7 COMPUTING THEORETICAL DENSITY OF CHROMIUM

Calculate the theoretical density of chromium (based on its crystal structure). Hint: use the data in Table 2.2. Chromium has an atomic radius of 0.125 nm, and its atomic weight is 52 g/mol. Compare the theoretical density with the experimentally calculated density.

SOLUTION

By reference to Table 2.2, chromium crystallizes as BCC lattice with $n = 2$;

$A = 52$ g/mol; $R = 0.125$ x 10^{-9} m $= 12.5$ x 10^{-9} cm

By reference to Figure E-2.1, $a = (4R)/\sqrt{3}$

Volume of BCC unit cell $= V_{uc} = a^3 = \left(\dfrac{4R}{\sqrt{3}}\right)^3 = \left(\dfrac{4\times12.5\times10^{-9}}{\sqrt{3}}\right)^3 = 24\times10^{-24}$ cm^3

By using Equation 2.6,

$$\rho = \frac{nA}{V_{uc}N_A} = \frac{2\times52}{24\times10^{-24}\times 6.02 \times 10^{23}} = 7.198\,\text{g/cm}^3$$

The calculated density based on the crystal structure ($\rho = 7.198$ g/cm^3) is well in agreement with the experimentally calculated value as available in the literature (($\rho = 7.19$ g/cm^3).

EXAMPLE 2.8 DETERMINING THE MILLER INDICES FOR A CRYSTALLOGRAPHIC DIRECTION

Define the crystallographic direction for the vector **d** (in Figure E-2.8) using Miller indices.

SOLUTION

By reference to subsection 2.4.1, we can find the Miler indices as follows:

1. Locate two arbitrary points on direction **d**: P1 and P2.
2. The coordinates of the two points are: P1(0,0,1) and P2(0, ½, 0).

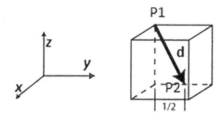

FIGURE E-2.8 Direction vector d in a crystal lattice.

3. Perform subtraction: P2(0, ½, 0) − P1(0, 0, 1) = (0, ½, −1).
4. Fraction to clear: 2 (0, ½, −1) = (0, 1, −2).
5. Miller indices: [0, 1, $\overline{2}$].

EXAMPLE 2.9 DETERMINATION OF MILLER INDICES FOR CRYSTALLOGRAPHIC PLANES

Determine the *Miller indices* for the planes shown in Figures E-2.9a–c.

SOLUTION

By reference to the procedure (steps 1–4) outlined in subsection 2.4.2, we can determine the Miller indices for the planes shown in Figures E-2.9(a–c), as follows.

FIGURE E-2.9(a)

x y z
1. Intercepts ∞ 1 ∞
2. Reciprocals 1/∞ 1/1 1/∞
3. Reduction 0 1 0
4. Miller indices (010)

FIGURE E-2.9(b)

x y z
1. Intercepts 1 1 ∞
2. Reciprocals 1/1 1/1 1/∞
3. Reduction 1 1 0
4. Miller indices (110)

FIGURE E-2.9(c)

x y z
1. Intercepts ∞ ∞ −1
2. Reciprocals 1/∞ 1/∞ 1/−1

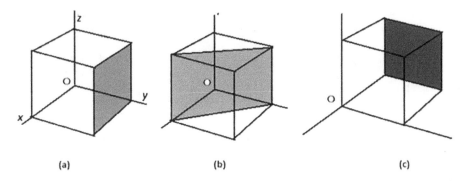

(a)   (b)   (c)

FIGURE E-2.9 Crystallographic planes with unknown Miller indices.

3. Reduction 0 0 −1
4. Miller indices (00 $\bar{1}$)

EXAMPLE 2.10 SKETCHING CRYSTALLOGRAPHIC PLANES WHEN THE MILLER INDICES ARE KNOWN

Sketch the crystallographic planes having the following Miller indices: (100), (110), and (111).

SOLUTION

(100) Plane

x y z
Reduction 1 0 0
Reciprocals 1/1 1/0 1/0
Intercepts 1 ∞ ∞

The (100) plane is sketched in Figure 2.10 (left)
(110) Plane

x y z
Reduction 1 1 0
Reciprocals 1/1 1/1 1/0
Intercepts 1 1 ∞

The (110) plane is sketched in Figure 2.10 (center)
(111) Plane

x y z
Reduction 1 1 1
Reciprocals 1/1 1/1 1/1
Intercepts 1 1 1

The (111) plane is sketched in Figure 2.10 (right)

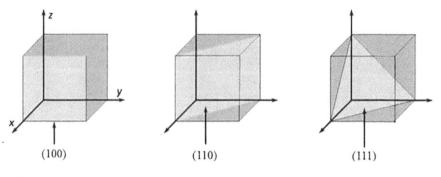

(100)   (110)   (111)

FIGURE E-2.10

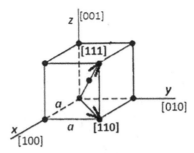

FIGURE E-2.11   The [100] and [111] directions in BCC unit cell.

EXAMPLE 2.11 COMPUTING LINEAR ATOMIC DENSITY

Calculate the linear atomic density along the following directions in a BCC unit cell:
(a) [110], and (b) [111].

SOLUTION

First, we need to indicate the [100] and [111] directions in BCC unit cell as shown
in Figure E-2.11.

A. By reference to Figure E-2.11,
Number of atoms centered in the [110] direction $= N_{linear} =$ (2 corner atoms x ½) = 1

By reference to Figure E-2.1, $a = (4/\sqrt{3})R$

Length of the line-section in the [110] direction $= l = \sqrt{2}\,a = 4\dfrac{\sqrt{2}}{\sqrt{3}}R$
By using Equation 2.7,

$$\rho_L = \frac{N_{linear}}{l} = \frac{1}{4\dfrac{\sqrt{2}}{\sqrt{3}}R} = \frac{\sqrt{3}}{4\sqrt{2}R}$$

B. From Figure E-2.11,
Number of atoms contained in [111] direction $= N_{linear} =$ (2 corner atoms x ½) +
(1 atom) = 2
By reference to Figure E-2.1,
Length of the line-section in the [111] direction $= l = 4R$
By using Equation 2.7,

$$\rho_L = \frac{N_{linear}}{l} = \frac{2}{4R} = \frac{1}{2R}$$

The linear atomic densities in the [110] and [111] directions are $\dfrac{\sqrt{3}}{4\sqrt{2}R}$ and $\dfrac{1}{2R}$,
respectively.

EXAMPLE 2.12 CALCULATING THE PLANAR ATOMIC DENSITY

Determine the planar atomic density of (110) plane in an FCC unit cell.

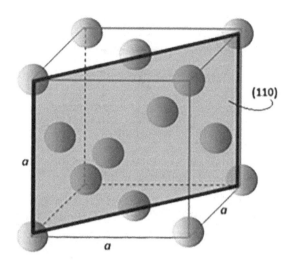

FIGURE E-2.12   The (110) plane in FCC unit cell.

SOLUTION

The (110) plane in FCC unit cell is shown in Figure E-2.12.
By reference to Figure E-2.12,

Number of atoms centered on the plane = N_{planar} = (4 corner atoms x ¼) + (2 face
  atoms x ½) = 2

Area of the (110) plane = $A = (a)(\sqrt{2}\, a)$

But Equation 2.4 yields: $\sqrt{2}\, a = 4R$, and $a = \sqrt{8}\, R$

Area of the (110) plane = $A = (\sqrt{8}\, R)(4R) = 8\sqrt{2}R^2$

By using Equation 2.8,

$$\rho_P = \frac{N_{planar}}{A} = \frac{2}{8\sqrt{2}R^2} = \frac{1}{4\sqrt{2}R^2}$$

EXAMPLE 2.13 CALCULATING INTER-PLANAR SPACING USING BRAGG'S LAW

Monochromatic radiation having a wavelength of 0.154 nm is used in the XRD for
a metal sample. The XRD pattern shows an intensity peak at a diffraction angle of
64.4°. Calculate the inter-planar spacing for the first order of reflection.

SOLUTION

$\lambda = 0.154$ nm, $n = 1$, $2\theta = 64.4°$, $\theta = 32.2°$, $d = ?$

By using Bragg's law (Equation 2.10),

$$2\, d \sin \theta = n\, \lambda$$

$$d = \frac{n\, \lambda}{2 \sin \theta} = \frac{1 \times 0.154}{2 \sin 32.2} = 0.1445\,\text{nm}$$

The inter-planar spacing = 0.1445 nanometer.

EXAMPLE 2.14 HOW IS AN XRD PATTERN HELPFUL IN COMPUTING LATTICE PARAMETER?

According to the XRD pattern for powdered lead, a high-intensity peak results at diffraction angle of 36° when the Bragg diffraction condition is satisfied by the plane (200) for the first order reflection. Calculate the lattice parameter of lead, if the wavelength of the monochromatic radiation, used in the diffractometer, is 0.1542 nm.

SOLUTION

$2\theta = 36°$, $\theta = 18°$, $n = 1$, $\lambda = 0.1542$ nm, $(hkl) = (200)$, h = 2, k = 0, l = 0, a = ?

By using Equation 2.10, we obtain: $d = 0.2495$ nm
   By using Equation 2.12,

$$a = d \sqrt{h^2 + k^2 + l^2} = 0.2495 \times \sqrt{2^2 + 0^2 + 0^2} = 0.499\,\text{nm}$$

The lattice parameter of lead = 0.499 nm.

EXAMPLE 2.15 DETERMINING INTER-PLANAR SPACING FOR A SPECIFIED SET OF PLANES

Aluminum has an atomic radius of 0.1431 nm. Calculate the inter-planar spacing for the (110) set of planes.

SOLUTION

Aluminum crystallizes as FCC lattice. R = 0.1431 nm, $(hkl) = (110)$, d = ?
By Equation 2.4,

$$\sqrt{2}\, a = 4\, R$$
$$\sqrt{2}\, a = 4 \times 0.1431 = 0.5724\,\text{nm}$$
$$a = 0.4048\,\text{nm}$$

By using Equation 2.11,

$$d = \frac{a}{\sqrt{h^2 + k^2 + l^2}} = \frac{0.4048}{\sqrt{1^2 + 1^2 + 0^2}} = 0.2863\,\text{nm}$$

The inter-planar spacing for the (110) set of planes in aluminum = 0.2863 nanometer.

EXAMPLE 2.16 DETERMINING INDEX NUMBERS FOR VARIOUS PEAKS FOR CUBIC LATTICES

An XRD pattern is obtained from a cubic crystal by using radiation of wavelength 0.154 nm. The Bragg angles for the first seven peaks in the XRD pattern are: 21.66°, 31.47°, 39.74°, 47.58°, 55.63°, 64.71°, and 77.59°. (a) Specify the first seven planes for which XRD reflections are allowed, and (b) determine the index numbers for the planes for each of the three cubic lattices.

SOLUTION

The cubic crystal structure may be a simple cubic (SC), a BCC, or an FCC.
(a) By reference to Table 2.3,

For SC, the reflection is allowed for the planes: (100), (110), (111), (200), (210), (211), and (220).
For BCC, the reflection is allowed for planes: (110), (200), (211), (220), (310), (222), and (321).
For FCC, the reflection is allowed for planes: (111), (200), (220), (311), (222), (400), and (331).
(b) By using Equation 2.14 and by reference to Table 2.4,
For SC, the index numbers for the reflecting planes are: 1, 2, 3, 4, 5, 6, and 8.
For BCC, the index numbers for the reflecting planes are: 2, 4, 6, 8, 10, 12, and 14.
For FCC, the index numbers for the reflecting planes are: 3, 4, 8, 11, 12, 16, and 19.

EXAMPLE 2.17 DETERMINATION OF CRYSTAL STRUCTURE WITH THE AID OF XRD PATTERN'S DATA

By using the data in Example 2.15, calculate the lattice parameters for the first two and seventh Bragg reflections for each of the cubic crystal structures, assuming first order reflection. Based on your calculations, determine the cubic crystal structure of the investigated material.

SOLUTION

$\lambda = 0.154$ nm; $n = 1$, $\theta = 21.66°$, $31.47°$, $39.74°$, $47.58°$, $55.63°$, $64.71°$, and $77.59°$

By reference to the data in Example 2.15,
For SC, the first two and the seventh reflecting planes are: (100), (110), and (220), respectively.

For the plane (100), $\theta = 21.66°$, $a = \dfrac{n\lambda \sqrt{h^2 + k^2 + l^2}}{2\sin\theta} =$
$\dfrac{(1)(0.154)\sqrt{1^2 + 0^2 + 0^2}}{2\sin 21.66} = 0.2086$ nm

For the plane (110), $\theta = 31.47°$, $a = \dfrac{n\lambda \sqrt{h^2 + k^2 + l^2}}{2 \sin \theta} = \dfrac{(1)(0.154)\sqrt{1^2 + 1^2 + 0^2}}{2 \sin 31.47}$

$= 0.2085$ nm

For the plane (220), $\theta = 77.59°$, $a = \dfrac{n\lambda \sqrt{h^2 + k^2 + l^2}}{2 \sin \theta}$

$= \dfrac{(1)(0.154)\sqrt{2^2 + 2^2 + 0^2}}{2 \sin 77.59} = 0.2229$ nm

For BCC, the first two and the seventh reflecting planes are: (110), (200), and (321), respectively.

For the plane (110), $\theta = 21.66°$, $a = \dfrac{n\lambda \sqrt{h^2 + k^2 + l^2}}{2 \sin \theta} = \dfrac{(1)(0.154)\sqrt{1^2 + 1 + 0^2}}{2 \sin 21.66}$

$= 0.2949$ nm

For the plane (200), $\theta = 31.47°$, $a = \dfrac{n\lambda \sqrt{h^2 + k^2 + l^2}}{2 \sin \theta} = \dfrac{(1)(0.154)\sqrt{2^2 + 0^2 + 0^2}}{2 \sin 31.47}$

$= 0.2949$ nm

For the plane (321), $\theta = 77.59°$, $a = \dfrac{n\lambda \sqrt{h^2 + k^2 + l^2}}{2 \sin \theta} = \dfrac{(1)(0.154)\sqrt{3^2 + 2^2 + 1^2}}{2 \sin 77.59}$

$= 0.2949$ nm

For FCC, the first two reflecting and the seventh reflecting planes are: (111), (200), and (331).

For the plane (111), $\theta = 21.66°$, $a = \dfrac{n\lambda \sqrt{h^2 + k^2 + l^2}}{2 \sin \theta} = \dfrac{(1)(0.154)\sqrt{1^2 + 1 + 1}}{2 \sin 21.66}$

$= 0.361$ nm

For the plane (200), $\theta = 31.47°$, $a = \dfrac{n\lambda \sqrt{h^2 + k^2 + l^2}}{2 \sin \theta} = \dfrac{(1)(0.154)\sqrt{2^2 + 0^2 + 0^2}}{2 \sin 31.47}$

$= 0.295$ nm

For the plane (331), $\theta = 77.59°$, $a = \dfrac{n\lambda \sqrt{h^2 + k^2 + l^2}}{2 \sin \theta} = \dfrac{(1)(0.154)\sqrt{3^2 + 3^2 + 1^2}}{2 \sin 77.59}$

$= 0.343$ nm

A look at the previous data indicates that the "*a*" values for FCC widely vary; this variation renders FCC to be rejected. For SC, the "*a*" values for the first two reflecting planes for SC are consistent, but the *a* value for the seventh plane deviates; this means we should also reject the SC lattice. The "*a*" values of the BCC lattice are perfectly consistent (*a* = 0.2949 nm).

Hence, the cubic crystal structure is BCC; and the lattice parameter = 0.2949 nm.

EXAMPLE 2.18 COMPUTING CRYSTALLITES SIZE USING SCHERRER'S FORMULA

An XRD pattern from a polycrystalline material showed a peak intensity at $2\theta=28.8°$. The pattern shows $2\theta\text{High}=44.6°$ and $2\theta\text{Low}=44.4°$. Calculate the crystallites size, if the wavelength of the radiation is 0.154 nm.

SOLUTION

λ = 0.154 nm, K = 0.9 (say), $2\theta=28.8°$, $\theta = 14.4°$

B = 2θHigh – 2θLow = 44.6–44.4 = 0.2° = ($\pi/180$) x 0.2 radians = 0.0035 rad.

By using Equation 2.15,

$$t=\frac{K\lambda}{B\cos\theta}=\frac{0.9\times0.154}{0.0035\times\cos14.4}=40.8\,\text{nm}$$

QUESTIONS AND PROBLEMS

2.1. Distinguish amorphous and crystalline solids with the aid of diagrams and examples.

2.2. Draw diagrams showing the seven crystal systems indicating the relationships between their edges and co-axial angles. Which two crystal systems are commonly found in metals?

2.3. Justify the following statements: (a) the coordination number of HCP structure is 12. (b) FCC and HCP lattices are closed packed structures whereas BCC is not.

2.4. Explain at least two industrial applications of crystallographic directions and planes.

2.5. Describe the X-ray diffraction technique with the aid of a diagram; and hence deduce Bragg's law.

2.6. Explain the solidification process mentioning the various stages in solidification of a metal with the aid of diagrams.

2.7. Justify the following statement: "rapid cooling results in a fine-grained microstructure whereas slow cooling results in a coarse-grained microstructure."

P2.8.   Calculate the atomic packing factor (APF) for HCP lattice justifying the values/expressions for the number of atoms per unit cell, the volume of atom, and the volume of unit cell for HCP.

P2.9.   Iron (at ambient temperature) has an atomic radius of 0.124 nm, and its atomic mass is 55.8 a.m.u. Compute the lattice parameter of the unit cell and the density of iron.

P2.10. Sketch the crystallographic planes with the Miller Indices: (101), (001), and ($1\,\bar{1}\,1$).

P2.11. Give real-life examples for the following crystallographic directions: [001] and [100].

P2.12. Sketch the following crystallographic directions: [021], [111], and [210].

P2.13. Copper (FCC) has: atomic mass = 63.5 g/mol; atomic radius = 0.1278 nm. Calculate the theoretical density of copper.

P2.14. Calculate the linear atomic density along [110] and [111] directions in an FCC unit cell.

P2.15. Calculate the following crystal structure properties for simple cubic lattice: (a) number of atoms per unit cell, (b) coordination number (CN), and (c) APF.

P2.16. In an XRD pattern for powdered metal, a high-intensity peak results at a Bragg angle of 77.59° when the Bragg diffraction condition is satisfied by the plane (321) for the first order reflection. Calculate the lattice parameter of the metal, if the wavelength of the monochromatic radiation, used in the diffractometer, is 0.1542 nm.

P2.17. An XRD pattern from a polycrystalline material showed a peak intensity at $2\theta=23.3°$. The pattern shows a peak width of 0.3°. Calculate the crystallites size, if the wavelength of the radiation is 0.154 nm.

P2.18. Determine the planar atomic density of (110) plane in a BCC unit cell.

REFERENCES

Barret, C.S. & Massalski, T. (1980) *Structure of Metals*. 3rd ed. Pergamon Press, Oxford.

Douglas, B. & Ho, S-H. (2006) *Structure and Chemistry of Crystalline Solids*. Springer-Verlag, New York.

Helliwell, J.R. (2015) *Perspectives in Crystallography*. CRC Press, Boca Raton, FL.

Langford, J.I. & Wilson, A.J.C. (1978) Scherrer after sixty years: A survey and some new results in the determination of crystallite size. *Journal of Applied Crystallography*, 11, 102–113.

Rohrer, G.S. (2001) *Structure and Bonding in Crystalline Materials*. 1st ed. Cambridge University Press, Cambridge.

Tisza, M. (2001) *Physical Metallurgy for Engineers*. ASM International, Materials Park, OH.

3 Crystal Imperfections and Deformation

3.1 REAL CRYSTALS AND CRYSTAL DEFECTS

Perfect and Real Crystals. Theoretically, crystals are defined as three-dimensional perfectly ordered arrangements of atoms or ions. However, there are defects or imperfections in real crystals. *Crystal imperfections* refer to the missing of atoms/ ions or misalignment of unit cell in an otherwise perfect crystal. Based on the inter-atomic bonding forces that exist in a metal, one can predict the elastic modulus of the metal. But, the actual elastic strength of most materials (real crystals) is far below their predicted strength values (Hertzberg, 1996). This discrepancy in the theoretical and actual strengths is due to crystal imperfections (see subsection 3.3.3). Crystal defects also influence electrical properties of solids. However, there are some beneficial effects of crystal imperfection. For example, many metal forming operations involving plastic deformation are due to crystal imperfections. Another example is an alloy (having point crystal defects); which may have higher strength as compared to pure metal.

Crystal Imperfections and Classification. Crystal imperfections or defects can be classified into the following four groups: (1) point defects, (2) line defects, (3) surface defects, and (4) volume defects. The various types of crystal defects in each group are illustrated in Figure 3.1. A *point defect* occurs at a single lattice point (e.g. a vacancy). A *line defect* occurs along a row of atoms (e.g. a *dislocation*); or as a planar defect occurring over a two-dimensional surface in the crystal. A volume defect is generally a *void* in the crystal. In general, crystal imperfections occur either as *point defects* or *line defects*; they are discussed in detail in the following sections.

3.2 POINT DEFECTS AND ALLOY FORMATION

3.2.1 WHAT IS A POINT DEFECT?

A *point defect* is a defect that exists at a specific lattice point in the crystal. In general, there are three types of point defects: (a) *vacancy*, (b) *interstitial point defect*, and (c) substitutional *point defect* (see Figure 3.2). Sometimes, there is a combination of the two types of the point defects (e.g. *Frenkel defect*). When a point defect exists, there is a *strain* in the crystal in the immediate vicinity of the point defect. The presence of point defects can be inferred through optical spectroscopy (Nordlund and Averback, 2005).

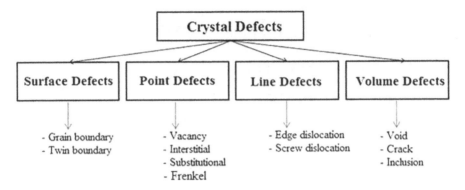

FIGURE 3.1   Classification of crystal defects.

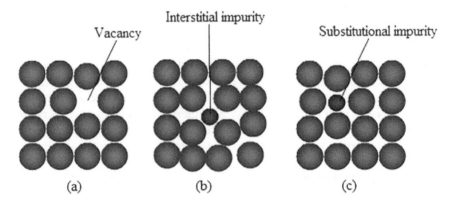

FIGURE 3.2   Point defects; (a) vacancy, (b) interstitial, (c) substitutional.

3.2.2   VACANCY

General. A missing atom in a crystal lattice is called a *vacancy* (Figure 3.2a). For most metals, one lattice site out of 10,000 will be empty at a temperature just below the melting temperature (Callister, 2007). The existence of vacancies in a crystal lattice control many industrial processes of technological importance; these processes include: semiconductor processing, diffusion in case-carburizing of steel gears, and the like.

Mathematical Modeling of Vacancy. The concentration of vacancies in a crystal is a function of temperature as well as the number of atomic sites per unit volume of the metal lattice. The number of atomic sites per cubic meter of the metal lattice (N) can be determined by:

$$N = \frac{N_A \rho}{A} \tag{3.1}$$

where N_A is the Avogadro's number; ρ is the density of the metal; and A is the atomic weight. The concentration of vacancies can be determined by the Arrhenius-type relationship as follows:

$$n_v = N \exp\left(-\frac{Q_v}{KT}\right) \tag{3.2}$$

where n_v is the number of vacancies at equilibrium per cubic meter of the lattice; N is the number of atom sites per cubic meter of the lattice; Q_v is the activation energy for vacancy formation, eV; k is Boltzmann's constant (= 8.62 x 10^{-5} eV/atom-K), and T is the temperature in Kelvin (K). Another important parameter is the *vacancy fraction*; which is defined as the ratio of the number of vacancies to the number of atom sites in the lattice (Cai and Nix, 2016). Numerically, vacancy fraction can be expressed by rewriting Equation 3.2 as follows:

$$\text{Vacancy fraction} = \frac{n_v}{N} = \exp\left(-\frac{Q_v}{kT}\right) \tag{3.3}$$

The significances of Equations 3.1–3.3 are illustrated in Examples 3.1–3.4.

3.2.3 INTERSTITIAL AND SUBSTITUTIONAL POINT DEFECTS

An *interstitial point defect* occurs when an atoms occupies the space between regular lattice sites (Figure 3.2b). A *self-interstitial* point defect occurs when an atom of the host crystal occupies an interstitial position; whereas an *impurity interstitial* refers to the point defect when a different kind of atom occupies the interstitial position. A substitutional point defect occurs when an impurity atom occupies a regular lattice site, in replacement of the host atom (Figure 3.2c). When there are a large number of impurity atoms, the material is called a solid solution (e.g. an alloy).

3.2.4 SOLID-SOLUTION ALLOYS

A solid-solution alloy is one in which the minor component is randomly dispersed in the crystal structure of the major component. There are two classes of solid solutions, (a) substitutional solid solutions and (b) interstitial solid solutions; which are explained in the following paragraphs as well as in Chapter 6 (see section 6.2).

Substitutional Solid Solutions. In a substitutional solid solution, atoms of the minor component (*solute*) are substituted for the atoms of the major component (*solvent*) on the lattice positions normally occupied by the solvent atoms (see Figure 3.2c). A typical example of substitutional solid solution is a copper-nickel (Cu-Ni) alloy.

Interstitial Solid Solutions. In an interstitial solid solution, the solute atoms occupy the interstitial positions (holes between the atoms) in the crystal lattice of the solute (Figure 3.2b). An example of interstitial solid solution is an iron-carbon (Fe-C) alloy.

3.2.5 ALLOY FORMATION—HUME-ROTHERY RULES

The formation of a solid-solution alloy is governed by certain rules. In general, there is a solubility limit i.e. a limit to the maximum amount of solute that can be added to the solvent without changing the lattice structure to a more complex form. There are only a few binary (two component) alloy systems where the solubility limit is 100%. These alloy systems are often referred to as *isomorphous* alloy systems. An example is the copper-nickel (Cu-Ni) alloy system; where both Cu and Ni have the FCC crystal structure. Besides crystal structure, there are other requirements in order to have extensive solid solubility. These conditions of extensive solid solubility are best described by Hume-Rothery Rules, as follows (Hume-Rothery, 1969):

A. *Unfavorable Atomic Size Factor. If the difference in atomic sizes of solute and host metal is more than* 15%, then the solute will have a low solubility in the host metal.

B. *Electronegativity Factor.* If a solute differs largely in electronegativity (or electro-positivity) as compared to the host metal, then the solubility of the solute in the solvent is limited, and it is more likely that a compound will be formed.

C. *Valence Factor.* A metal with a lower valence is more likely to dissolve in a host metal having a higher valence, than *vice versa.*

3.3   DISLOCATIONS

3.3.1   WHAT IS A DISLOCATION?

A dislocation is a *line defect* that occurs along a row of atoms. Dislocations cause lattice distortions centered around a line in a crystalline solid. They can be observed in crystalline materials by using transmission electron microscopy (TEM) techniques (see Figure 3.3). In reality, all crystalline materials contain some dislocations that

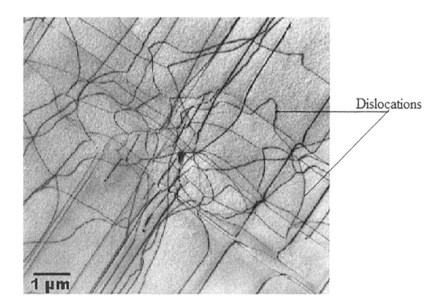

Dislocations

1 µm

FIGURE 3.3   TEM micrograph showing dislocations in a silicon crystal.

were introduced during solidification, plastic deformation, or as a consequence of thermal stresses that result from rapid cooling (Hull and Bacon, 2011). Dislocations weaken the crystal structure along a one-dimensional space. An important parameter in dislocation is *dislocation density*; which is defined as the number of dislocations in a unit volume of a crystalline material (see Chapter 9, Section 9.4). The dislocation density can be measured by using the X-ray diffraction technique (Shintani and Murata, 2011).

3.3.2 EDGE DISLOCATION AND SCREW DISLOCATION

When a crystal is subjected to shear stress, there are atomic displacements due to dislocation motion. The direction and amount of the atomic displacement caused by the dislocation is represented by a vector—called *Burgers vector* (see Figure 3.4). A dislocation in a crystalline solid may be one of the following three types: (a) edge dislocation, (b) screw dislocations, or (c) mixed dislocation.

An *edge dislocation* is formed in a crystal by the insertion of an extra half-plane of atoms (Figure 3.4a). An *edge* dislocation has its Burgers vector ($\vec{b}$) perpendicular to the dislocation line; which runs along the bottom of the extra half-plane (Figure 3.4a). When a crystal is subjected to a shear stress, an edge dislocation moves in the direction of the Burgers vector. A *screw dislocation* has the Burgers vector ($\vec{b}$) parallel to the dislocation line (Figure 3.4b). When a crystal is subjected to a shear stress, a screw dislocation moves in a direction perpendicular to the Burgers vector. A combination of edge and screw dislocations results in *mixed dislocation*.

3.3.3 DISLOCATION MOVEMENT—DEFORMATION BY SLIP

In section 3.1, we learned that the actual strength of a real crystal is much lower than that of a perfect one. This discrepancy can be explained on the basis of dislocation motion. Plastic deformation of a metal involves the motion of a large number of dislocations; which are line defects. In the presence of a dislocation, a much lower shear stress is required to move the dislocation to produce plastic deformation. This

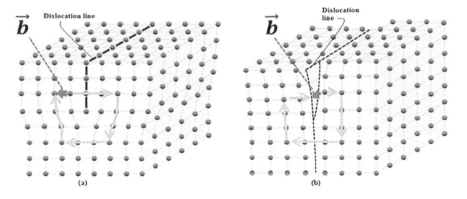

FIGURE 3.4 Two types of dislocations; (a) edge dislocation, (b) screw dislocation.

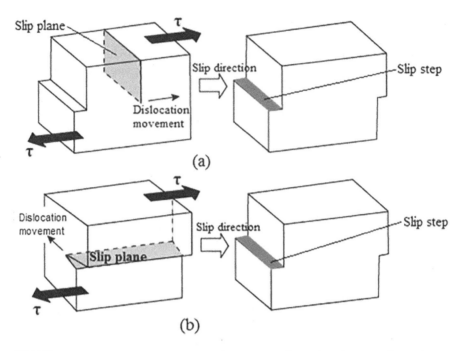

FIGURE 3.5   Plastic deformation by slip with: (a) edge dislocation, (b) screw dislocation.

is why the actual strength of a real crystal is much lower than that of a perfect crystal (having no defect).

When a shear stress (τ) is applied to a crystalline solid, individual atoms move in a direction parallel to Burgers vector (slip direction) (see Figure 3.4 and Figure 3.5). When the shear force is increased, the atoms will continue to slip to the right (see Figure 3.5). A row of the atoms find their way back into a proper spot in the lattice; and another row of the atoms will slip out of position forming a slip step (Figure 3.5). The process by which plastic deformation results by dislocation movements, is called *slip*; which occurs along slip planes. A *slip plane* is the crystallographic plane along which the dislocation line traverses under the action of a shear stress. In an edge dislocation, the dislocation moves (in its slip plane) along the slip direction (Figure 3.5a). On the other hand, in screw dislocation, the dislocation moves (in its slip plane) in a direction perpendicular to the slip direction (Figure 3.5b).

3.4   SLIP SYSTEMS

We have learned in the preceding section that plastic deformation of a solid results by *slip*; which occurs across *slip planes*. A *slip plane* is a crystallographic plane with the greatest atomic packing (see Figure 3.5). The slip across a slip plane occurs in a specified direction, called *slip direction;* which is the direction of the greatest atomic density. A combination of *slip planes* and *slip directions* in a crystal form a *slip system*. A slip system for a BCC unit cell is shown in Figure 3.6.

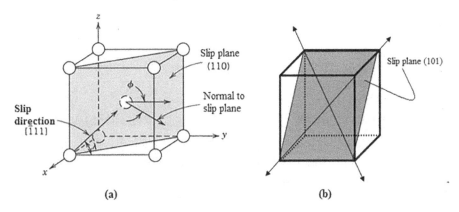

FIGURE 3.6 Slip system in BCC lattice; (a) slip plane (110) and slip direction $[\bar{1}11]$, (b) Slip plane (101) with family of slip directions $<\bar{1}11>$.

TABLE 3.1
Slip Systems in FCC, BCC, and HCP Crystals

Crystal	Slip plane	Slip direction	No. of slip planes	No. of slip directions per plane	No. of slip systems	Slip system
FCC	{111}	<110>	4	3	12	{111}<111>
BCC	{110}	<111>	6	2	12	{110}<111>
HCP	{0001}	[1000]	1	3	3	{0001}[1000]

If we compare (100) and (110) planes in a BCC unit cell, the former contains four atoms whereas the latter contains five atoms; it means that (110) is the plane of the greatest atomic packing (slip plane). It can be shown from Figure 3.6a that the BCC unit cell has six slip planes: (110), (101), (011), ($\bar{1}\bar{1}0$), ($\bar{1}0\bar{1}$), and (0$\bar{1}\bar{1}$); these slip planes may be represented by a family of slip planes: {110}. Each slip plane in the BCC unit cell has two slip directions (Figure 3.6b); the family of BCC slip directions can be represented by <111> (Kelly and Knowels, 2012; Jackson, 1991). For a crystal structure, the number of slip systems can be determined by:

Number of slip systems = Number of slip planes x Number of slip directions (3.4)

Thus, for BCC, there are 12 slip systems (6 x 2 = 12); these 12 slip systems can be represented by {110}<111>. Similarly, we can determine slip systems for FCC and HCP crystals. The slip systems for FCC, BCC, and HCP crystal structures are presented in Table 3.1.

The data in Table 3.1 indicates that both FCC and BCC metals have 12 slip systems; which is much higher as compared to the slip systems in HCP metals. This crystallographic feature explains why metals that crystallize in FCC and BCC systems have excellent formability whereas HCP metals have poor formability. This mechanical behavior can be observed in industrial practice, as follows. Mild steel

(BCC) and aluminum (FCC) can be readily cold worked (plastically deformed at room temperature) to desired shapes. But, zinc (HCP) is difficult to be cold worked due to lesser number of available slip systems: three slip systems (see Table 3.1). Owing to the poor formability, zinc components are usually manufactured by casting (die-casting) techniques.

3.5   DEFORMATION IN SINGLE CRYSTALS—SCHMID'S LAW

It is explained is subsection 3.3.3 that plastic deformation by slip occurs as a result of dislocation motion across slip planes when shear stresses are applied along a slip plane in a slip direction. Even if an applied stress is pure tensile (or compressive), there exist shear components at parallel or perpendicular alignments to the applied stress direction. These shear components are called *resolved shear stresses*.

It is interesting to determine how the applied stress is resolved onto the slip system. Consider a load, F, is applied along the tensile axis of a single crystal (Figure 3.7). This force per unit area is the tensile stress, σ. It is possible to resolve the stress into shear stress (along the slip plane) and normal to the slip plane, as shown in Figure 3.7.

In order to derive an expression for the resolved shear stress, we first compute the tensile and shear forces with reference to Figure 3.7. The tensile force (F) can be related to tensile stress (σ) by:

$$F = \sigma A_0 \tag{3.5}$$

By considering the shear force (F_s) in the slip direction, we can write:

$$F_s = F \cos \lambda \tag{3.6}$$

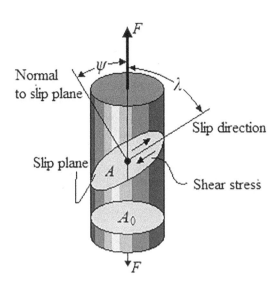

FIGURE 3.7   Resolution of tensile stress ($\sigma = F/A_0$) into shear stress ($\tau_r = F_s/A$) under tensile loading.

where F is the tensile force, and λ is the angle between the tensile axis and the slip direction.

By combining Equations 3.5 and 3.6, we obtain:

$$F_s = \sigma A_0 \cos \lambda \tag{3.7}$$

By considering the angle between the tensile axis and normal to the slip plane (ψ), we obtain:

$$A_0 = A \cos \psi \tag{3.8}$$

By combining Equations 3.7 and 3.8, we obtain:

$$F_s = \sigma A (\cos \psi)(\cos \lambda)$$
$$\frac{F_s}{A} = \sigma (\cos \psi)(\cos \lambda)$$

But the shear force divided by slip-plane area (F_s / A) is the resolved shear stress (τ_r). Thus,

$$\tau_r = \sigma (\cos \psi)(\cos \lambda) \tag{3.9}$$

where σ is the tensile stress, ψ is the angle between the tensile axis and normal to slip plane; and λ is the angle between the tensile axis and the slip direction (see Examples 3.5–3.6).

In general, for a cubic system, the angle (θ) between any two directions $\vec{a}$ and $\vec{b}$ represented by $[u_1 v_1 w_1]$ and $[u_2 v_2 w_2]$ can be obtained from the scalar product of the two vectors, as follows:

$$\theta = \cos^{-1} \frac{a\,b}{|a||b|^\infty} = \cos^{-1} \frac{u_1 u_2 + v_1 v_2 + w_1 w_2}{\sqrt{u_1^2 + v_1^2 + w_1^2}\sqrt{u_2^2 + v_2^2 + w_2^2}} \tag{3.10}$$

The significance of Equation 3.10 is explained in Example 3.6.

A single crystal of a metal has a number of different slip systems that are capable to cause plastic deformation. The resolved shear stress is normally different for each slip system; however, there is one slip system that is the most favorably oriented to initiate slip. The minimum shear stress required to initiate slip in a single crystal under an applied tensile or compressive stress is called the *critical resolved shear stress*, τ_{crss}. The *critical resolved shear stress* refers to the maximum value of the resolved shear stress to result in plastic deformation of single crystal by yielding. Hence, Equation 3.9 can be rewritten as:

$$\tau_{crss} = \sigma_y [(\cos \psi)(\cos \lambda)]_{max} \tag{3.11}$$

where σ_{ys} is the yield strength of the single crystal. Equation 3.11 is known as **Schmid's law**. The usefulness of Schmid's law is illustrated in Examples 3.7–3.8.

It has been experimentally shown that for single crystals of several metals, the critical resolved shear stress (τ_{crss}) is related to the dislocation density (ρ_D) by (Callister, 2007):

$$\tau_{crss} = \tau_0 + A\sqrt{\rho_D} \qquad (3.12)$$

where τ_0 and A are constants (see Example 3.9).

3.6 PLASTIC DEFORMATION—COLD WORKING/ROLLING

All metal-forming manufacturing processes involve plastic deformation. In the early stages of plastic deformation, slip is essentially on primary glide (slip) planes. As deformation proceeds, cross slip takes place. The cold-worked structure forms high dislocation density regions that soon develop into networks. The grain size decreases with strain at low deformation but soon reaches a fixed size. This is why cold working will decrease ductility (see Chapter 7).

In industrial practice, most metals are worked at ambient temperature (cold worked) by cold rolling. Rolling is a process by which the metal stock is introduced between rollers and then compressed and squeezed. Figure 3.8 illustrates *flat rolling*

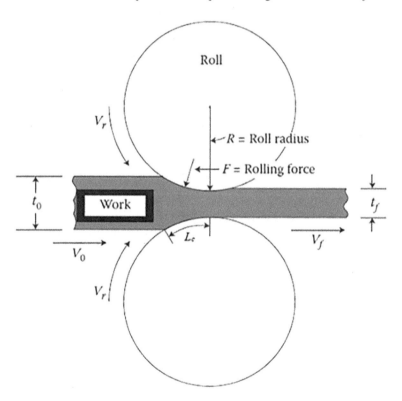

FIGURE 3.8 Flat rolling of metallic work (V_r = roll speed, L_c is contact length, V_0 = work speed before rolling; V_f is work speed after rolling).

i.e. rolling of a metal strip with rectangular cross sections in which the width is greater than the thickness. The strip is reduced in thickness from t_0 to t_f with width of the strip assumed to be constant during rolling (Figure 3.8).

True strain (ε), in the work (strip) is given by:

$$\varepsilon = In\left(\frac{t_0}{t_f}\right) \tag{3.13}$$

where t_0 is the original thickness, mm; and t_f is the final thickness, mm (see Example 3.10).

The amount of thickness reduction is expressed as draft, d, as follows:

$$d = t_0 - t_f \tag{3.14}$$

The amount of deformation in rolling is measured by percent reduction, as follows:

$$\% \text{ reduction} = \frac{d}{t_0} \times 100 \tag{3.15}$$

where d is the draft, mm; and t_0 is the original thickness of the strip, mm (see Example 3.11).

In general, the amount of deformation is expressed as percent cold work (% CW) in terms of the cross-sectional areas before and after cold work (see Chapter 6, section 6.4).

3.7 CALCULATIONS—EXAMPLES ON CRYSTAL IMPERFECTIONS AND DEFORMATION

EXAMPLE 3.1 COMPUTING THE NUMBER OF ATOMIC SITES PER UNIT VOLUME IN A LATTICE

The atomic weight and density (at 500°C) for aluminum is 27 g/mol and 2.62 g/cm³, respectively. Calculate the number of atomic sites per cubic meter of the aluminum lattice.

SOLUTION

$N_A = 6.02 \times 10^{23}$; $\rho = 2.62$ g/cm³; $A = 27$ g/mol, $N = ?$

By using Equation 3.1,

$$N = \frac{N_A \rho}{A} = \frac{6.02 \times 10^{23} \times 2.62}{27} = 5.84 \times 10^{22} \text{ atoms/cm}^3$$

$N = 5.84 \times 10^{22}$ atoms/cm³ x 10^6 cm³/m³ = 5.84×10^{28} atoms/m³

The number of atomic sites per unit volume of Al lattice at 500°C = 5.84 x 10^{28} atoms/m^3.

EXAMPLE 3.2 COMPUTING THE EQUILIBRIUM VACANCIES CONCENTRATION PER UNIT VOLUME

By using the data in Example 3.1, calculate the equilibrium number of vacancies per cubic meter for aluminum at 500°C. The activation energy for vacancy formation in aluminum at 500°C is 0.75 eV/atom.

SOLUTION

N = 5.84 x 10^{28} atoms/m^3, Q_v = 0.75 eV/atom, k = 8.62 x10^{-5} eV/atom-K, T = 500°C = 773 K

By using Equation 3.2,

$$n_v = N\exp(-\frac{Q_v}{kT}) = 5.84\times10^{28}\exp(-\frac{0.75}{8.62\times773\times10^{-5}})$$
$$= 5.84\times10^{28}(e^{-11.25}) = 7.59\times10^{23}$$

The equilibrium number of vacancies per unit volume = 7.59 x 10^{23} vacancies/m^3.

EXAMPLE 3.3 CALCULATING THE VACANCY FRACTION IN A CRYSTAL LATTICE

The equilibrium concentration of vacancies for copper at 1,000°C is 2.2 x 10^{25} vacancies/m^3. The atomic weight and density (at 1,000°C) for copper are 63.5 g/mol and 8.4 g/cm^3, respectively. Compute the vacancy fraction for copper at 1,000°C. Give a meaning to your answer.

SOLUTION

By using Equation 3.1,

$$N = \frac{N_A\rho}{A} = \frac{6.02\times10^{23}\times8.4}{63.5} = 7.96\times10^{22} \text{ atom/cm}^3 / m^3$$
$$= 7.96\times10^{28} \text{ atom/m}^3$$

By using Equation 3.3,

$$\text{Vacancy fraction} = \frac{n_v}{N} = \frac{2.2\times10^{25}}{7.96\times10^{28}} = 2.76\times10^{-4}$$

It means that there are around three vacancies in each 10,000 atomic sites.

EXAMPLE 3.4 CALCULATING THE ACTIVATION ENERGY FOR VACANCY FORMATION

By using the data in Example 3.3, calculate the activation energy for vacancy formation in copper at 1,000°C.

SOLUTION

N = 7.96 x 10^{28} atoms/m^3, n_v = 2.2 x 10^{25} vacancies/m^3, T = 1,000°C = 1,273 K
By taking natural logarithm of both sides of Equation 3.2,

$$In\, n_v = In\, N - \frac{Q_v}{kT}$$

$$Q_v = -kT\, In\frac{n_v}{N} = -8.62\times10^{-5}\times1273\, In\frac{2.2\times10^{25}}{7.96\times10^{28}}$$
$$= -10973.2\times10^{-5}\,(-8.194) = 0.899$$

The activation energy for vacancy formation for copper at 1,000°C = 0.9 eV/atom.

EXAMPLE 3.5 CALCULATING THE RESOLVED SHEAR STRESS WHEN THE TENSILE STRESS IS KNOWN

A tensile stress of 1.5 MPa is applied to a cubic single crystal. The angle between the tensile axis and slip direction is 49°, and the angle between the tensile axis and normal to slip plane is 57.6°. Calculate the resolved shear stress.

SOLUTION

σ = 1.5 MPa, λ = 49°, ψ = 57.6°, τ_r = ?
By using Equation 3.9,

$\tau_r = \sigma(\cos \psi)(\cos \lambda)$ = 1.5 (cos 57.6°) (cos 49°) = 1.5 x 0.536 x 0.656 = 0.5273 MPa

The resolved shear stress = 527.3 kPa.

EXAMPLE 3.6 CALCULATING THE RESOLVED SHEAR STRESS

A tensile stress of 50 MPa is applied to a BCC iron single crystal such that the direction of the tensile axis is [010]. The slip plane and the slip direction are as shown in Figure 3.6a. Calculate the resolved shear stress.

SOLUTION

σ = 50 MPa, Directions for λ: [010] and [$\bar{1}$11], Directions for ψ: [010] and [110]
For determining the angle λ, [$u_1 v_1 w_1$] = [010], and [$u_2 v_2 w_2$] = [$\bar{1}$11]

By using Equation 3.10 taking $\theta = \lambda$,

$$\lambda = \cos^{-1} \frac{u_1 u_2 + v_1 v_2 + w_1 w_2}{\left|\sqrt{u_1^2 + v_1^2 + w_1^2}\right|\left|\sqrt{u_2^2 + v_2^2 + w_2^2}\right|}$$

$$= \cos^{-1} \frac{(0)(-1) + (1)(1) + (0)(1)}{\left|\sqrt{0^2 + 1^2 + 0^2}\right|\left|\sqrt{(-1)^2 + 1^2 + 1^2}\right|} = 54.7°$$

For determining the angle ψ, $[u_1 v_1 w_1] = [010]$, and $[u_2 v_2 w_2] = [110]$
By using Equation 3.10 taking $\theta = \psi$,

$$\psi = \cos^{-1} \frac{u_1 u_2 + v_1 v_2 + w_1 w_2}{\left|\sqrt{u_1^2 + v_1^2 + w_1^2}\right|\left|\sqrt{u_2^2 + v_2^2 + w_2^2}\right|}$$

$$= \cos^{-1} \frac{(0)(1) + (1)(1) + (0)(0)}{\left|\sqrt{0^2 + 1^2 + 0^2}\right|\left|\sqrt{1^2 + 1^2 + 0^2}\right|} = 45°$$

By using Equation 3.9,
$\tau_r = \sigma (\cos \psi)(\cos \lambda) = (50)(\cos 45°)(\cos 54.7°) = 50 \times 0.707 \times 0.578 = 20.4$
The resolved shear stress = 20.4 MPa.

EXAMPLE 3.7 CALCULATING THE YIELD STRENGTH OF A SINGLE-CRYSTAL WHEN τ_{CRSS} IS KNOWN

The *critical resolved shear stress* for BCC iron single crystal is 30 MPa. The angle between the tensile axis and normal to slip plane is 54.7°, and the angle between the tensile axis and the slip direction is 45°. Calculate the yield strength of BCC iron single crystal.

SOLUTION

$\tau_{crss} = 30$ MPa, $\psi = 54.7°$, $\lambda = 45°$, $\sigma_y = ?$
By rearranging the terms in Equation 3.11,

$$\sigma_{ys} = \frac{\tau_{crss}}{(\cos \psi)(\cos \lambda)} = \frac{\tau_{crss}}{(\cos 54.7)(\cos 45)} = \frac{30}{0.578 \times 0.707} = 73.41$$

The yield strength of BCC iron single crystal = 73.41 MPa.

EXAMPLE 3.8 COMPUTING THE YIELD STRENGTH WHEN THE ANGLES ARE UNKNOWN

The critical resolved shear stress for an FCC crystal is 0.5 MPa. The tensile axis direction is [010], and the slip direction is [$\bar{1}$10]. Compute the yield strength of the crystal if yielding occurs across the slip plane (111).

SOLUTION

τ_{crss} = 0.5 MPa, Directions for λ: [010] and $[\bar{1}10]$, Directions for ψ: [010] and
  [111]

For determining the angle λ, $[u_1 v_1 w_1] = [010]$, and $[u_2 v_2 w_2] = [\bar{1}10]$
By using Equation 3.10 taking θ = λ,

$$\lambda = \cos^{-1} \frac{u_1 u_2 + v_1 v_2 + w_1 w_2}{\left|\sqrt{u_1^2 + v_1^2 + w_1^2}\right|\left|\sqrt{u_2^2 + v_2^2 + w_2^2}\right|}$$

$$= \cos^{-1} \frac{(0)(-1)+(1)(1)+(0)(0)}{\left|\sqrt{0^2 + 1^2 + 0^2}\right|\left|\sqrt{(-1)^2 + 1^2 + 0^2}\right|} = 45°$$

For determining the angle ψ, $[u_1 v_1 w_1] = [010]$, and $[u_2 v_2 w_2] = [111]$
By using Equation 3.10 taking θ = ψ,

$$\psi = \cos^{-1} \frac{u_1 u_2 + v_1 v_2 + w_1 w_2}{\left|\sqrt{u_1^2 + v_1^2 + w_1^2}\right|\left|\sqrt{u_2^2 + v_2^2 + w_2^2}\right|}$$

$$= \cos^{-1} \frac{(0)(1)+(1)(1)+(0)(1)}{\left|\sqrt{0^2 + 1^2 + 0^2}\right|\left|\sqrt{1^2 + 1^2 + 1^2}\right|} = 54.7°$$

By rearranging the terms in Equation 3.11,

$$\sigma_{ys} = \frac{\tau_{crss}}{(\cos\psi)(\cos\lambda)} = \frac{\tau_{crss}}{(\cos 54.7)(\cos 45)} = \frac{0.5}{0.578 \times 0.707} = 1.22$$

The yield strength of the FCC crystal =1.22 MPa.

EXAMPLE 3.9 CALCULATING THE τ_{CRSS} WHEN THE DISLOCATION DENSITY IS KNOWN

The value of τ_0 for single-crystal copper is 0.069 MPa. Also, for copper, the critical resolved shear stress is 2.10 MPa at a dislocation density of 10^5 mm^{-2}. Calculate the values of (a) A, and (b) the critical resolved shear stress for copper at a dislocation density of 10^4 mm^{-2}.

SOLUTION

(a)   Taking τ_0=0.069 MPa, τ_{crss} = 2.10 MPa, and ρ_D = 10^5 mm^{-2} in Equation 3.12,

$$\tau_{crss} = \tau_0 + A\sqrt{\rho_D}$$
$$2.10 = 0.069 + A\sqrt{10^5}$$
$$A = 6.35 \times 10^{-3} \text{ MPa-mm}$$

(b)   By using Equation 3.12,

$$\tau_{crss} = \tau_0 + A\sqrt{\rho_D} = 0.069 + (6.35 \times 10^{-3})\sqrt{10^4} = 0.069 + 0.635 = 0.7$$

At a dislocation density of 10^4 mm^{-2} in copper, the critical resolved shear stress = 0.7 MPa.

EXAMPLE 3.10 COMPUTING THE TRUE STRAIN DUE TO COLD ROLLING

A 45-mm thick aluminum plate was rolled to 38-mm thickness and then again rolled down to 30-mm thickness. Calculate the true strain in the metal.

SOLUTION

t_0 = 45 mm, t_f = 30 mm, ε = ?
By using Equation 3.13,

$$\varepsilon = In\left(\frac{t_0}{t_f}\right) = In\left(\frac{45}{30}\right) = In\,1.5 = 0.405$$

True strain in the metal = 0.405.

EXAMPLE 3.11 COMPUTING THE DRAFT AND % REDUCTION IN ROLLING

By using the data in Example 3.10, calculate: (a) the draft, and (b) % reduction.

SOLUTION

(a) By using Equation 3.14,

$d = t_0 - t_f = 45 - 30 = 15$ mm

Draft = 15 mm
(b) By using Equation 3.15,

$$\%\ \text{reduction} = \frac{d}{t_0} \times 100 = \frac{15}{45} \times 100 = 33$$

QUESTIONS AND PROBLEMS

3.1. Why is the (actual) strength of a real crystalline material much lower than the theoretically predicted strength of otherwise perfect crystal?
3.2. Draw a classification chart of crystal imperfections. Which defect is the cause of plastic deformation in a metal?
3.3. Sketch the various point defects; and explain one of them.

3.4.   Define the terms *dislocation* and *Burgers vector*. Draw sketches for edge dislocation and screw dislocation indicating Burgers vector for each type of dislocation.

3.5.   Explain deformation by slip with the aid of diagrams.

3.6.   (a) Define the following terms: (i) slip plane, (ii) slip direction, (iii) slip system. (b) Draw a sketch showing slip plane and slip directions in an FCC unit cell.

3.7.   Why do FCC and BCC metals have better formability as compared to HCP metals?

3.8.   Differentiate between the terms: resolved shear stress and the critical resolved shear stress.

P3.9. A tensile stress of 5 kPa is applied in a cubic crystal such that the tensile axis direction is [432]. Calculate the resolved shear stress on the slip plane (11 $\bar{1}$) in the slip direction: [011].

P3.10. The equilibrium number of vacancies in silver at 800°C is 3.6 x 10^{23} m^{-3}. The atomic weight and density (at 800°C) of silver are 107.9 g/mol and 9.5 g/cm^3, respectively. Calculate the activation energy for vacancy formation at 800°C for silver.

P3.11. By using the data in P3.10, compute the vacancy fraction for silver at 800°C. Give a meaning to your answer.

P3.12. By using the data in Example 3.9, calculate the critical resolved shear stress for a single crystal of copper at a dislocation density of 10^6 mm^{-2}.

P3.13. The critical resolved shear stress for zinc is 0.91 MPa. The angle between the tensile axis and the normal to the slip plane is 65°. Three possible slip directions make angles of 78°, 48°, and 30° with the tensile axis. Calculate the yield strength of zinc.

P3.14. A 40-mm-thick aluminum plate was rolled to 30-mm thickness and then again rolled down to 15-mm thickness. Calculate the true strain in the metal.

P3.15. By using the data in P3.14, calculate: (a) the draft, and (b) % reduction.

P3.16. A tensile stress of 0.8 MPa is applied to an FCC metal single crystal such that the tensile axis direction is [010], and the slip direction is [1 $\bar{1}$ 0] . The slip plane is (111). Calculate the resolved shear stress.

REFERENCES

Cai, W. & Nix, W.D. (2016) *Imperfections in Crystalline Solids*. Cambridge University Press, Cambridge.

Callister, W.D. (2007) *Materials Science and Engineering: An Introduction*. John Wiley & Sons Inc, Hoboken, NJ.

Hertzberg, R.W. (1996) *Deformation and Fracture Mechanics of Engineering Materials*. John Wiley & Sons Inc, Hoboken, NJ.

Hull, D. & Bacon, D.J. (2011) *Introduction to Dislocations*. 5th ed. Elsevier Science Publications Inc, New York.

Hume-Rothery, W. (1969) *Atomic Theory for Students of Metallurgy (fifth reprint)*. The Institute of Metals, London.

Jackson, A.G. (1991) *Handbook of Crystallography*. Springer, Inc, Verlag, New York.

Kelly, J. & Knowels, K.M. (2012) *Crystallography and Crystal Defects*. John Wiley & Sons, Inc, Hoboken, NJ.

Nordlund, K. & Averback, R. (2005) Point defects in metals. In: Yip, S. (eds) *Handbook of Materials Modeling*. Springer, Dordrecht.

Shintani, T. & Murata, Y. (2011) Evaluation of the dislocation density and dislocation character in cold rolled type 304 steel determined by profile analysis of X-ray diffraction. *Acta Materiala*, 59(11), 4314–4322.

4 Diffusion and Applications

4.1 INTRODUCTION TO DIFFUSION AND ITS APPLICATIONS

4.1.1 WHAT IS DIFFUSION?

Diffusion is a mass transfer phenomenon that involves the movement of one atomic specie into another. A thorough understanding of diffusion in materials is crucial for materials development in engineering (Memrer, 2007). Although diffusion occurs in all three states (gaseous, liquid, and solid), this chapter/text focuses on the solid-state diffusion. *Diffusion* is essentially a solid-state process where dissimilar materials attempt to achieve equilibrium as a result of a driving force due to a concentration gradient. It means that diffusion involves the movement of particles in a solid/metal from a high-concentration region to a low-concentration region, resulting in the uniform distribution of the substance (see Figure 4.1). This movement/exchange of atomic species occurs because of the point defects in metallic solids (Pichler, 2004). In diffusion, atomic vacancies and other small-scale point defects allow atoms to exchange places. In order for the atoms to have sufficient energy for exchange of positions, high temperatures are required.

4.1.2 INDUSTRIAL APPLICATIONS OF DIFFUSION

Solid-state diffusion is widely applied in microelectronics, biomedical, and other engineering sectors. In modern engineering applications, materials are often chosen for products and/or processes that are carried out at high temperatures. An example of industrial process involving diffusion is the carburization of steels for case hardening of automotive components (e.g. gears, crank-shafts, etc.). In carburizing, carbon is added to the steel surface by exposing the steel part to a carbon-rich atmosphere at a temperature in the range of 900–1,100°C; thereby allowing carbon atoms to diffuse into the steel (see Chapter 16, section 16.1). Another example of diffusion is the thermal barrier coating (TBC) of gas-turbine (GT) blades that operate at a temperature as high as 1,500°C (Huda, 2017). In TBC, a GT blade is often coated by depositing (*Al*) on the surface of the Ni-base superalloy blade via dissociation of AlF_3 gas.

The controlled diffusion of dopants (beneficial impurities) into silicon (Si) wafers is a technologically important process in the fabrication of microelectronic/integrated circuits (ICs) (Huda and Bulpett, 2012). The process of junction formation at the transition from *p* to *n*-type semiconducting material (or vice versa) is typically accomplished by diffusing the appropriate dopant impurities (e.g. phosphorus P, boron

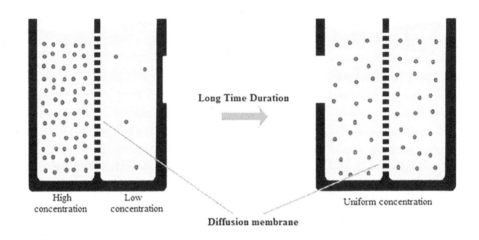

High   Low
concentration   concentration

Diffusion membrane

Long Time Duration

Uniform concentration

FIGURE 4.1   Diffusion in a solid due to concentration gradient.

B, etc.), into the intrinsic silicon (Si) wafer. The performance of such ICs strongly depends on the spatial distribution of the impurity atoms in the diffusion zone (Sedra and Smith, 2014). A notable biomedical application of diffusion is in a heart-lung machine that enables surgeons to operate on a human heart. Here, the controlled diffusions of oxygen (O) and carbon dioxide (CO_2) through a Si-based rubber membrane play an important role in the functioning of the machine (Schaffer *et al.*, 1999).

4.2   FACTORS AFFECTING RATE OF DIFFUSION

In an industrial process involving diffusion, the extent of diffusion per unit time (rate of diffusion) is technologically important. The rate of diffusion depends on four factors: (1) temperature, (2) concentration gradient, (3) diffusion distance, and (4) diffusing and host materials. The effects of each factor is briefly explained in the following paragraphs.

Temperature. Temperature has the most pronounced effect on the rate of diffusion. Increasing temperature results in an increase of diffusion rate by adding energy to the diffusing atoms. An increase of temperature results in atomic vibrations with higher amplitude thereby assisting diffusion i.e. speeding up diffusion (see section 4.7).

Concentration Gradient. The rate of diffusion strongly depends on the difference between the concentrations across the diffusion membrane (see Figure 4.1). For example, diffusion through a membrane will occur rapidly if there is a high concentration of a gas on one side and none or very low concentration of the gas on the other side of the membrane (see section 4.4).

Diffusion Distance. The rate of diffusion varies inversely as the distance through which atoms are diffusing. This is why a gas diffuses through a thin wall (membrane) much faster than it would diffuse through a thick wall.

Diffusing and Host Material. The diffusion kinetics also depends on both the diffusing material as well as the host materials. Diffusing materials made up of small-sized atoms (e.g. H, C, N, O) move faster than diffusing materials made up of larger-size atoms (e.g. Cu, Fe).

4.3 MECHANISMS AND TYPES OF DIFFUSION

4.3.1 Diffusion Mechanisms

General. Diffusion involves step-wise migration of atoms from one lattice site to another site. During diffusion, atoms are in constant motion by rapidly changing their positions in a solid material. In order for an atom to diffuse, two conditions must be satisfied: (a) there must be a vacant adjacent site, and (b) the atom must have sufficient vibrational energy to break bonds with its neighbor atoms thereby causing lattice distortion. There are two basic diffusion mechanisms in solids: (1) vacancy diffusion, and (2) interstitial diffusion.

Vacancy Diffusion. Vacancy diffusion occurs by the motion of vacancies in a lattice (see Figure 4.2). In Figure 4.2, the diffusion of a particular lattice atom by a vacancy mechanism seems to be the movements of vacancies, but it is something different. Vacancy diffusion is actually the interchange of an atom from a normal lattice position to an adjacent vacant lattice site (vacancy).

Interstitial Diffusion. Interstitial diffusion involves jumping of atoms from one interstitial site to another without permanently displacing any other atoms in the crystal lattice. Interstitial diffusion occurs when the size of the diffusing (interstitial) atoms are very small as compared to the matrix atoms' size (see Figure 4.3). For example, in wrought iron, α-ferrite solid solution is formed when small-sized carbon atoms diffuse into BCC lattice of larger-sized iron atoms.

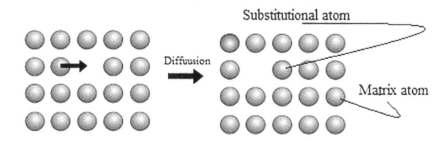

FIGURE 4.2 Vacancy diffusion.

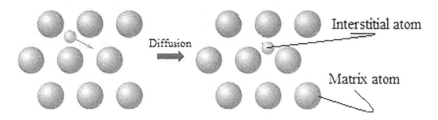

FIGURE 4.3 Interstitial diffusion.

4.3.2 TYPES OF DIFFUSION

In general, there are two types of diffusion: (a) steady-state diffusion, and (b) non-steady state diffusion. The main difference between steady-state diffusion and non-steady state diffusion is that **steady-state diffusion occurs at a constant rate whereas the rate of non-steady state diffusion is a function of time**. Both of these types can be quantitatively explained by Fick's laws; which were developed by Adolf Fick in the nineteenth century.

4.4 STEADY-STATE DIFFUSION—FICK'S FIRST LAW

Steady-State Diffusion. In steady-state diffusion, it is important to know how fast diffusion occurs. The rate of diffusion or the rate of mass transfer is generally expressed as *diffusion flux*. The *diffusion flux (J)* may be defined as the rate of mass transfer through and perpendicular to a unit cross-sectional area of solid (see Figure 4.4a). Numerically,

$$J = \frac{1}{A}\frac{dM}{dt} \qquad (4.1)$$

where J is the diffusion flux, kg/m^2-s or atoms/m^2-s; A is the area of the solid through which diffusion is occurring, m^2; and dM is the mass transferring or diffusing through the area A in time-duration dt, kg or atoms (see Example 4.1).

The concentration of a diffusive specie y (in atoms per unit volume) in a host metal (X) can be computed by:

$$C_{atom} = \frac{(wt\% \, y)(\rho_X)(N_A)}{A_y} \qquad (4.2)$$

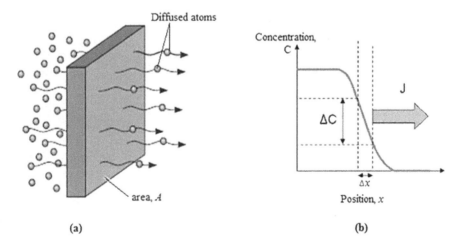

FIGURE 4.4 (a) Steady-state diffusion across a sheet, (b) variation of C with x for the diffusion situation in (a).

where C_{atom} is the concentration of the diffusive specie y, atoms/cm^3; ρ_x is the density of the host metal X, g/cm^3; A_y is the atomic weight of the diffusive specie y, g/mol; and N_A is the Avogadro's number (see Example 4.2).

Fick's First Law of Diffusion. *Fick's First Law* states that "diffusion flux is proportional to the concentration gradient." Mathematically, *Fick's First Law of diffusion* can be expressed as:

$$J = -D\frac{dC}{dx} \qquad (4.3)$$

where the constant of proportionality D is the diffusion coefficient, m^2/s; and dC/dt is the concentration gradient i.e. the rate of concentration C with respect to the position x (see Figure 4.4b). The diffusion coefficient, D, indicates the ease with which a specie can diffuse in some medium (see Example 4.3). The negative sign in Equation 4.3 indicates that the direction of diffusion J is opposite to the concentration gradient (from high to low concentration).

Applications of Fick's First Law. Diffusion in hydrogen storage tanks plays an important role. Hydrogen (H_2) gas can be stored in a steel tank at some initial pressure and temperature. The hydrogen concentration on the steel surface may vary with pressure. If the density of steel, and the diffusion coefficient of hydrogen in steel at the specified temperature are known, the rate of pressure drop as a result of diffusion of hydrogen through the wall can be computed by the application of Fick's First Law. Another example of steady-state diffusion is found in the purification of hydrogen gas. Here, a thin sheet of palladium (Pd) metal is used as a diffusion membrane in the purification vessel such that one side of the sheet is exposed to the impure gas comprising H_2 and other gases. When the pressure on the other side of the vessel is reduced, the hydrogen selectively diffuses through the Pd sheet to the other side giving pure hydrogen gas.

4.5 NON-STEADY STATE DIFFUSION—FICK'S SECOND LAW

Non-steady state diffusion applies to many circumstances in metallurgy. During non-steady state diffusion, the concentration C is a function of both time t and position x. The equation that describes the one-dimensional non-steady state diffusion is known as Fick's Second Law. Fick's Second Law is based on the principle that "an increase in the concentration across a cross-section of unit area with time is equal to the difference of the diffusion flux entering the volume (J_{in}) and the flux exiting the volume (J_{out}) (see Figure 4.5)." Mathematically,

$$\frac{\Delta C}{\Delta t} = \frac{\Delta J}{\Delta x} = \frac{J_{in} - J_{out}}{\Delta x} \qquad (4.4)$$

By taking partial derivatives, Equations 4.3 and 4.4 can be combined as:

$$\frac{\partial C}{\partial t} = \frac{\partial J}{\partial x} = \frac{\partial}{\partial x}(D\frac{\partial C}{\partial x}) \qquad (4.5)$$

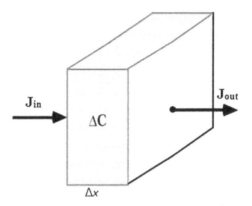

FIGURE 4.5 Diffusion flux in and out of a volume element.

In its simplest form, Equation 4.5 can be expressed as Fick's Second Law, as follows:

$$\frac{\partial C}{\partial t} = \frac{\partial}{\partial x}(D\frac{\partial C}{\partial x})$$

(4.6)

It may be noted that Equation 4.6 involves partial derivatives of $C(x, t)$ with respect to x and t. It means that we may take derivative with respect to one variable while treating the other variable as a constant. Also, in Equation 4.6, an assumption is made that the diffusion coefficient D itself does not depend on $C(x)$, or indirectly on x, but is only a function of time (t). Fick's Second Law is in essence an expression of the continuity of particles (or continuity of mass): it amounts to stating mathematically that the time rate of change in concentration in a small volume element is due to the sum total of particle fluxes into and out of the volume element.

In case the diffusion coefficient (D), is independent of concentration (C), Equation 4.6 can be simplified to the following expression:

$$\frac{\partial C}{\partial t} = D\frac{\partial^2 C}{\partial x^2}$$

(4.7)

By integration, it is possible to find a number of solutions to Equation 4.7; however, it is necessary to specify meaningful boundary conditions (Carslaw and Jeager, 1986). By considering diffusion for a semi-infinite solid (in which the surface concentration is constant), the following boundary conditions may be assumed (Callister, 2007):

For $t = 0$, $C_x = C_0$ at $0 \leq x \leq \infty$

For $t > 0$, $C_x = C_s$ (the constant surface concentration) at $x=0$

$C_x = C_0$ at $x=\infty$

By using the previous boundary conditions, Equation 4.7 can be solved to obtain (Huda, 2018):

$$C_x = C_s - (C_s - C_o)erf(\frac{x}{2\sqrt{Dt}})$$

(4.8)

where C_x is the concentration of the diffusive atoms at distance x from a surface; C_o is the initial bulk concentration, C_s is the surface concentration, erf is the Gaussian error function; and t is the time duration for diffusion (see Example 4.4). A partial list of error function (erf) values are given in Table 4.1.

In case of approximate value of depth of case x, Equation 4.8 can be simplified to:

$$C_x = C_s - (C_s - C_o)\left(\frac{x}{2\sqrt{Dt}}\right) \tag{4.9}$$

The significance of Equation 4.9 is illustrated in Example 4.5.

In some metallurgical situations, it is desired to achieve a specific concentration of solute in an alloy; this situation enables us to rewrite Equation 4.9 as (Huda, 2018):

$$\frac{x^2}{Dt} = \text{constant} \tag{4.10}$$

$$\text{or} \quad \frac{x_1^2}{D_1 t_1} = \frac{x_2^2}{D_2 t_2} \tag{4.11}$$

where x_1 is the case depth after carburizing for time-duration of t_1; and x_2 is the case depth after carburizing for t_2. If the materials are identical and their carburizing temperature are the same, the diffusion coefficients are also the same i.e. $D_1 = D_2 = D$. Thus, Equation 4.11 simplifies to:

or

$$\frac{x_1^2}{t_1} = \frac{x_2^2}{t_2} \tag{4.12}$$

It has been recently reported that the depth of case in gas carburizing of steel exhibits a time-temperature dependence such that (Schneider and Chatterjee, 2013):

Case depth $= D \sqrt{t}$ (4.13)

TABLE 4.1

Partial Listing of Error Function Values (Huda, 2018)

z	erf (z)	z	erf (z)	z	erf (z)
0	0	0.30	0.3286	0.65	0.6420
0.025	0.0282	0.35	0.3794	0.70	0.6778
0.05	0.0564	0.40	0.4284	0.75	0.7112
0.10	0.1125	0.45	0.4755	0.80	0.7421
0.15	0.1680	0.50	0.5205	0.85	0.7707
0.20	0.2227	0.55	0.5633	0.90	0.7970
0.25	0.2763	0.60	0.6039	0.95	0.8209

The technological importance of Equations 4.7–4.12 are illustrated in Examples 4.4–4.6. Fick's Second Law has significant industrial applications; which are explained in section 4.6.

4.6 APPLICATIONS OF FICK'S SECOND LAW OF DIFFUSION

A number of important industrial process are governed by Fick's Second Law of diffusion; these processes include: impurity diffusion in Si wafers in ICs fabrication; case-carburizing (surface hardening) of steel, the kinetics of decarburization of steel boiler plates; and the like.

Impurity Diffusion in Si Wafers in ICs Fabrication. It is learned in subsection 4.1.2 that the controlled diffusion of phosphorus (P), boron (B), or other dopants (beneficial impurities) into silicon (Si) wafers is a technologically important process in the fabrication of integrated circuits (ICs). There are mainly two types of physical mechanisms by which impurities can diffuse into the Si lattice: (a) vacancy diffusion, and (b) interstitial diffusion (see Figure 4.6).

It has been reported that diffusion of gallium (Ga) into silicon (Si) wafer occurs at 1,100°C; at this temperature the diffusivity coefficient of *Ga* in *Si* is 7×10^{-17} m²/s (Huda and Bulpett, 2012). If the initial bulk concentration (C_0), surface concentration (C_s), and concentration C_x are known, the depth of penetration x can be computed by using Equation 4.9 (see Example 4.7).

Case Carburizing of Steel. Surface hardening by case carburizing of steel is an important industrial process applied to some machine elements (e.g. gears, crankshafts, etc.). In this process, carbon is added to the steel surface by exposing the steel part to a carbon-rich atmosphere at a temperature in the range of 900–1,100°C; thereby allowing carbon atoms to diffuse into the steel part. Here, interstitial diffusion of carbon (C) atoms into the host iron/steel solid results in case carburized steel gear (Figure 4.7). In accordance with Fick's Second Law of diffusion, the diffusion in case carburizing will work only if the steel has initially a low carbon content since diffusion works on the differential concentration principle (see Equation 4.5).

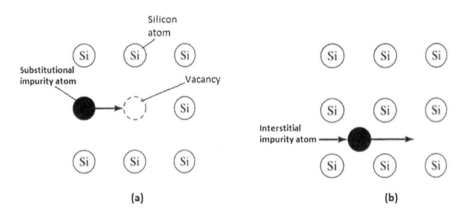

FIGURE 4.6 Impurity diffusion in Si wafer; (a) vacancy diffusion, (b) interstitial diffusion.

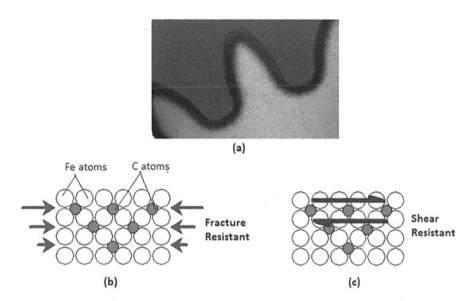

FIGURE 4.7 Interstitial diffusion of carbon into steel; (a) case carburized steel gear, (b) C atoms put the surface in compression, (c) C atoms lock crystal planes to resist shear.

4.7 THERMALLY ACTIVATED DIFFUSION—ARRHENIUS LAW

We have learned in section 4.1 that in diffusion, atomic vacancies and other small-scale point defects allow atoms to exchange places. In order for the atoms to have sufficient energy for exchange of positions, high temperatures are required. It means that diffusion is generally a thermally activated process. It is seen in section 4.4 that the diffusion coefficient, D, indicates the ease or speed with which a specie can diffuse in some mediums. It has been experimentally shown that diffusion coefficient (D) increases with increasing temperature according to the following Arrhenius-type mathematical relationship (Maaza, 1993):

$$D = D_0 \exp\left(-\frac{Q_d}{RT}\right) \tag{4.14}$$

where D is the diffusion coefficient, m²/s; D_0 is the pre-exponential (a temperature independent) constant, m²/s; Q_d is the activation energy for diffusion, J/mol or eV/atom; R is the gas constant (=8.314 J/mol-K = 8.62 x 10^{-5} eV/atom-K), and T is the temperature in Kelvin (K). The activation energy for diffusion, Q_d, is the minimum energy required to produce the diffusive motion of one mole of atoms. A large Q_d in Equation 4.14 results in a smaller D (see Examples 4.8–4.10). The activation energies for diffusion and the pre-exponentials for some diffusing systems are listed in Table 4.2.

By taking natural logarithm of both sides of Equation 4.14, we obtain:

$$\ln D = \ln D_o - \frac{Q_d}{RT} \tag{4.15}$$

TABLE 4.2

Diffusion Data for Some Diffusing Systems

Host solid	Diffusing specie	D_0 (m²/s)	Q_d (eV/atom)	Temperature range (°C)
Silicon	Nickel	1.0×10^{-5}	1.9	550–900
Silicon	Iron	6.2×10^{-5}	0.87	1,100–1,250
Silicon	Sodium	1.6×10^{-5}	0.76	800–1,100
BCC iron	Iron	2.8×10^{-4}	2.6	750–900
FCC iron	Iron	5.0×10^{-5}	2.94	1,100–1,250
Aluminum	Copper	6.5×10^{-5}	1.41	500–700
Aluminum	Aluminum	2.3×10^{-4}	1.49	300–600
Copper	Zinc	2.4×10^{-5}	1.96	800–1050

or

$$logD = log\, D_o - \frac{Q_d}{2.3R}\left(\frac{1}{T}\right) \tag{4.16}$$

Equation 4.16 indicates that a graphical plot of *log D* versus (1/T) would yield a straight line having intercept of *log D_o* and slope as shown in Figure 4.8 (see Example 4.9).

Equation 4.15 can be transformed into a more practical form by considering the diffusion coefficients D_1 and D_2 at temperatures T_1 and T_2, respectively, as follows.

By using Equation 4.15 for the diffusion coefficient D_1,

$$In\, D_1 = In\, D_o - \frac{Q_d}{RT_1} \tag{4.17}$$

By using Equation 4.15 for the diffusion coefficient D_2,

$$In\, D_2 = In\, D_o - \frac{Q_d}{RT_2} \tag{4.18}$$

By performing the subtraction of Equations (4.18) – (4.17), we obtain:

$$\frac{Q_d}{RT_1} - \frac{Q_d}{RT_2} = In\, D_2 - In\, D_1$$

$$\frac{Q_d}{R}\left(\frac{1}{T_1} - \frac{1}{T_2}\right) = In\frac{D_2}{D_1} \tag{4.19}$$

The significance of Equation 4.19 is illustrated in Examples 4.9 and 4.10.

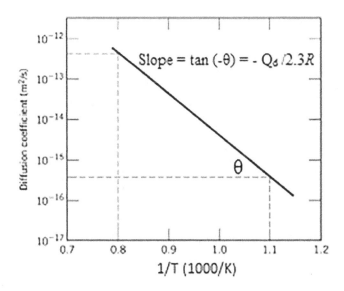

FIGURE 4.8   Graphical plot of *log D* versus (1/T) for a typical diffusing system.

4.8   CALCULATIONS—EXAMPLES ON DIFFUSION AND ITS APPLICATIONS

EXAMPLE 4.1   CALCULATING THE MASS TRANSFERRED WHEN THE DIFFUSION FLUX IS KNOWN

In purification of hydrogen gas, a sheet of palladium metal is used as a diffusion membrane in the purification vessel. Calculate the mass of H_2 that pass in 30 min through a 6-mm-thick sheet of palladium having an area of 0.25 m^2 at 600°C. The diffusion flux at the temperature is 4.5 x 10^{-6} kg/m^2-s.

SOLUTION

J = 4.5 x 10^{-6} kg/m^2-s; dx = 6 mm = 6 x 10^{-3} m; A = 0.25 m^2; dt = 30 min = 1800 s, dM = ?

By rearranging the terms in Equation 4.1,

dM = J A (dt) = 4.5 x 10^{-6} x 0.25 x 1800 = 2 x 10^{-3} kg = 2 g

EXAMPLE 4.2   COMPUTING THE CONCENTRATION OF THE DIFFUSIVE SPECIE IN ATOMS/VOLUME

A 1-mm-thick plate of wrought iron (BCC iron) is surface treated such that one side is in contact with a carbon-rich atmosphere that maintains the carbon concentration at the surface at 0.18 wt%. The other side is in contact with a decarburizing atmosphere that keeps the surface free of carbon. Calculate the concentration of carbon (in atoms per cubic meter) at the surface on the carburizing side. The density of BCC iron is 7.9 g/cm^3.

SOLUTION

Density of the host metal (BCC iron) $= \rho_X = 7.9$ g/cm^3; $N_A = 6.02 \times 10^{23}$ atoms/mol
Atomic weight of the diffusive specie (carbon) $= A_y = 12$ g/mol;
Wt% y = wt% carbon = 0.18%
By using Equation 4.2,

$$C_{atom} = \frac{(wt\%y)(\rho_X)(N_A)}{A_y} = \frac{(0.18\%)(7.9)}{12} \times 6.02 \times 10^{23}$$
$$= \frac{(0.0018)(7.9)}{12} \times 6.02 \times 10^{23}$$

$C_{atom} = 7.13 \times 10^{20}$ atoms/cm^3 = 7.13 $\times$ 10^{20} x 10^6 atoms/m^3 = 7.13 $\times$ 10^{26} atoms/m^3

The concentration of carbon at the surface at the carburizing side = 7.13 x 10^{26} atoms/m^3.

EXAMPLE 4.3 COMPUTING THE DIFFUSION FLUX WHEN THE DIFFUSION COEFFICIENT IS KNOWN

By using the data in Example 4.2, calculate the diffusion flux of carbon through the plate under the steady-state diffusion condition at a temperature of 727°C. The diffusion coefficient of carbon in BCC iron at 727°C is 8.7 x 10^{-11} m^2/s.

SOLUTION

The concentration of carbon at the surface at the carburizing side = C_1 = 7.13 x 10^{26} atoms/m^3.
The concentration of carbon at the surface at the decarburizing side = C_2 = 0.

$dC = C_1 - C_2 = 7.13 \times 10^{26} - 0 = 7.13 \times 10^{26}$ atoms/m^3

$dx = 1$ mm = 0.001 m
$D = 8.7 \times 10^{-11}$ m^2/s; Diffusion flux $= J = ?$
By using Fick's First Law (Equation 4.3),

$$J = -D\frac{dC}{dx} = 8.7 \times 10^{-11} \times \frac{7.13 \times 10^{26}}{0.001} = 6.2 \times 10^{19} \text{ atom/m}^2\text{-s}$$

The diffusion flux = 6.2 x 10^{19} atoms/m^2-s.

EXAMPLE 4.4 CALCULATING THE CARBURIZING TIME FOR STEEL FOR KNOWN CASE COMPOSITION

A component made of 1015 steel is to be case-carburized at 1,000°C. The component design requires a carbon content of 0.8% at the surface and a carbon content of 0.3% at a depth of 0.7 mm from the surface. The diffusivity of carbon in BCC iron at 1,000°C is 3.11 x 10⁻¹¹ m²/s. Compute the time required to carburize the component to achieve the design requirements.

SOLUTION

Diffusivity $= D = 3.11$ x 10^{-11} m²/s
Surface carbon concentration $= C_s = 0.8\%$
Initial bulk carbon concentration $= C_o = 0.15\%$ (The AISI-1015 steel contains 0.15% C)
Concentration of diffusive carbon atoms at distance x from the surface $= C_x = 0.3\%$; $x = 0.7$ mm

By the application of Fick's Second Law of diffusion (by using Equation 4.8),

$$C_x = C_s - (C_s - C_o)\,erf\,(\frac{x}{2\sqrt{Dt}})$$

$$0.3 = 0.8 - (0.8 - 0.15)\,erf\,(\frac{0.0007}{2\sqrt{3.11 x 10^{-11} t}})$$

$erf\,(62.7/\sqrt{t}) = 0.7692$
Taking $Z = 62.7/\sqrt{t}$, we get erf $Z = 0.7692$

By reference to Table 4.1, we can develop a new table as follows:

erf Z	0.7421	0.7692	0.7707
Z	0.80	X	0.85

By using the interpolation mathematical technique,

$$\frac{0.7692 - 0.7421}{0.7707 - 0.7421} = \frac{x - 0.80}{0.85 - 0.80}$$

$x = 0.847 = Z$
or $Z = 62.7/\sqrt{t} = 0.847$
$t = 5480$ s $= 1.5$ h

The time required to carburize the component $= 1.5$ hours.

**EXAMPLE 4.5 ESTIMATING THE DEPTH OF CASE IN
DE-CARBURIZATION OF STEEL**

An AISI-1017 boiler-plate steel was exposed to air at a high temperature of 927°C
due to malfunctioning of its heat exchanger. Estimate the depth from the surface
of the boiler plate at which the concentration of carbon decreases to one-half of its
original content after exposure to the hot oxidizing environment for 20 hours. The
concentration of carbon at the steel surface was 0.02%. The diffusivity coefficient of
carbon in steel at 927°C is 1.28 x 10⁻¹¹ m²/s.

SOLUTION

D = 1.28 x 10⁻¹¹ m²/s; Time = t = 20 h = 72,000 s; Surface carbon
concentration = C_s = 0.02%

Initial bulk carbon concentration = C_o = 0.17% (AISI-1017 steel contains 0.17% C)

Concentration of diffusive C atoms at distance x from the surface = C_x = (½)
(0.17%) = 0.085%

By using Equation 4.9,

$$C_x = C_s - (C_s - C_o)(\frac{x}{2\sqrt{Dt}})$$

$$0.085 = 0.02 - (0.02 - 0.17)(\frac{x}{2\sqrt{1.28 \times 10^{-11} \times 72000}})$$

or x = 3.03 mm

Hence, the depth from the surface of the boiler plate = 3 mm (approx.).

**EXAMPLE 4.6 COMPUTING CARBURIZING TIME FOR
SPECIFIED COMPOSITION AT CASE DEPTH**

A carburizing heat treatment of a steel alloy for a duration of 11 hours raises the carbon concentration to 0.5% at a depth of 2.2 mm from the surface. Estimate the time
required to achieve the same concentration at a depth of 4 mm from the surface for
an identical steel at the same carburizing temperature.

SOLUTION

x_1 = 2.2 mm when t_1 = 11 h; x_2 = 4 mm when t_2 = ?

By using Equation 4.12,

$$\frac{x_1^2}{t_1} = \frac{x_2^2}{t_2}$$

$$\frac{(2.2)^2}{11} = \frac{4^2}{t_2}$$

$$t_2 = 36.36 h$$

The required carburization time = 36.36 hours.

EXAMPLE 4.7 CALCULATING THE DIFFUSION COEFFICIENT FOR A DIFFUSING SYSTEM

By reference to the diffusion data in Table 4.2, calculate the diffusion coefficient for zinc in copper at 550°C.

SOLUTION

$D_0 = 2.4 \times 10^{-5}$ m²/s, $Q_d = 1.96$ eV/atom, $R = 8.62 \times 10^{-5}$ eV/atom-K, T = 550+273= 823K, D = ?

By using Equation 4.14,

$$D = D_0 \exp\left(-\frac{Q_d}{RT}\right) = 2.4 \times 10^{-5} \exp\left(-\frac{1.96}{8.62 \times 10^{-5} \times 823}\right)$$
$$= 2.4 \times 10^{-5} \exp(-27.62) = 2.43 \times 10^{-17}$$

The diffusion coefficient for zinc in copper at 550°C = 2.43 x 10⁻¹⁷ m²/s.

EXAMPLE 4.8 CALCULATING THE TEMPERATURE WHEN Q_D, D_0, AND D ARE KNOWN

By using the data in Table 4.2, calculate the temperature at which the diffusion coefficient of nickel in silicon is 1.0 x 10⁻¹⁵ m²/s.

SOLUTION

$D_0 = 1.0 \times 10^{-5}$ m²/s and $Q_d = 1.9$ eV/atom = 183390 J/mol, $D = 1.0 \times 10^{-15}$ m²/s, T = ?

By rearranging the terms in Equation 4.15, we obtain:

$$T = \frac{Q_d}{R(In\, D_0 - In\, D)}$$
$$T = \frac{183390}{8.314(In\, 0.00001 - In\, 10^{-15})} = \frac{183390}{8.314(-11.51 + 34.5)} = 959.58\,K$$
$$T = 959.58 - 273 = 686.6\,{}^{\circ}C$$

At 686.6°C, the diffusion coefficient of nickel in silicon is 1.0 x 10⁻¹⁵ m²/s.

EXAMPLE 4.9 COMPUTING THE TEMPERATURE WHEN Q_D IS KNOWN BUT D_0 IS UNKNOWN

The activation energy for diffusion for copper in silicon is 41.5 kJ/mol. The diffusion coefficient of Cu in Si at 350°C is 15.7 x 10⁻¹¹ m²/s. At what temperature is the diffusion coefficient 7.8 x 10⁻¹¹ m²/s?

SOLUTION

$Q_d = 41,500$ J/mol, $D_1 = 15.7 \times 10^{-11}$ m²/s at $T_1 = 350 + 273 = 623K$, $D_2 = 7.8 \times 10^{-11}$ m²/s at $T_2 = ?$

By using Equation 4.19,

$$\frac{Q_d}{R}\left(\frac{1}{T_1}-\frac{1}{T_2}\right)=In\frac{D_2}{D_1}$$

$$\frac{41500}{8.314}\left(\frac{1}{623}-\frac{1}{T_2}\right)=In\frac{7.8\times10^{-11}}{15.7\times10^{-11}}$$

$$4991.6\left(\frac{1}{623}-\frac{1}{T_2}\right)=-0.6995$$

$$T_2=573K=300\,°C$$

EXAMPLE 4.10 CALCULATING THE ACTIVATION ENERGY (Q_D) AND D_0 FOR A DIFFUSION SYSTEM

The diffusion coefficients of lithium (Li) in silicon (Si) are 10^{-9} m²/s and 10^{-10} m²/s at 1,100°C and 700°C, respectively. Calculate the values of the activation energy (Q_d) and the pre-exponential constant (D_0) for diffusion of Li in silicon.

SOLUTION

$D_1 = 10^{-9}$ m²/s at $T_1 = 1,100°C = 1,100 + 273 = 1,373$ K
$D_2 = 10^{-10}$ m²/s at $T_2 = 700°C = 700 + 273 = 973$ K
By using Equation 4.19,

$$\frac{Q_d}{R}\left(\frac{1}{T_1}-\frac{1}{T_2}\right)=In\frac{D_2}{D_1}$$

$$\frac{Q_d}{8.314}\left(\frac{1}{1373}-\frac{1}{973}\right)=In\frac{10^{-10}}{10^{-9}}$$

$$\frac{Q_d}{8.314}(0.000728-0.001)=-2.3$$

$$Q_d=\frac{2.3\times8.314}{0.000272}=70,302.2\,J/mol=0.72\,eV/atom$$

In order to find the value of D_0, we may rearrange the terms in Equation 4.17 as follows:

$$In\,D_o=In\,D_1+\frac{Q_d}{RT_1}$$

$$In\,D_o=In10^{-9}+\frac{70302.2}{8.314\times1373}=-20.723+6.16=-14.56$$

$$D_o=exp(-14.56)=4.75\times10^{-7}\,m^2/s$$

$D_o = 4.75$ x 10^{-7} m²/s; the activation energy for diffusion = 70,302.2 J/mol. = 0.72 eV/atom.

QUESTIONS AND PROBLEMS

4.1. Define the following terms: (a) diffusion, (b) diffusion coefficient, (c) activation energy for diffusion.

4.2. List the factors affecting the rate of diffusion; and briefly explain them.

4.3. Distinguish between the following: (a) vacancy diffusion and interstitial diffusion; (b) steady-state diffusion and non-steady state diffusion.

4.4. Explain the term *diffusion flux* with the aid of a diagram.

4.5. (a) State Fick's First Law of Diffusion. (b) What industrial application areas of Fick's First Law of Diffusion do you identify? Explain.

4.6. Explain Fick's Second Law of diffusion with the aid of a sketch.

4.7. How is Fick's Second Law helpful in explaining case carburizing of steel?

4.8. The activation energy for diffusion is related to temperature by the Arrhenius expression: $D = D_0 \exp(-\dfrac{Q_d}{RT})$. Obtain an expression relating the diffusion coefficients D_1 and D_2 at temperatures T_1 and T_2.

P4.9. Calculate the mass of hydrogen that pass in 45 min through a 4-mm-thick sheet of palladium having an area of 0.20 m² at 600°C during purification of hydrogen gas. The diffusion flux at the temperature is 4.5 x 10^{-6} kg/m²-s.

P4.10. A 2-mm-thick plate of wrought iron (BCC iron) is surface treated such that one side is in contact with a carbon-rich atmosphere that maintains the carbon concentration at the surface at 0.21 wt%. The other side is in contact with a decarburizing atmosphere that keeps the surface free of carbon. Calculate the concentration of carbon (in atoms per cubic meter) at the surface on the carburizing side. The density of BCC iron is 7.9 g/cm³.

P4.11. By using the data in P4.10, calculate the diffusion flux of carbon through the plate under steady-state diffusion condition at a temperature of 500°C. The diffusion coefficient of carbon in BCC iron at 500°C is 2.4 x 10^{-12} m²/s.

P4.12. A component made of 1016 steel is to be case-carburized at 900°C. The component design requires a carbon content of 0.9% at the surface and a carbon content of 0.4% at a depth of 0.6 mm from the surface. The diffusivity of carbon in BCC iron at 900°C is 1.7 x 10^{-10} m²/s. Compute the time required to carburize the component to achieve the design requirements.

P4.13. An AISI-1018 boilerplate steel was exposed to air at a high temperature of 1,100°C due to a fault in its heat exchanger. Estimate the depth from the surface of the boilerplate at which the concentration of carbon decreases to one-half of its original content after exposure to the hot oxidizing environment for 15 hours. The concentration of carbon at the steel surface was 0.02%. The diffusivity coefficient of carbon in γ-iron at 1,100°C is 5.3 x 10^{-11} m²/s.

P4.14. The diffusion coefficients for copper in aluminum at 500°C and 600°C are 4.8 x 10^{-14} m²/s and 5.3 x 10^{-13} m²/s, respectively. Calculate the values of the activation energy (Q_d) and the pre-exponential constant (D_o) for diffusion of copper in aluminum.

P4.15. A carburizing heat treatment of a steel alloy for a duration of nine hours raises the carbon concentration to 0.45% at a depth of 2 mm from the surface. Estimate the time required to achieve the same concentration at a depth of 3.5 mm from the surface for an identical steel at the same carburizing temperature.

P4.16. The activation energy for diffusion for carbon in γ-iron is 148 kJ/mol. The diffusion coefficient for carbon in γ-iron at 900°C is 5.9 x 10^{-12} m²/s. At what temperature is the diffusion coefficient 5.3 x 10^{-11} m²/s?

REFERENCES

Callister, W.D. (2007) *Materials Science and Engineering: An Introduction*. John Wiley & Sons Inc, Hoboken, NJ.

Carslaw, H.S. & Jeager, J.C. (1986) *Conduction f Heat in Solids*. 2nd ed. Oxford University Press, Oxford.

Huda, Z. (2017) Energy-efficient gas-turbine blade-material technology—*A review. Materiali in Tehnologije (Materials and Technology)*, 51(3), 355–361.

Huda, Z. (2018) *Manufacturing: Mathematical Models, Problems, and Solutions*. CRC Press Inc., Boca Raton, FL.

Huda, Z. & Bulpettt, R. (2012) *Materials Science and Design for Engineers*. Trans Tech Publications, Zurich.

Maaza, M. (1993) Determination of diffusion coefficient d and activation energy Q_d of Nickel into Titanium in Ni-Ti Multilayers by Grazing-Angle neutron reflectometry. *Journal of Applied Crystallography*, 26, 334–342.

Memrer, H. (2007) *Diffusion in Solids*. Springer-Verlag, Berlin.

Pichler, P. (2004) *Intrinsic Point Defects, Impurities, and Their Diffusion in Silicon*. Springer-Verlag, Wien.

Schaffer, J.P., Saxene, A., Antolovich, S.D., Sanders, T.H. & Warner, S.B. (1999) *The Science and Design of Engineering Materials*. McGraw Hill Inc, Boston.

Schneider, M.J. & Chatterjee, B.M.S. (2013) *Introduction to Surface Hardening of Steels*. ASM International, Materials Park, OH.

Sedra, A.S. & Smith, K.C. (2014) *Microelectronic Circuits*. 7th ed. Oxford University Press, Oxford.

Part II

Physical Metallurgy—
Microstructural Developments

5 Metallography and Material Characterization

5.1 EVOLUTION OF GRAINED MICROSTRUCTURE

The study of grained microstructure is of great technological importance since many materials properties are strongly dependent on the microstructural design features. The mechanism of solidification of metals forms the basis of the evolution of grained microstructure in a polycrystalline material. In metallurgical practice, single crystals are rarely produced only under strictly controlled conditions of cooling leading to solidification. The expense of producing a single crystal is only justified for advanced applications; which include gas-turbine blades, solar cells, piezoelectric materials, etc. In general, when a metal begins to solidify, multiple crystals begin to grow in the liquid resulting in the formation of a polycrystalline (more than one crystal) solid.

The solidification of a polycrystalline metal involves the following four stages: (I) *nucleation*: the formation of stable nuclei in the melt (Figure 5.1a); (II) *crystal growth:* growth of nuclei into crystals (Figure 5.1b); (III) *formation of randomly oriented crystals* (Figure 5.1c); and (IV) grained microstructure with grain boundaries (Figure 5.1d) (see also Figure 5.6).

Nucleation refers to the moment a crystal begins to grow; and the point where it occurs is called the nucleation point. At the solidification temperature, atoms of molten metal begin to bond together at the nucleation points and start to form crystals. The final sizes of the individual crystals depend on the number of nucleation points. The crystals grow by the progressive bonding of atoms until they impinge upon adjacent growing crystals. If the rate of solidification is high (fast cooling), many nucleation points are formed thereby resulting in fine grain size in the metal. On the other hand, slow cooling results in a coarse grain-sized material. In most steels, the cast solidified structure is dendritic: tree-like.

In a polycrystalline metal, a crystal is usually referred to as a *grain*. A *grain* may be considered as a crystal without smooth faces because its growth was obstructed by contact with another grain or a boundary surface during solidification. The interface formed between two grains is called a *grain boundary* (see Figure 5.1d). In general, *grains* or *crystals* are too small to be seen by the naked eye; however, sometimes they are large enough to be visible under the unaided eye (e.g. zinc crystals in a galvanized steel) (Millward *et al.*, 2013).

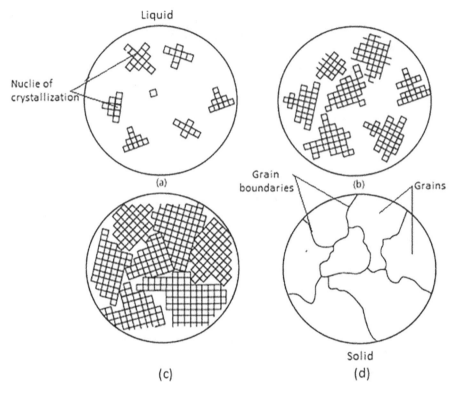

FIGURE 5.1 Evolution of grained microstructure in a polycrystalline metal; (a) nucleation of crystals, (b) crystal growth, (c) randomly oriented grains form as crystals grow together, (d) grained microstructure as seen under a microscope.

5.2 METALLOGRAPHY—METALLOGRAPHIC EXAMINATION OF MICROSTRUCTURE

5.2.1 METALLOGRAPHY AND ITS IMPORTANCE

Metallography is the branch of physical metallurgy that involves the study of the microstructure of a metallic material. The study and analysis of microstructure helps us to determine if the material has been processed correctly. Metallographic examination is, therefore, a critical step for determining product reliability and for determining why a material failed. In order to study a material's microstructure under a microscope, it is first important to prepare a metallographic specimen; the latter is described in the following subsection.

5.2.2 METALLOGRAPHIC SPECIMEN PREPARATION

The preparation of a metallographic specimen involves the following seven steps: (I) sectioning, (II) rough grinding, (III) mounting, (IV) fine grinding,

(V) polishing, (VI) etching, and (VII) microscopy. These steps in the preparation of a metallographic specimen are discussed one-by-one in the following paragraphs.

Sectioning. Sectioning involves cutting-off a small piece from a bulk metallic piece by use of a cut-off machine. One must ensure that the cut-off machine's cutter is harder and stronger than the specimen so as to avoid damage to the machine and the specimen. When sectioning a specimen from a larger piece of material, care must be taken to ensure that the specimen contains all the information required to investigate a feature of interest. It is important to ensure cooling by using cutting fluid during sectioning. For low deformation sectioning, it is appropriate to use a low-speed diamond saw for subsequent microscopic examination.

Rough grinding. Rough grinding involves the removal of rust/surface deposits or leveling of irregular surfaces. This objective can be achieved either by filing or by using a belt/disc grinder.

Mounting. Mounting of specimen provides a safe means of holding it and protecting its edges from rounding. In general, a mounting press is used to mount rigid metals; and the process is called *compression molding*. Specimens can be hot mounted (at 150°C) using a mounting press that involves use of a thermosetting plastic—Bakelite. Alternatively, soft or large-sized specimen can be cold mounted by using a cold-setting resin (e.g. epoxy, acrylic, or polyester resin). The edges of the metallographic specimen mount should be rounded to avoid the damage to grinding and polishing discs (see Figure 5.2).

Fine Grinding. Fine grinding involves the use of a series of abrasive papers with grits of decreasing coarseness. The purpose of fine grinding is to remove

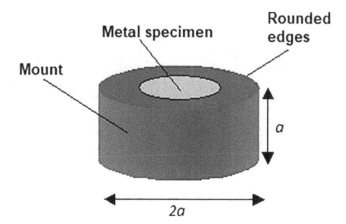

FIGURE 5.2   A mounted metallographic specimen.

the surface layers damaged by sectioning and to systematically abrade the metallographic specimen. The fine-grinding operation may be performed either on a manual metallographic grinder or a rotating wheel grinding machine. Mounted specimens are ground with rotating discs of abrasive paper (e.g. wet silicon carbide [emery] paper). The coarseness of the paper is indicated by a grit number: the number of SiC grains per square inch. For example, 600-grit paper is finer than 400-grit paper. The grinding procedure involves several stages, using a finer paper (higher grit number) each time. Each grinding stage removes the scratches from the previous coarser paper; which can be easily achieved by orienting the specimen perpendicular to the previous scratches. Between each grade, the specimen is washed thoroughly with soapy water to prevent contamination from coarser grit present on the specimen surface. Typically, the finest grade of paper used is the 1200, and fine grinding is complete once the only scratches left on the specimen are from this grade. In a typical metallographic grinding operation, the specimen is ground using grit number sequence: 180, 400, 600, 800, 1000, and 1200. Figure 5.3(a–b) shows two of the progression of the specimen when ground with progressively finer papers.

Polishing. Polishing involves the removal of the remaining scratches on the ground specimen so as to produce a smooth lustrous surface required for microscopic examination. *Polishing* is accomplished by using a polishing machine with rotating *polishing discs* that are covered with soft cloth impregnated with abrasive particles (alumina powder suspension or diamond paste) and an oily lubricant or water lubricant. In the case of polishing with diamond paste, particles of different grades are used: a coarser polish (with diamond particles 6 µm in diameter), and then a finer polish (with diamond particles 1 µm in diameter) to produce a smooth surface. At each polishing stage, the specimen should be washed thoroughly with warm soapy water to prevent contamination of the next disc. An alcohol rinse should be used in the final polishing stage; drying can be accelerated by using a hot air drier. The polished specimens reflects light like a mirror (see Figure 5.4).

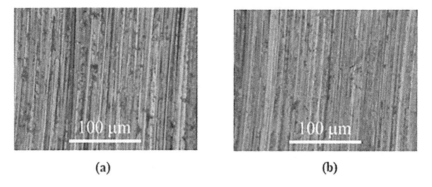

(a)   (b)

FIGURE 5.3   Copper specimen ground with (a) 800-grit paper; and (b) 1200-grit paper.

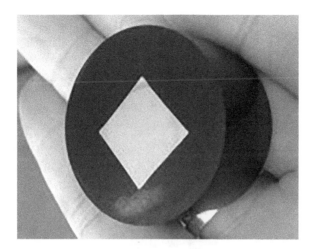

FIGURE 5.4 Polished metallographic specimen.

Polishing by using rotating discs usually leaves a layer of disturbed material on the surface of the specimen. In some cases, electro-polishing can be used to obtain an undisturbed surface; this is particularly important where high resolution SEM analysis is undertaken.

Etching. Metallographic etching involves revealing the microstructure of the metal through selective chemical attack. In alloys with more than one phase, etching creates contrast between different regions through differences in the reflectivity of the different phases. In pure metals or solid-solution alloys, the rate of etching is affected by crystallographic orientation resulting in contrast between grains. The etching reagent preferentially etches high-energy sites such as grain boundaries; which results in a surface relief that enables different crystal orientations, grain boundaries, phases, and precipitates to be easily distinguished. For etching carbon steels, *nital* (2% nitric acid-98% alcohol) solution is used. For etching stainless steel or copper alloys, a saturated aqueous solution of ferric chloride (Fe_3Cl) containing a few drops of *HCl* is used. *Metallographic etching* is applied using a grit-free cotton bud wiped over the surface a few times; the specimen should then immediately be washed in alcohol and dried.

Leveling and Microscopy. The metallographic etched specimen to be examined under a microscope should be perfectly flat and leveled; the *leveling* is accomplished by using a specimen leveling press (Figure 5.5). The leveling device presses the mounted specimen into plasticine on a microscope slide, making it leveled.

The etched and leveled metallographic specimen is finally observed under an optical microscope. The optical micrograph for a properly etched metallographic specimen is shown in Figure 5.6. A detailed account of microscopy is given in Sect. 5.3. In this section, qualitative aspects of metallography are discussed; the quantitative metallography is given in section 5.5.

FIGURE 5.5 A metallographic specimen leveler.

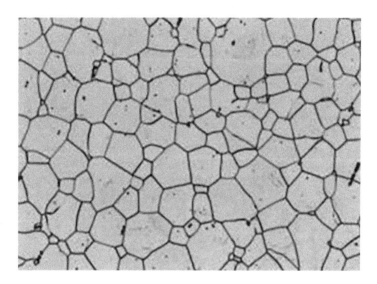

FIGURE 5.6 The optical micrograph of an etched metallographic specimen showing the microstructure of a single-phase metal/alloy.

5.3 MICROSCOPY

Microscopy involves the use of microscopes to examine grained microstructures, phases, and other microstructural details in a metallographic specimen. There are mainly two types of microscope: (a) optical microscope (OM) or light microscope (LM), and (b) electron microscope; these microscopes are explained in the following subsections.

5.3.1   Optical Microscopy (OM)

5.3.1.1   The OM Equipment

Optical microscopy (or light microscopy) refers to the inspection of a metallographic specimen at a moderately high magnification (in the range of 100×–1,000×) by using an instrument known as an optical or metallurgical microscope. A basic metallurgical or optical microscope (OM) has the following six parts: (a) a lamp to illuminate the specimen; (b) a nose piece to hold four to five objective lenses used in changing the viewing magnification; (c) an aperture diaphragm to adjust the resolution and contrast; (d) a field diaphragm to adjust the field of view; and (e) an eyepiece lens to magnify the objective image (usually by 10X); and (f) a stage for manipulating the specimen. In addition to the previously mentioned basic features, modern metallurgical microscopes are equipped with a computerized imaging system, including a USB digital camera (see Figure 5.7).

5.3.1.2   Imaging in Optical Microscopy

During metallographic examination, the specimen is positioned perpendicularly to the axis of the objective lens of the microscope; this specimen positioning is ensured by *leveling*, as explained in the preceding section. Light is then incident on the specimen, which reflects some light back to the lens. The image seen in the optical microscope (OM) depends not only on how the specimen is illuminated and positioned, but also on the characteristics of the specimen. An OM's overall magnification is a combination of the magnifications of eyepiece and the objective lens. For example, an OM with 10X eyepiece and a 50X objective has an overall magnification of 500X. There are however, limits to the amount of overall magnification that can be reached before poor resolution results; which occurs when the image continues to be enlarged, but no additional detail is resolved. This is often the case when higher magnification eyepieces are used. The spatial resolution of a well-designed OM is mainly

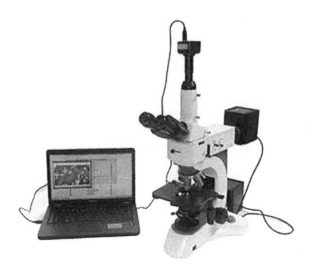

FIGURE 5.7   A modern metallurgical microscope with computerized image-capturing system.

determined by the objective lens. It means that for good resolution, the magnification of the objective should be high (from 50X to 100X) keeping the eyepiece at 10X.

5.3.1.3   Resolution of an Optical Microscope

The *resolution* of a microscope is the minimum resolvable distance i.e. the minimum distance to resolve fine detail in the object being observed. A metallurgical microscope's resolution strongly depends on the numerical aperture (*NA*) of its objective lens. The *NA* of the objective is given by:

$$NA = n \sin \alpha \tag{5.1}$$

where *n* is the refractive index between the lens and the object, and α is the half-angle aperture (the half inclination of the lens to the objective points), deg. (see Example 5.1).

The spatial resolution of an *OM* is given by the Rayleigh's equation (Martin, 1966):

$$\Delta r_0 = \frac{0.62\,\lambda}{NA} = \frac{0.62\,\lambda}{n\sin\alpha} \tag{5.2}$$

where Δr_0 is the resolution of the *OM*, nm; λ is the wavelength of the light source, nm; and *NA* is the numerical aperture of the objective (see Example 5.2). Equation 5.2 suggests that for good resolution, the *NA* of the microscope should be reasonably high. Since there are practical limitations on achieving excellent resolution, the resolution of a conventional optical microscope cannot be less than 200 nm.

5.3.2   ELECTRON MICROSCOPY

In the preceding subsection, we learned that there is a limitation on the resolution of an optical microscope due to the large wavelength of the visible light waves. This difficulty in microscopic resolution is overcome by the development of electron microscopes. Since the wavelength of electronic wave (beam) is much shorter than the visible light waves, the resolution of an electron microscope (using electronic beam) in theory can reach as low as 0.1 nm (Smith, 2008). A number of electron microscopic techniques have been developed; these include scanning electron microscopy (*SEM*), electron probe micro-analysis (*EPMA*), transmission electron microscopy (*TEM*), high resolution TEM (*HRTEM*), and the like. Since SEM and TEM are commonly used in microstructural characterization, they are discussed in the following subsections.

5.3.2.1   Scanning Electron Microscopy

In a scanning electron microscope (*SEM*), a focused beam of electrons is "scanned" across the surface of the sample so as to create an image from signals resulting from the interaction between the electron beam and the sample; these signals include: high-energy backscattered electrons (*BSE*), low-energy secondary electrons, X-rays, and the like (see Figure 5.8). In a SEM, the short wavelength of the electron beam

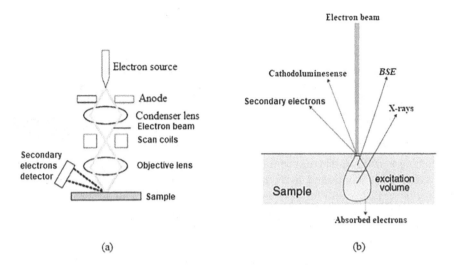

FIGURE 5.8 Basic components of SEM (a), electron-sample interaction (b).

used to form the image can resolve details smaller than 2 nm, allowing image magnifications of over 500,000X.

If an electron is accelerated from rest through a potential difference V, then the energy acquired by the electrons will appear as kinetic energy, as follows:

$$\tfrac{1}{2}\, m_e\, v^2 = e\, V \tag{5.3}$$

where m_e is the mass of an electron, kg; v is the velocity of electron, m/s; e is the electronic charge, Coul; and V is the accelerating voltage, volts. By rearranging the terms in Equation 5.3, we can express the velocity of electrons in terms of the electronic charge and mass, as follows:

$$v = \sqrt{\frac{2\,e\,V}{m_e}} \tag{5.4}$$

Equation 5.4 enables us to compute the velocity of electrons for a given V (see Example 5.3).

According to L. de Broglie, the wavelength of a fast-moving particle (here, the electron beam) can be calculated by:

$$\lambda = \frac{h}{m_e\, v} \tag{5.5}$$

where h is the Planck's constant ($h = 6.63 \times 10^{-34}$ J-s); $m_e = 9.1 \times 10^{-31}$ kg; $e = 1.6 \times 10^{-19}$ Coul

By combining Equations 5.4–5.5, we obtain:

$$\lambda = \sqrt{\frac{1.5}{V}} \tag{5.6}$$

where λ is the wavelength, nm; and V is the accelerating voltage, volts (see Example 5.4). The high resolution (< 0.2 nm) achieved in SEM is attributed to the short wavelength of the electron beam; which in turn strongly depends on the accelerating voltage. By reference to Equations 5.2 and 5.6, we can relate the accelerating voltage to the resolution of a microscope as follows. The higher the accelerating voltage, the smaller the wavelength of the electron beam and better the achievable resolution of the electron microscope.

5.3.2.2 Transmission Electron Microscopy

Working Principle. The transmission electron microscope (TEM) involves the use of a thin foil specimen and an electron beam to scrutinize objects at very fine resolutions. In a TEM, a thin-foil specimen is irradiated with a high-energy (in the range of 100–200 keV) electron beam that is focused by magnetic lenses. The illumination aperture and the size of the illuminated area are varied using condenser lenses that are provided before the specimen. The electron intensity distribution of the beam after interaction with the specimen is imaged onto a fluorescent screen by the objective lens and the post-objective lens system (see Figure 5.9). Images are generally recorded by a digital charge-coupled device (CCD) camera.

The TEM has its primary uses in metallurgy (or the study of metals and minerals), especially in terms of developing images of crystals and metals at the molecular level—allowing metallurgists to study material's structure and interactions, and

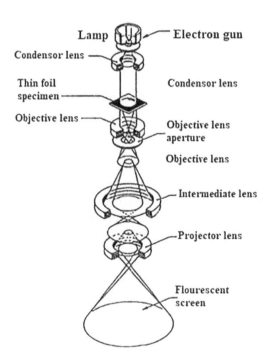

FIGURE 5.9 Components and working principle of TEM.

identify flaws. TEM micrographs enable us to observe structures and phases in materials at nano levels (Fultz and Howe, 2013). The dark areas of the image in a TEM micrograph correspond to areas on the specimen where fewer electrons were able to pass through; whereas the lighter areas are where more electrons did pass through; the varying amounts of electrons in these areas enable the user to see structures and gradients.

Resolution in TEM. In order to obtain an expression for resolution, the value for λ in Equation 5.6 can then be substituted into Equation 5.2. Since the angle α is usually very small ($\sim 10^{-2}$ *radians*) (a likely figure for TEM), the value of α approaches that of sin α, so we replace it. Since the refractive index, n, is unity (1), we eliminate it. Thus, combination of Equations 52 and 5.6 for TEM yields:

$$\Delta r_0 = \frac{0.76}{\alpha \sqrt{V}} \qquad (5.7)$$

where Δr_0 is the resolution, nm; α is the half aperture angle, radians; and V is the accelerating voltage, volts (see Example 5.5).

5.4 MATERIAL ANALYSIS AND SPECIFICATION OF COMPOSITION OF AN ALLOY

5.4.1 MATERIAL ANALYSIS/CHARACTERIZATION TOOLS

There are a number of modern material-characterization techniques/tools available to determine the chemical analysis of alloys; these tools include energy dispersive X-ray spectroscopy (*EDS*), electron probe micro-analyzer (*EPMA*), and the like. Since the *EDS* technique is commonly used in material analysis, it is briefly discussed in the following paragraph.

The energy dispersive X-ray spectroscopy (*EDS*) involves the use of the X-ray spectrum emitted by a solid sample bombarded with a focused beam of electrons to obtain a localized chemical analysis i.e. the elemental distribution in a specified region in an alloy. By scanning the beam in a television-like raster and displaying the intensity of a selected X-ray line, element distribution images or "maps" can be produced (Agarwal, 1991; Grieken and Markowicz, 2002). Although, a scanning electron microscope (SEM) is primarily designed for producing electron images (see Figure 5.8), it can also be used for elemental mapping, and even point analysis, provided an X-ray spectrometer is added. In *EDS*, the relative intensity of an X-ray line (peak) is approximately proportional to the mass concentration of the element concerned (see Figure 5.10).

The main outcome from the EDS measurement is the value of the relative intensity of the spectral lines (*K-ratio*) measured after the optimization of the shape of the peaks. For an alloy, the concentration of the ith element in the specimen (C_i) is calculated by (Goldstein *et al.*, 2003):

$$\frac{c_i}{c_{(i)}} = \text{ZAF}\left(\frac{I_i}{I_{(i)}}\right) = (\text{ZAF})(K - ratio) \qquad (5.8)$$

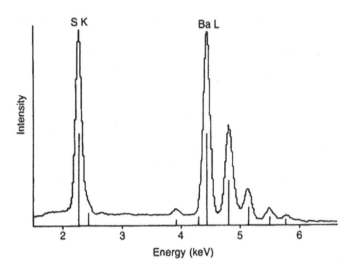

FIGURE 5.10 Line markers for S K and Ba L lines in the ED spectrum of barite (S = sulfur; Ba = barium; K = first atomic orbit, L = second atomic orbit).

where $\dfrac{c_i}{c_{(i)}}$ is the relative concentration of the element in specimen and in standard; $\dfrac{I_i}{I_{(i)}}$ is relative intensity of spectral lines (= K-ratio); Z is the mean atomic number correction factor; A is the absorption correction factor; and F is the fluorescence correction factor (see Example 5.6).

5.4.2 Specification of Composition in Atom Percent

It is often necessary to express the composition (or concentration) of an alloy in terms of atom percent (at%). The atom percent of an element in alloy is the ratio of the number of moles of the element to the total number of moles of all the elements in the alloy. Considering a binary alloy comprising of hypothetical elements X and Y, we may express the at% of each element by:

$$C'_Y = \frac{n_{mX}}{n_{mX} + n_{mY}} \times 100 \tag{5.9}$$

$$C'_Y = \frac{n_{mY}}{n_{mX} + n_{mY}} \times 100 \tag{5.10}$$

where C'_X and C'_Y are the at% of elements X and Y, respectively, in the alloy; n_{mX} is the number of moles in some specified mass of element X; and n_{mY} is the number of moles in some specified mass of element Y.

The number of moles of the element X, n_{mX}, can be calculated by:

$$n_{mX} = \frac{m_X}{A_X} \tag{5.11}$$

where m_X is the mass of element X in grams, and A_X is the atomic weight of X. The number of moles of the element Y can similarly be calculated (see Example 5.7).

If the weight percent of elements in an alloy is given, the atom percent can be computed by:

$$C'_X = \frac{C_X A_Y}{C_X A_Y + C_Y A_X} \times 100 \tag{5.13}$$

$$C'_Y = \frac{C_Y A_X}{C_X A_Y + C_Y A_X} \times 100 \tag{5.14}$$

where C'_X and C'_Y are the at% of elements X and Y, respectively, in the alloy; and C_X and C_Y are the wt% X and wt% Y, respectively, in the alloy. It may be noted that for a ternary alloy, Equations 5.13 and 5.14 may be generalized accordingly (see Example 5.8).

5.4.3 SPECIFICATION OF AVERAGE DENSITY OF AN ALLOY

It is sometimes required to determine the density of an alloy when the composition in wt% of an alloy is known. The density of a hypothetical X-Y-Z alloy can be determined by:

$$\rho_{ave} = \frac{100}{\dfrac{C_X}{\rho_X} + \dfrac{C_Y}{\rho_Y} + \dfrac{C_Z}{\rho_Z}} \tag{5.15}$$

where ρ_{ave} is the average (approximate) density of the alloy comprising of elements X, Y, and Z; C_X, C_Y, and C_Z are the elemental concentrations wt% X, wt% Y, and wt% Z, respectively; and ρ_X, ρ_Y, and ρ_Z are the densities of metals X, Y, and Z, respectively (see Example 5.9).

5.5 QUANTITATIVE METALLOGRAPHY— GRAIN SIZE MEASUREMENT

Quantitative metallography involves the quantitative analysis of microstructures by various quantitative metallographic techniques; which include: grain size measurement; the determination of volume fractions of phases or microconstituents, and the amount and distribution of phases (Colpaert, 2018). These microstructural parameters can directly affect the mechanical properties. There are two main approaches in quantitative metallography: (a) manual quantitative metallography, and (b) computer-aided quantitative metallography.

5.5.1 Manual Quantitative Metallography

Manual quantitative metallographic techniques are traditionally in use by metallurgists for analyzing the microstructure of a metal. These techniques mainly include: (a) *chart method*, and (b) *counting method*; which are explained in the following paragraphs.

5.5.1.1 The Chart Method of Manual Quantitative Metallography

In the standard *chart method*, the sample is viewed while continuously being referred to a standard chart; which contains micrographs at the same magnification as the sample with varying percentages of the parameter(s) in question. This process enables the metallurgist to directly compare the sample with standard samples; and hence the most representative microstructure is determined. The *chart method* is especially useful in process environments where large number of samples is routinely analyzed. It is a quick, inexpensive method of analysis.

5.5.1.2 The Counting Method of Manual Quantitative Metallography

The *counting method* involves direct measurement/counting of a metallographic parameter by the operator. A common example of manual quantitative metallography by *counting method* is the grain size measurement. The first step in the grain size measurement is the determination of the magnification of the image. If there is a magnification bar on the micrograph, the magnification (M) can be determined by:

$$M = \frac{Measured\ length\ of\ the\ bar}{Length\ shown\ on\ the\ bar} \tag{5.16}$$

Equation 5.16 can also be used to draw the magnification bar when a microscope's magnification is known (see Example 5.10).

There are mainly three techniques of grain size determinations: (a) *line intercept method*, (b) *ASTM* grain size number, and (c) *Jeffries planimetric* method.

Line Intercept Method of Grain Size Measurement. The *line intercept method*, also called *the average grain intercept (AGI) method*, is a manual quantitative metallographic technique used to quantify the grain size for a metal by drawing a set of randomly positioned line segments (each of length *l*) on the micrograph, and counting the number of times each line segment intersects a grain boundary. Then, the average line length intersected is determined by:

$$N_l = \frac{l}{\overline{N}} \tag{5.17}$$

where N_l is the average line length intersected, mm; l is the length of each line segment, mm; and N is the average number of grain boundaries intersected (see Example 5.11).

The average grain size or grain diameter is determined by:

$$d = \frac{N_l}{Magnification} \tag{5.18}$$

where d is the average grain size of the metal, mm.

ASTM Grain Size Number. ASTM refers to the American Society of Testing of Materials. The ASTM grain size number for a given photomicrograph can be determined by using:

$$N = 2^{G-1} \qquad (5.19)$$

where N is the number of grains observed in an area of 1 in² on a photomicrograph taken at a magnification of 100X, and G is the ASTM grain size number. It is evident in Equation 5.19 that the ASTM grain size number increases as the number of grains per unit area increases. By rearranging the terms in Equation 5.19, we obtain:

$$G = \frac{\log N}{\log 2} + 1 \qquad (5.20)$$

Equation 5.20 enables us to conveniently calculate the ASTM grain size number (G) when N is known (see Example 5.12).

The ASTM grain size number can be used to compute the average grain size of a metal, as follows:

$$d = \sqrt{\frac{\frac{25.4^2}{2^{G-1}}}{100^2}} \times 1000 = \sqrt{\frac{25.4^2}{2^{G-1}}} \times 10 \qquad (5.21)$$

where d is the average grain size of the metal, μm; and G is the ASTM grain size number (see Example 5.13).

Jeffries Planimetric Method. The *Jeffries method* is a counting method of metallography; which involves drawing a circle onto a photomicrograph and measuring the grain size in mm.

The number of grains per mm², N_A, is computed by:

$$N_A = \frac{M^2 \left(n_1 + \frac{n_2}{2} \right)}{A_t} \qquad (5.22)$$

where M is the magnification of the photomicrograph, A_t is the test area (area of the inscribed circle), mm²; n_1 is the number of (full) grains in the inscribed area; and n_2 is the number of grains intersecting the perimeter of the test area. The ASTM grain size number (G) is computed by:

$$G = (3.322 \log N_A) - 2.95 \qquad (5.23)$$

where N_A is the number of grains per mm² (see Example 5.14).

5.5.2 COMPUTER-AIDED QUANTITATIVE METALLOGRAPHY— COMPUTERIZED IMAGE ANALYSIS

Like optical microscopy, computerized image analysis utilizes the reflective properties of the sample to allow the phases/boundaries of the microstructure to be distinguished. In the computer-aided metallographic technique, the micrograph is divided into discrete optical units—called *pixels*. The dedicated computer (the computer linked to the microscope) is used to record *pixel* information; which is then used for quantitative metallographic analysis. A number of commercial image-analysis software is available; which allows us to determine various microstructural parameters; these parameters include: area, perimeter, length, width, grain size, and area fraction of phases present in the microstructure.

5.6 CALCULATIONS—EXAMPLES ON METALLOGRAPHY AND MATERIALS CHARACTERIZATION

EXAMPLE 5.1 COMPUTING NUMERICAL APERTURE OF THE OBJECTIVE LENS OF AN *OM*

The refractive index between the lens and the object of a metallurgical microscope is 1.63. The half-angle aperture is 72°. Calculate the numerical aperture of the objective lens of the *OM*.

SOLUTION

$n = 1.63$, $\alpha = 72°$, NA = ?
By using Equation 5.1,

NA = $n \sin \alpha$ = 1.63 x sin 72° = 1.63 x 0.95 = 1.54

EXAMPLE 5.2 COMPUTING RESOLUTION OF A METALLURGICAL MICROSCOPE

By using the data in Example 5.1, calculate the resolution of the microscope if the visible light from the illumination system has a wavelength of 550 nm.

SOLUTION

$\lambda = 550$ nm; $NA = 1.54$, Δr_0 = ?
By using Equation 5.2,

$$\Delta r_0 = \frac{0.62\,\lambda}{NA} = \frac{0.62 \times 550}{1.54} = 221.4\,\text{nm}$$

The resolution of the metallurgical microscope is 221.4 nm or 0.22 μm.

EXAMPLE 5.3 COMPUTING THE VELOCITY OF ELECTRONS IN SEM WHEN THE VOLTAGE IS KNOWN

In an experiment, an accelerating voltage of 80 kV was applied to a SEM. Calculate the velocity of electrons.

SOLUTION

$V = 80$ kV $= 80,000$ V; $m_e = 9.1$ x 10^{-31} kg; $e = 1.6$ x 10^{-19} Coul; $v = ?$
By using Equation 5.4, we obtain:

$$v = \sqrt{\frac{2\,eV}{m_e}} = \sqrt{\frac{2 \times 1.6 \times 10^{-19} \times 80,000}{9.1 \times 10^{-31}}} = 1.67 \text{ x } 10^8 \text{ m/s}$$

The velocity of electrons $= 1.67$ x 10^8 m/s.

EXAMPLE 5.4 COMPUTING THE WAVELENGTH OF AN ELECTRON BEAM WHEN THE VOLTAGE IS KNOWN

The voltage applied to accelerate electron beam in a SEM is 10 kV. What is the wavelength of the electron beam?

SOLUTION

$V = 10$ kV $= 10,000$ volts, $\lambda = ?$
By using Equation 5.6,

$$\lambda = \sqrt{\frac{1.5}{V}} = \sqrt{\frac{1.5}{10,000}} = 0.0122 \text{ nm}$$

The wavelength of the electronic beam $= 0.0122$ nm or 12.2 pm.

EXAMPLE 5.5 CALCULATING THE RESOLUTION OF TEM

Calculate the resolution of the TEM if the accelerating voltage is 100 kV. (Take a reasonable value for the half aperture angle).

SOLUTION

$V = 100$ kV $= 100,000$ V; $\alpha = 10^{-2}$ radians ; $\Delta r_0 = ?$
By using Equation 5.7,

$$\Delta r_0 = \frac{0.76}{\alpha \sqrt{V}} = \frac{0.76}{0.01\sqrt{100000}} = 0.24 \text{ nm}$$

The resolution of TEM $= 0.24$ nm.

EXAMPLE 5.6 COMPUTING RELATIVE CONCENTRATION OF ELEMENT IN AN ALLOY USING EDS DATA

The dispersion of the energy spectrum of X-rays from the selected point (micro-area) of an alloy sample resulted in the EDS data, shown in Table E-5.6.

TABLE E-5.6
The EDS Data for the Investigated Alloy

Element	K-ratio	Z	A	F
Fe K	0.71	1.002	1.019	0.992
Cr K	0.21	0.999	1.006	0.879
Ni K	0.08	0.990	1.065	1.000

Calculate relative concentration of each element in specimen; and identify the alloy. (The standard was 100% pure so the value for the standard may be taken as 1.)

SOLUTION

By using Equation 5.8 for Fe K,

$$\frac{c_{Fe}}{c_{(Fe)}} = (ZAF)(K - ratio) \ = \ 1.002 \times 1.019 \times 0.992 \times 0.71$$
$$= \ 0.719 \sim 72\%$$

By using Equation 5.8 for Cr K,

$$\frac{c_{Cr}}{c_{(Cr)}} = (ZAF)(K - ratio) = 0.999 \times 1.006 \times 0.879 \times 0.21$$
$$= 0.185 \sim 18.5\%$$

By using Equation 5.8 for Ni K,

$$\frac{c_{Cr}}{c_{(Cr)}} = (ZAF)(K - ratio) = 0.990 \times 1.065 \times 1.000 \times 0.08$$
$$= 0.084 \sim 8.5\%$$

By taking the standard concentrations for each element to be unity, we obtain:

Wt% Fe = 72, wt% Cr = 18.5, and wt% Ni = 8.5

Hence, we identify the alloy to be the 18-8 stainless steel (containing 18% Cr and 8% Ni).

EXAMPLE 5.7 COMPUTING THE ATOM PERCENT OF AN ALLOY WHEN THE WEIGHTS ARE KNOWN

Calculate the composition, in atom percent, of alloy that contains 32 g copper and 48 g zinc. The atomic weights of copper and zinc are 63.5 and 65.4 g/mol, respectively.

SOLUTION

$m_{Cu} = 32$ g, $m_{Zn} = 48$ g; $A_{Cu} = 63.5$ g/mol, $A_{Zn} = 65.4$ g/mol

By using Equation 5.11,

$$n_{mCu} = \frac{m_{Cu}}{A_{Cu}} = \frac{32}{63.5} = 0.504 \, \text{g-moles}$$

$$n_{mZn} = \frac{m_{Zn}}{A_{Zn}} = \frac{48}{65.4} = 0.734 \, \text{g-moles}$$

By using Equation 5.9,

$$C'_{Cu} = \frac{n_{mCu}}{n_{mCu+n_{mZn}}} \times 100 = \frac{0.504}{0.504+0.734} \times 100 = 40.7 \, \text{at\%}$$

By using Equation 5.10,

$$C'_{Zn} = \frac{n_{mZn}}{n_{mCu+n_{mZn}}} \times 100 = \frac{0.734}{0.504+0.734} \times 100 = 59.3 \, \text{at\%}$$

Hence, the concentrations of copper and zinc are 40.7 at% and 59.3 at%, respectively.

EXAMPLE 5.8 COMPUTING COMPOSITION IN ATOM PERCENT WHEN THE WEIGHTS % ARE KNOWN

Calculate the atom percent of each element in the 60Cu-40Ni alloy.

SOLUTION

Wt% Cu = C_{Cu} = 60, wt% Ni = C_{Ni} = 40, A_{Cu} = 63.5 g/mol, A_{Ni} = 58.7 g/mol

By using Equations 5.13 and 5.14,

$$C'_{Cu} = \frac{C_{Cu}A_{Ni}}{C_{Cu}A_{Ni} + C_{Ni}A_{Cu}} \times 100 = \frac{60 \times 58.7}{(60 \times 58.7) + (40 \times 63.5)} \times 100 = 58 \, \text{at\%}$$

$$C'_{Ni} = \frac{C_{Ni}A_{Cu}}{C_{Cu}A_{Ni} + C_{Ni}A_{Cu}} \times 100 = \frac{40 \times 63.5}{(60 \times 58.7) + (40 \times 63.5)} \times 100 = 42 \, \text{at\%}$$

Hence, the concentrations of copper and nickel are 58 at% and 42 at%, respectively.

EXAMPLE 5.9 COMPUTING THE DENSITY OF AN ALLOY WHEN WT% OF ELEMENTS ARE KNOWN

Determine the approximate density of a leaded brass that has a composition of 64.7 wt% Cu, 33.5 wt% Zn, and 1.8 wt% Pb.

SOLUTION

Wt% Cu = 64.7, wt% Zn = 33.5, wt% Pb = 1.8, ρ_{Cu} = 8.94 g/cm³, ρ_{Zn} = 7.13, ρ_{Pb} = 11.35 g/cm³

By using Equation 5.15,

$$\rho_{ave} = \frac{100}{\dfrac{C_X}{\rho_X} + \dfrac{C_Y}{\rho_Y} + \dfrac{C_Z}{\rho_Z}} = \frac{100}{\dfrac{C_{Cu}}{\rho_{Cu}} + \dfrac{C_{Zn}}{\rho_{Zn}} + \dfrac{C_{Pb}}{\rho_{Pb}}} = \frac{100}{\dfrac{64.7}{8.94} + \dfrac{33.5}{7.13} + \dfrac{1.8}{11.35}}$$
$$= 8.28 \ g \ / \ cm^3$$

The average density of the leaded brass = 8.28 g/cm³.

EXAMPLE 5.10 DRAWING MAGNIFICATION BAR ON A MICROGRAPH

Draw magnification bar on the micrograph in Figure 5.6 given that the magnification is 100X.

SOLUTION

Magnification = M = 100. Let us draw a bar with measured length = 2 cm. By rearranging the terms in Equation 5.18, we can write:

$$\text{Length to be shown on the magnification bar} = \frac{Measured \ length \ of \ the \ bar}{M}$$
$$= \frac{2 \ cm}{100} = 200 \ \mu m$$

Hence, we draw the magnification bar within the micrograph as shown in Figure E-5.10.

EXAMPLE 5.11 MEASURING GRAIN SIZE BY THE LINE INTERCEPT METHOD

Determine the average grain size of the metal whose micrograph is shown in Figure E-5.10. (Use the line intercept method of grain size measurement.)

SOLUTION

First we need to determine the magnification by referring to the magnification bar in Figure E-5.10.
By using Equation 5.16,

$$M = \frac{Measured \ length \ of \ the \ bar}{Length \ shown \ on \ the \ bar} = \frac{2 \ cm}{200 \ \mu m} = 100$$

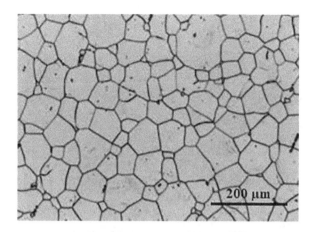

FIGURE E-5.10   The optical micrograph showing magnification bar.

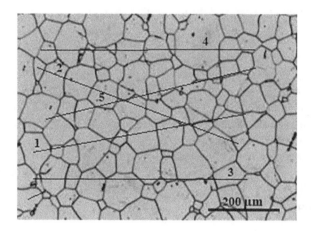

FIGURE E-5.11   The optical micrograph after drawing line segments.

TABLE E-5.11

Data Showing the Number of Grain Boundaries Intersected by Each Line Segment

Line #	1	2	3	4	5
No. of grain boundaries intersected	13	11	12	12	15

By drawing a set of randomly positioned line segments (each of length 60 mm) on the micrograph, and counting the number of grain boundaries intersected by each line segment, we obtain Figure E-5.11 and the data in Table E-5.11.

$$\text{The average number of grain boundaries intersected} = \bar{N} = \frac{13+11+12+12+15}{5}$$

$$=12.6$$

By using Equation 5.17,

$$N_l = \frac{l}{\overline{N}} = \frac{60\,mm}{12.6} = 4.76 \text{ mm}$$

By using Equation 5.18,

$$d = \frac{N_l}{Magnification} = \frac{4.76}{100} = 0.0467\,mm$$

The average grain size of the metal = 0.0467 mm = 46.7 μm (see this book's cover design image).

EXAMPLE 5.12 CALCULATING THE ASTM GRAIN SIZE NUMBER

Calculate the ASTM grain size number of the metal whose micrograph is shown in Figure E-5.10.

SOLUTION

In order to calculate the ASTM grain size, we need to first determine the number of grains observed in an area of 1 in² on the micrograph (Figure E-5.10). Hence, we specify 1 in² area on the micrograph as shown in Figure E-5.12. It is known that the magnification is 100X (see Example 5.10).

By reference to Figure E-5.12,

The number of grains observed in an area of 1 in² on the micrograph at 100X = N = 28

By using Equation 5.20,

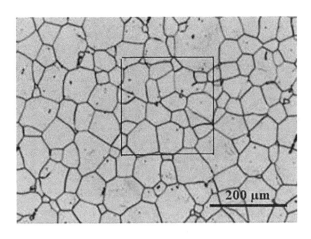

FIGURE E-5.12   Drawn one square inch area on the micrograph for finding *N*.

$$G = \frac{\log N}{\log 2} + 1 = \frac{\log 28}{\log 2} + 1 = 5.8$$

The ASTM grain size number of the metal = 5.8.

EXAMPLE 5.13 COMPUTING THE AV. GRAIN SIZE OF A METAL WHEN ASTM GRAIN SIZE IS KNOWN

By using the data in Example 5.12, compute the average grain size of the metal. Compare your result with the one obtained by line intercept method illustrated in Example 5.11.

SOLUTION

$G = 5.8$, $d = ?$ µm
By using Equation 5.21

$$d = \sqrt{\frac{25.4^2}{2^{G-1}}} \times 10 = \sqrt{\frac{25.4^2}{2^{5.45-1}}} \times 10 = 48 \, \mu m$$

The calculated average grain size of the metal ($d = 48$ µm) is in close agreement with that obtained in Example 5.11 ($d = 46.7$ µm).

EXAMPLE 5.14 COMPUTING THE ASTM GRAIN SIZE NUMBER BY USING *JEFFRIES* METHOD

Calculate the ASTM grain size number for the photomicrograph of the metal at a magnification of 100X, as shown in Figure 5.6, by using the *Jeffries planimetric* method. Compare your result with the one obtained in Example 5.12.

SOLUTION

In order to cover the maximum area of the microstructure in Figure 5.6, we draw a circle having a diameter of 54 mm; this corresponds to an inscribed area of 2,290 mm² (see Figure E-5.14).

By counting the number of grains within the inscribed circle (n_1) and the number of grain boundaries intersecting the circle (n_2) in Figure E-5.14, we obtain:
M = 100; A_t = 2,290 mm², n_1 = 48, and n_2 = 28
By using Equation 5.22,

$$N_A = \frac{M^2 \left(n_1 + \frac{n_2}{2} \right)}{A_t} = \frac{100^2 \left(48 + \frac{28}{2} \right)}{2290} = 270.7$$

By using Equation 5.23,

$$G = (3.322 \log 270.7) - 2.95 = 5.13$$

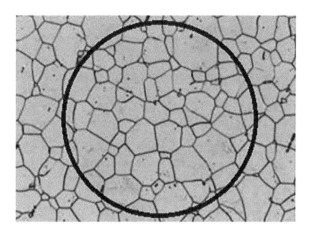

FIGURE E-5.14   The inscribed circle in the photomicrograph shown in Figure 5.6.

By comparing the calculated ASTM grain size number of the metal ($G = 5.13$) with the one obtained in Example 5.12 ($G = 5.8$), we find that the two G values are in close agreement.

Questions and Problems

5.1. Give the meanings of the following abbreviations as used in material characterization.
(a) *SEM*, (b) *BSE*, (c) *ASTM*, (d) *TEM*, (e) *EDS*, (f) *OM*, (g) *EPMA*, (h) *HRTEM*.

5.2. Encircle the correct answers for the following multiple choice questions (MCQs).
(a) Which operation results in directional scratches onto the metallographic specimen? (i) polishing, (ii) etching, (iii) grinding, (iv) sectioning, (v) mounting.
(b) Which operation enables us to observe grained microstructure of the specimen? (i) polishing, (ii) etching, (iii) grinding, (iv) sectioning, (v) mounting.
(c) Which operation requires use of fluid for cooling? (i) polishing, (ii) etching, (iii) grinding, (iv) sectioning, (v) mounting.
(d) Which operation involves the use of Bakelite for metallographic specimen preparation? (i) polishing, (ii) etching, (iii) grinding, (iv) sectioning, (v) mounting.
(e) Which techniques enable us to determine the chemical composition of an alloy? (i) TEM, (ii) SEM, (iii) OM, (iv) quantitative metallography, (v) EDS.
(f) Which technique involves penetration of an electron beam through the sample? (i) TEM, (ii) EDS, (iii) OM, (iv) quantitative metallography, (v) SEM.

(g) Which technique has low microscopic resolution? (i) TEM, (ii) EDS, (iii) OM, (iv) quantitative metallography, (v) SEM.

(h) Which technique enables us to measure grain size? (i) mounting, (ii) EDS, (iii) OM, (iv) quantitative metallography.

(*i*) Which technique involves use of X-ray to determine the chemical composition? (i) TEM, (ii) EDS, (iii) OM, (iv) quantitative metallography.

(j) For which metal/alloy, is 2% nitric acid-98% alcohol a suitable metallographic etchant? (i) nickel, (ii) copper, (iii) carbon steel, (iv) stainless steel.

5.3. Explain the evolution of grained microstructure of a polycrystalline material with the aid of sketches. Give two examples of applications of single crystals.

5.4. (a) List the main steps in the preparation of a metallographic specimen. (b) Explain the leveling and etching of a metallographic specimen.

5.5. What is meant by the resolution of a microscope? On what factors does resolution depend?

5.6. What are the advantages of electron microscopy over OM?

5.7. Explain the working principle of TEM with the aid of sketch.

5.8. Describe the EDS technique of material characterization.

P5.9. Determine the average grain size, in μm, for the metallographic specimen whose microstructure is shown in Figure P5.9 using line intercept method.

P5.10. By using the micrograph in Figure 5.6, calculate the ASTM grain size number of the metal by two (manual metallography) methods; and compare your results.

P5.11. Calculate the average grain size of a sample of annealed aluminum given that the ASTM grain size number for the metal's sample is 6.5.

P5.12. The refractive index between the lens and the object of a metallurgical microscope is 1.6. The half-angle aperture is 70°. Calculate the resolution of the microscope if the visible light from the illumination system has a wavelength of 540 nm.

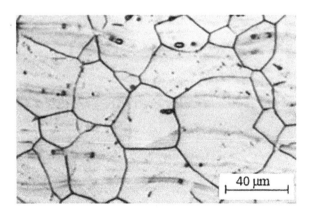

FIGURE P5.9 Optical micrograph for the metallographic specimen.

P5.13. In an experiment, an accelerating voltage of 100 kV was applied to a SEM. Calculate the velocity of electrons.

P5.14. Calculate the composition, in atom %, of the alloy that contains 40 g copper and 36 g tin. The atomic weights of copper and tin are 63.5 and 118.7 g/mol, respectively.

P5.15. The voltage applied to an accelerate electron beam in a SEM is 15 kV. What is the wavelength of the electron beam?

P5.16. Calculate the atom percent of each element in the 70-30 brass.

P5.17. Calculate the approximate density of 18-8 stainless steel, given that densities of iron, chromium, and nickel are 7.8, 7.2, and 8.9 g/cm³, respectively.

REFERENCES

Agarwal, B.K. (1991) *X-ray Spectroscopy*. 2nd ed. Springer-Verlag, Berlin.

Colpaert, H. (2018) *Metallography of Steels*. ASM International, Materials Park, OH.

Fultz, B. & Howe, J. (2013) *Transmission Electron Microscopy and Diffractometry of Materials*. 4th ed. Springer-Verlag, Berlin.

Goldstein, J., Newbury, D.E., Joy, D.C., Lyman, C.E., Echlin, P., Lifshin, E., Sawyer, L. & Michael, J.R. (2003) *Scanning Electron Microscopy and X-ray Microanalysis*. Springer, Germany.

Grieken, R.E.V. & Markowicz, A.A. (2002) *Handbook of X-ray Spectrometry*. Marcel Dekker, Inc, New York.

Martin, L.C. (1966) *The Theory of the Microscope*. Blackie Publishers, London.

Millward, D., Ahmet, K. & Attfield, J. (2013) *Construction and the Built Environment*. 2nd ed. Routledge Publishers, London.

Smith, D.J. (2008) Ultimate resolution in the electron microscope? *Materials Today*, 11, 30–38.

6 Phase Diagrams

6.1 THE BASIS OF PHASE DIAGRAMS

6.1.1 GIBBS'S PHASE RULE

It is important to be first familiar with *Gibbs's phase rule*, since it forms the basis of phase diagrams. The *Gibbs's phase rule* expresses the possible number of degrees of freedom (F) in a closed system at equilibrium, in terms of the maximum number of equilibrium phases (P) and the number of system components (C), as follows (Hillert, 2007):

$$P + F = C + 2 \tag{6.1}$$

The degrees of freedom, F, can be calculate by:

$$F = C - P + 2 \tag{6.2}$$

The number of degrees of freedom, F, for a system at equilibrium is the number of intensive variables (often taken as the pressure, temperature, and composition fraction) that may be arbitrarily specified without changing the number of phases. If $F=0$, an *invariant equilibrium* is defined (i.e. equilibrium can be attained only for a single set of values of all the state variables). If $F=1$, an *univariant equilibrium* is defined (i.e. the set of stable phases depends on the value of one stable variable only).

6.1.2 WHAT IS A PHASE DIAGRAM?

A phase diagrams, also called an equilibrium diagram, is a graphical representation of the stabilities of various phases (at equilibrium) at given compositions, temperatures, and/or pressures (Okamoto *et al.*, 2016; Rhines, 1956). A *phase* is the part of a system with distinct chemical and physical properties; and which can be physically separated from the system. The phase diagram of a one-component (H_2O) system is shown in Figure 6.1. The H_2O phase diagram can be explained by applying Gibbs's phase rule (see Example 6.1).

6.2 PHASE TRANSFORMATION REACTIONS AND CLASSIFICATION OF PHASE DIAGRAMS

6.2.1 PHASE TRANSFORMATION REACTIONS

A triple point in a phase diagrams represents a phase transformation involving three phases (see Figure 6.1). There are a number of triple-point phase transformation reactions; however the following four reactions are generally encountered in

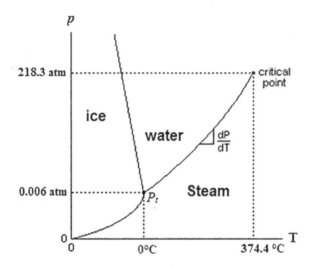

FIGURE 6.1   Phase diagram for the one-component (H_2O) system.

alloy systems: (a) eutectic, (b) eutectoid, (c) peritectic, and (d) peritectoid reactions (Hillert, 2007). These phase transformation reactions are briefly explained in the following paragraphs.

Eutectic reaction. The *eutectic reaction* involves transformation (on cooling) of a liquid phase into two different solid phases, as follows:

$$\text{Liquid} \rightarrow \text{Solid}_1 + \text{Solid}_2 \tag{6.3}$$
$$\text{or } L \rightarrow \alpha + \beta \tag{6.4}$$

The *eutectic reaction* occurs at a specific composition (eutectic composition) and at a particular temperature (eutectic temperature) (see section 6.4).

Eutectoid reaction. The *eutectoid reaction* involves transformation (on cooling) of a solid phase into two different solid phases, as follows:

$$\text{Solid}_1 \rightarrow \text{Solid}_2 + \text{Solid}_3 \tag{6.5}$$

A typical example of eutectoid reaction is the transformation of the austenite phase (γ) to ferrite (α) and cementite (Fe_3C) phases when steel of eutectoid composition (0.8% C) is cooled below the eutectic temperature (723°C); this eutectoid reaction is represented as:

$$\text{or } \gamma \rightarrow \alpha + Fe_3C \tag{6.6}$$

Peritectic reaction. The *peritectic reaction* involves transformation (on cooling) of a solid phase and a liquid phase into another solid phase, as follows:

$$\text{Solid}_1 + \text{liquid} \rightarrow \text{Solid}_2 \tag{6.7}$$

Peritectoid reaction. The *peritectoid reaction* involves transformation (on cooling) of two different solid phases into a third solid phase, as follows:

$$Solid_1 + Solid_2 \rightarrow Solid_3 \tag{6.8}$$

6.2.2 CLASSIFICATION OF PHASE DIAGRAMS

Having defined the phase transformation reactions in the preceding subsections, the classification of phase diagrams is now presented. Based on the number of components, phase diagrams may be grouped into three categories:

1.   One-component or unary phase diagrams (e.g. H_2O system),
2.   Two-components or binary phase diagrams (e.g. Cu-Ni system), and
3.   Three-components or ternary phase diagrams (e.g. Fe-Ni-Cr system).

Based on phase transformation reactions, phase diagrams may be classified into four types:

1.   Phase diagrams involving eutectic reactions (e.g. Pb-Sn system, Al-Si system, etc.),
2.   Phase diagrams involving peritectic/eutectoid reactions (e.g. Fe-Fe$_3$C system),
3.   Phase diagrams involving peritectoid reactions (e.g. Cu-Al system), and
4.   Complex phase diagrams (e.g. Al-Ni system).

6.2.3 UNARY PHASE DIAGRAMS

Unary phase diagrams are one-component phase diagrams (see subsection 6.1.2). Examples of unary phase diagrams include H_2O pressure-temperature phase diagram, SiO_2 phase diagram, Fe phase diagram, and the like.

6.3 BINARY PHASE DIAGRAMS

6.3.1 BASICS OF BINARY PHASE DIAGRAMS

Definition. Most materials are composed of at least two components. A phase diagram based on two components, is called a *binary phase diagram* (Okamoto, 2002; Massalski *et al.*, 1990).

   Classification of Binary Phase Diagrams. In general, binary phase diagrams may be classified into the following types: (a) isomorphous binary phase diagrams, (b) binary phase diagrams involving eutectic reactions, (c) binary phase diagrams involving peritectic/eutectoid reactions, and (d) binary complex phase diagrams. The isomorphous binary phase diagrams are discussed in this section; the other types of binary phase diagrams are discussed in the subsequent sections.

6.3.2 ISOMORPHOUS BINARY PHASE DIAGRAMS

Basics. A material exhibiting complete solid solubility and liquid solubility of its constituent for all compositions, is called *isomorphous alloy/material system.* Owing to 100% solid solubility, there exists only a single type of crystal structure for all compositions in an isomorphous material system. Examples of isomorphous material systems include: Cu-Ni alloy, Sb-Bi alloy, NiO-MgO ceramic system, and the like (McHale, 2010).

The binary phase diagram of a typical isomorphous Cu-Ni alloy is shown Figure 6.2. It is evident in Figure 6.2 that there are three different phase regions: the α (FCC structure) region, the liquid region, and the α + liquid (α + L) region; these regions are separated by two phase boundaries: liquidus and solidus.

The liquidus and solidus lines enable us to determine the temperature of liquid and solid phases for a given overall composition of the alloy. The *liquidus* gives the temperature of the liquid phase at which, for a given composition, the first solid forms. The *solidus* gives the temperature of the solid phase at which, for a given composition, the first liquid forms (or during cooling, the temperature of the two-phase mixture at which the last liquid solidifies). The liquidus and solidus lines also enable

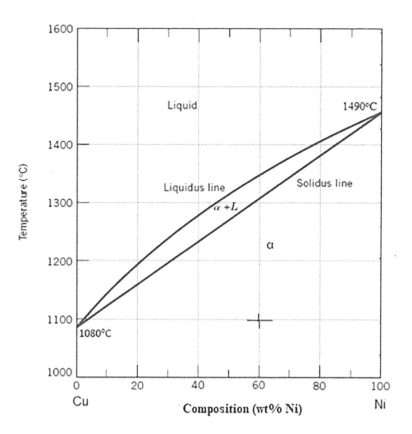

FIGURE 6.2 Copper-nickel phase diagram.

us to determine compositions of the first solid to form and the last liquid to solidify, respectively (see Example 6.3). Since solidification of an alloy occurs over a range of temperatures, the cooling of Cu-Ni alloy in the two-phase ($\alpha + L$) region results in a *cored structure*; which indicates non-homogeneity in the composition of the alloy. This problem can be overcome by a homogenization annealing heat treatment; which involves diffusion of Ni atoms from Ni-rich regions to Cu-rich regions thereby homogenizing the composition.

6.3.3 MATHEMATICAL MODELS FOR BINARY SYSTEMS

Mathematical Modeling for Composition. It is important to mathematically model the various ways to specify composition of an alloy/material system. If an alloy is composed of two metals, A and B, and the composition is given in weight %, the conversion to atomic % is done as follows:

$$at.\, \%A = \frac{\dfrac{wt\, \%A}{at.\, wt\, A}}{\dfrac{wt\, \%A}{at.\, wt\, A} + \dfrac{wt\, \%B}{at.\, wt\, B}} \times 100 \tag{6.9}$$

$$at.\, \%B = \frac{\dfrac{wt\, \%B}{at.wt\, B}}{\dfrac{wt\,\%A}{at.wt\, A} + \dfrac{wt\,\%B}{at.wt\, B}} \times 100 \tag{6.10}$$

In ceramic systems, the components are usually compounds; and the composition of a ceramic is generally expressed as mole fraction, N. The mole fraction of a component A in a two-component system, N_A, is given by:

$$N_A = \frac{n_A}{n_A + n_B} \tag{6.11}$$

where n_A and n_B are the number of moles of A and B in the material, respectively. The sum of mole fractions of all components is unity i.e.

$$N_A + N_B = 1 \tag{6.12}$$

Mathematical Modeling for Amounts of Phases—Lever Rule. The *lever rule*, also called *inverse lever rule*, states that: "the relative amount (weight fraction) of a particular phase is equal to the ratio of the tie line portion which lies between the overall composition and the phase boundary curve away from the phase of interest, to the total tie line length." The *rule* enables us to determine the weight fraction of each phase present in a two-phase field in the phase diagram.

The *inverse lever rule* can be mathematically expressed as follows:

$$W_L = \frac{C_\alpha - C_o}{C_\alpha - C_L} \tag{6.13}$$

where W_L is the weight fraction of the liquid; C_o is the composition of the alloy/material under consideration; C_α is the composition of the α (solid); and C_L is the composition of the liquid.

Similarly, the weight fraction of the solid phase (W_s) can be calculated as follows:

$$W_s = \frac{C_o - C_L}{C_\alpha - C_L} \tag{6.14}$$

The significance of Equation 6.13 through Equation 6.14 is illustrated in Example 6.6.

6.4 BINARY PHASE DIAGRAMS INVOLVING EUTECTIC REACTIONS

The term "*eutectic*" is a Greek word; which means "*easy to melt.*" An alloy of eutectic composition has the lowest melting temperature as compared to other compositions of the alloy. The term *eutectic reaction* has been introduced in section 6.2. There are two main types of binary phase diagrams involving a eutectic reaction: (a) eutectic phase diagrams with complete insolubility in solid state, and (b) eutectic diagrams with partial solid solubility.

6.4.1 EUTECTIC BINARY PHASE DIAGRAMS WITH COMPLETE INSOLUBILITY IN SOLID STATE

Eutectic phase diagrams with complete insolubility in solid state describe behavior of a material system, two components of which are completely soluble in the liquid state and entirely insoluble in the solid state. This diagram has two liquidus curves,

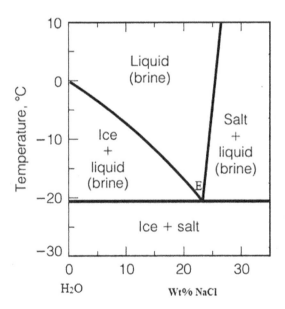

FIGURE 6.3 The H_2O-NaCl phase diagram (E = eutectic point).

starting from the freezing points of the two metals and intersecting in a minimum point—*eutectic point*. A notable examples of this type of phase diagram is H_2O-NaCl system (Figure 6.3).

Figure 6.3 indicates that the freezing temperature (T_f) of pure water (H_2O) is 0°C; however, the T_f is lowered by adding NaCl (salt); T_f of brine (solution of water and salt) become the minimum at the eutectic temperature (−21°C) and the composition: H_2O-23wt%NaCl. This principle is applied by sprinkling salt on icy roads or sideways during winter. The sprinkled salt causes ice (at a temperature > −21°C) to melt because brine has a lower freezing point than pure water.

6.4.2 EUTECTIC BINARY PHASE DIAGRAMS WITH PARTIAL SOLID SOLUBILITY

Basics. *Eutectic binary phase diagrams with partial solid solubility* refer to the eutectic binary systems in which the solubility of each component in each of the solid phases is limited. Examples of this type of phase diagrams include the systems: Pb-Sn; Bi-Sn; Cu-Ag; and the like. Figure 6.4 shows the lead-tin (Pb-Sn) phase diagram; here E refers to the eutectic point.

It is evident in Figure 6.4 that the maximum solid solubility of Sn in Pb is 18.3 wt%.; and that of Pb in Sn is 97.8 wt%. The line BC refers to the *solvus line*; which indicates that the solid solubility (of the solute in the solvent) decreases with decreasing temperature (see Example 6.11). Figure 6.4 shows that the lead-tin alloy has a eutectic temperature of 183°C; and that the 61.9%Sn-39.1%Pb alloy (eutectic composition alloy) will undergo eutectic reaction (see section 6.2) at the point E. A lead-tin

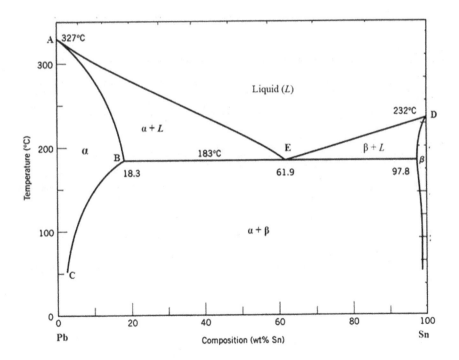

FIGURE 6.4 A lead-tin phase diagram (E represents the eutectic point).

alloy containing wt% Sn < 61.9 is called a *hypoeutectic alloy*; the one containing wt% Sn > 61.9 is called a *hypereutectic alloy*. The 60%Sn-40%Pb alloy is attractive for use as a solder owing to its low melting temperature (eutectic temperature). However, lead is mildly toxic thereby rendering the use of 60%Sn-40%Pb as a solder unsafe. This is why materials technologists have developed a safe lead-free eutectic solder alloy based on bismuth-tin (Bi-Sn) system (see Example 6.12).

Mathematical Modeling. It is evident in Figure 6.4 that two phases (α and β) are stable below the eutectic temperature for almost all compositions of Pb-Sn alloys. For a given composition (C_0) the weight fraction of each solid phase can be calculated by applying the *lever rule*, as follows:

$$W_\alpha = \frac{C_\beta - C_o}{C_\beta - C_\alpha} \tag{6.15}$$

$$W_\beta = \frac{C_0 - C_\alpha}{C_\beta - C_\alpha} \tag{6.16}$$

where W_α and W_β are the weight fractions of the solid α and the solid β, respectively; C_α is the composition (wt% Sn in Pb) of α; and C_β is the composition of the solid β (see Example 6.9).

The densities of the two solid phases (ρ_α and ρ_β, respectively) can be calculated by:

$$\rho_\alpha = \frac{100}{\dfrac{wt\ \%\ Sn\ in\ \alpha}{\rho_{Sn}} + \dfrac{wt\ \%\ Pb\ in\ \alpha}{\rho_{Pb}}} \tag{6.17}$$

$$\rho_\beta = \frac{100}{\dfrac{wt\ \%\ Sn\ in\ \beta}{\rho_{Sn}} + \dfrac{wt\ \%\ Pb\ in\ \beta}{\rho_{Pb}}} \tag{6.18}$$

Once the weight fractions and densities of α and β have been determined by using Equation 6.15 through Equation 6.18, the volume fractions of the two solid phases can be computed by (Callister, 2007):

$$V_\alpha = \frac{\dfrac{W_\alpha}{\rho_\alpha}}{\dfrac{W_\alpha}{\rho_\alpha} + \dfrac{W_\beta}{\rho_\beta}} \tag{6.19}$$

$$V_\beta = \frac{\dfrac{W_\beta}{\rho_\beta}}{\dfrac{W_\alpha}{\rho_\alpha} + \dfrac{W_\beta}{\rho_\beta}} \tag{6.20}$$

where V_α and V_β are the volume fractions of α and β, respectively.

The significance of Equation 6.17 through Equation 6.20 is illustrated in Example 6.10.

6.5 BINARY PHASE DIAGRAMS INVOLVING PERITECTIC/EUTECTOID REACTIONS

The *peritectic reaction* involves transformation (on cooling) of a solid phase and a liquid phase into another solid phase (see Equation 6.7). Examples of peritectic phase diagrams include Fe-Fe$_3$C system; Cu-Zn system; and the like. The Fe-Fe$_3$C phase diagram involving peritectic reaction (and eutectic and eutectoid reactions) is shown in Figure 6.5. The *eutectoid reaction* involves transformation (on cooling) of a solid phase into two different solid phases. It is evident in Figure 6.5 that there occurs a eutectoid reaction when a Fe-0.76wt%C alloy is cooled from a higher temperature to 727°C (see Equation 6.6). A detailed discussion of eutectoid phase transformation is given in Chapter 7 (section 7.5).

A look at the top region of Figure 6.5 reveals that cooling of the Fe-0.16 wt.% C (peritectic composition) alloy from a higher temperature to 1,499°C (peritectic temperature) would result in the following peritectic reaction:

$$Liq + \delta \rightarrow \gamma \qquad (6.21)$$

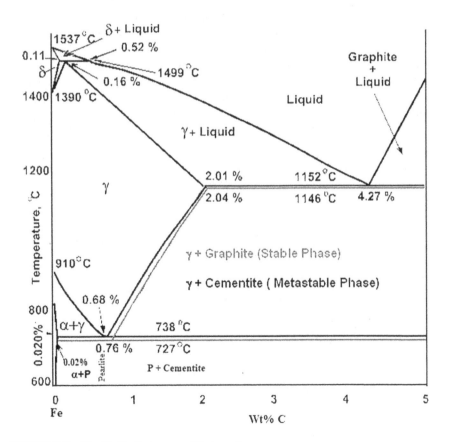

FIGURE 6.5 The Fe-Fe$_3$C phase equilibrium diagram.

where δ is the solid solution of carbon in BCC (δ) iron and γ is the austenite (solid solution of carbon in FCC iron). The maximum solid solubility of carbon in δ iron is 0.11 wt%. The quantitative analysis of peritectic phase diagram is illustrated in Example 6.12.

6.6 BINARY COMPLEX PHASE DIAGRAMS

In general, most phase diagrams of practical systems are complex (Okamoto *et al.*, 2016). The complex phase diagrams involve repetitive points corresponding to phase transformation reactions. Typical examples of complex phase diagrams include Cu-Zn system, Al-Ni system (see Figure E-6.15), and the like. The identifications of phase transformation reaction points in a complex phase diagram are illustrated in Example 6.15.

6.7 TERNARY PHASE DIAGRAMS

Basics. Ternary phase diagrams are the phase diagrams based on three-components systems; they enable us to compare three components at once. In order to view all three compositions at the same time, a triangular plot is set up with an element at each of the vertexes with the temperature and pressure stated (Huda and Bulpett, 2012). For example, the phase diagram for a stainless steel would show iron (Fe), nickel (Ni), and chromium (Cr) compositions (see Figure 6.6).

Method of Determining Compositions in a Ternary Phase Diagram. In order to calculate the composition of a ternary alloy/material represented by any point in a ternary phase diagram (see Figure E-6.16(A)), two lines are drawn through the point, parallel to any two sides of the triangle. The intersection of these two lines with the third side divides this side into three segments (see Figure E-6.16(B)). The lengths of

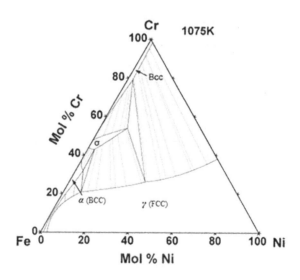

FIGURE 6.6   Fe-Cr-Ni (ternary) phase diagram.

the individual line segments (so obtained) are proportional to the relative amounts of the three components (see Example 6.16).

6.8   CALCULATIONS—EXAMPLES IN PHASE DIAGRAMS

EXAMPLE 6.1 USING GIBBS'S PHASE RULE TO DETERMINE TYPE OF AN EQUILIBRIUM

Water can *exist* in *three* different *states*; as a solid, liquid, or gas (see Figure 6.1). Use Gibbs's phase rule to determine whether the triple point (P_t) in Figure 6.1 refers to an *invariant equilibrium* or an *univariant equilibrium*.

SOLUTION

The system in Figure 6.1 is entirely composed of H_2O, so there is only one component present i.e. C=1. The three phases (liquid water, solid ice, and vapor/steam) indicate P=3. There is only one point (triple point) on the phase diagram (Figure 6.1) where all three phases coexist in equilibrium.

By using *Gibbs's phase rule* (Equation 6.2),

$$F = C—P + 2 = 1–3 + 2 = 0$$

Since *F=0*, the triple point (P_t) in Figure 6.1 refers to as an *invariant equilibrium*.

EXAMPLE 6.2 IDENTIFYING TRIPLE POINTS/COEXISTING PHASES IN A UNARY PHASE DIAGRAM

The iron phase diagram is shown in Figure E-6.2. Identify: (a) the triple points (indicating the temperature and pressure), and (b) the phases that coexist at equilibrium at each triple point.

SOLUTION

A.   In Figure E-6.2, there are three triple points: P_{t1}, P_{t2}, and P_{t3}. The triple point P_{t1} corresponds to T = 910°C, $p = 10^{-10}$ bar; the point P_{t2} corresponds to T = 1394°C, $p \cong 10^{-6}$ bar; and P_{t3} corresponds to T = 1538°C, $p \cong 10^{-5}$ bar.

B.   At the triple point P_{t1}, the phases α-iron, γ-iron, and vapor co-exist. At P_{t2}, three phases γ-iron, δ-iron, and vapor coexist. At the point P_{t3}, the phases δ-iron, liquid, and vapor coexist.

EXAMPLE 6.3 IDENTIFYING THE PHASE TRANSFORMATION TEMPERATURES AND COMPOSITIONS DURING COOLING OF AN ISOMORPHOUS ALLOY

A Cu-20%Ni alloy is cooled from 1,500°C to room temperature. Determine: (a) the temperature of the liquid at which first solid forms, and (b) the temperature of the

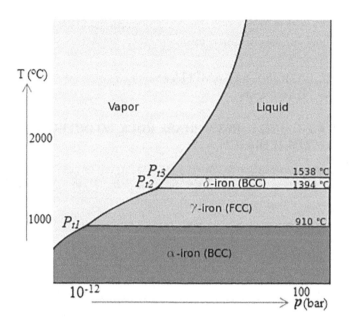

FIGURE E-6.2   Iron (Fe) phase diagram.

two phase mixture at which the last liquid solidifies, (c) the composition of the first solid to form, and (d) composition of the last liquid to solidify.

SOLUTION

We redraw the binary phase diagram in Figure 6.2 as Figure E-6.3.

In Figure E-6.3, the point A represents the Cu-20Ni alloy at 1,500°C. The line AC represents cooling of the alloy to complete solidification. Line BD is the tie line in the two-phase region:

A.   The temperature of the liquid at which first solid forms is 1,200°C (see line AB).
B.   The line AC intersects the solidus at C (1,160°C). Hence, the temperature of the two phase mixture at which the last liquid solidifies is 1,160°C.
C.   Point D gives the composition of the first solid to form; the composition is Cu-30%Ni.
D.   Point E gives the composition of the last liquid to solidify; the composition is Cu-16%Ni.

EXAMPLE 6.4   IDENTIFYING THE PHASE TRANSFORMATION TEMPERATURES AND COMPOSITIONS DURING COOLING OF ALBITE-ANORTHITE CERAMIC SYSTEM

Plagioclase is one of the most common minerals in the earth's crust. The phase diagram for the albite-anorthite ceramic system is shown in Figure E-6.4(A).

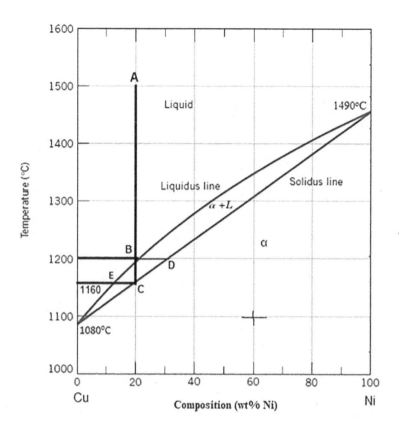

FIGURE E-6.3 The copper-nickel phase diagram showing cooling of Cu-20Ni alloy from 1,500°C.

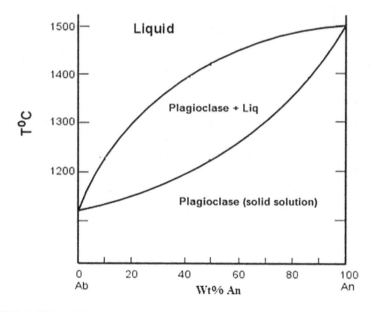

FIGURE E-6.4(A) Albite-anorthite phase diagram.

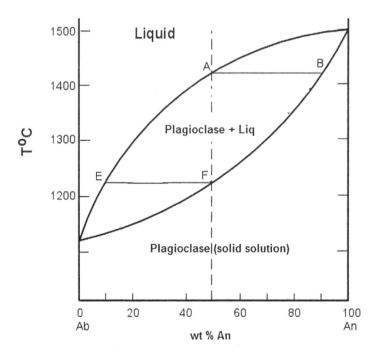

FIGURE E-6.4(B) Albite-anorthite phase diagram with drawn lines to determine desired temperatures and compositions.

A 50%Ab-50%An material is cooled from 1,500°C to room temperature. (a) Which type of phase diagram is shown in Figure E-10.5(A)? (b) What is the melting temperature of albite? (c) What is the melting temperature of anorthite? (d) At what temperature of the liquid does the first solid form? (e) At what temperature of the two-phase mixture does the last liquid solidify? (f) What is the composition of the first solid to form? (g) And what is the composition of the last liquid to solidify?

SOLUTION

A. The phase diagram (Figure E-6.4(A)) refers to an isomorphous binary phase diagram;
B. The melting temperature of pure albite $(Ab) = 1,118°C$;
C. The melting temperature of pure anorthite $(An) = 1,500°C$;

A dotted vertical line and two tie lines are drawn, as shown in Figure E-6.4(B).

D. The temperature of the liquid at which the first solid forms = 1,410°C;
E. The temperature of the two-phase mixture at which the last liquid solidifies = 1,220°C;
F. The composition of the first solid to form is 90%An-10%Ab;
G. The composition of the last liquid to solidify is 90%Ab-10%An.

EXAMPLE 6.5 DETERMINING THE ATOMIC PERCENT OF ELEMENTS IN A BINARY ALLOY

Determine the atomic percent of copper in the *Al-4wt%Cu* binary alloy.

SOLUTION

Wt% Cu = 4, at. wt. Cu = 63.5 g/mol, wt% Al = 96, at. wt. Al = 26.9

For the Al-4Cu alloy, Equation 6.9 can be rewritten as

$$\text{at. \% Cu} = \frac{\dfrac{wt\,\%\,Cu}{at.\,wt\,Cu}}{\dfrac{wt\,\%\,Cu}{at.\,wt\,Cu} + \dfrac{wt\,\%\,Al}{at.\,wt\,Al}} \times 100$$

$$\text{or} \quad \text{at. \% Cu} = \frac{\dfrac{4}{63.5}}{\dfrac{4}{63.5} + \dfrac{96}{26.9}} \times 100 = \frac{0.063}{0.063 + 3.56} \times 100 = 1.73$$

$$\text{or} \quad Cu = 1.73 \text{ at \%}$$

$$Al = 100 - 1.73 = 98.27 \text{ at \%}$$

EXAMPLE 6.6 CALCULATING THE WEIGHT FRACTION OF PHASES BY USING LEVER RULE

A 50%Ab-50%An ceramic is cooled from 1,500°C to complete solidification. Calculate the weigh fractions of the liquid and solid phases at a temperature of 1,400°C.

SOLUTION

By reference to the given composition (50Ab-50n) and the given temperature (1,400°C), we draw a dotted vertical line and a horizontal (tie) line, respectively as shown in Figure E-6.6.

By reference to Figure E-6.6 and by using Equation 6.13,

$$W_L = \frac{y}{x+y} = \frac{C_\alpha - C_o}{C_\alpha - C_L} = \frac{85 - 50}{85 - 40} = 0.78$$

By reference to Figure E-6.6 and by using Equation 6.14,

$$W_s = \frac{x}{x+y} = \frac{C_o - C_L}{C_\alpha - C_L} = \frac{50 - 40}{85 - 40} = 0.22$$

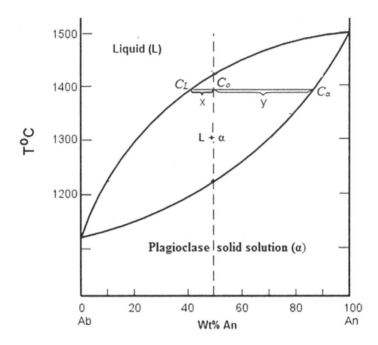

FIGURE E-6.6 *Ab-An* phase diagram showing the tie line for calculating the amounts of phases.

EXAMPLE 6.7 SCIENTIFIC AND TECHNOLOGICAL SIGNIFICANCE OF H₂O-NACL SYSTEM

Refer to the eutectic binary H_2O-NaCl phase diagram in Figure 6.3. (a) What are the compositions of solid and liquid when H_2O-10 wt% NaCl (brine) is cooled from 8°C to a temperature of −15°C? (b) Write the eutectic reaction at the point *E*. (c) Will it be effective to sprinkle salt on an icy road (to melt the ice) at a temperature of −25°C? Justify your answer.

SOLUTION

(a) For H_2O-10 wt% NaCl at −15°C, we may draw a tie line; the composition of solid (ice) is 100% H_2O and the composition of liquid (brine) is around H_2O-17 wt% NaCl.

(b) By using Equation 11.3, the following eutectic reaction will occur at *E* (see Figure 6.3):

$$Brine \rightarrow salt + ice$$

(c) No! It is evident in the H_2O-NaCl phase diagram that at a temperature below −21°C, there is always a mixture of solid salt and ice; it means that there would never be any liquid whatever proportions of salt and water we had.

EXAMPLE 6.8 DETERMINING PHASES AND PHASE COMPOSITIONS IN A PB-SN SYSTEM

Consider a Pb-30 wt% Sn alloy at 120°C. (a) Identify the phase(s). (b) What do α and β represent in Figure 6.3. (c) Determine the composition(s) of the identified phase(s).

SOLUTION

Figure 6.3 is redrawn to represent Pb-30 wt% Sn alloy at 120°C (point P) as shown in Figure E-6.8.

By drawing the tie line through P, we can determine the phases and their compositions.

A.   The phases present in the Pb-30 wt% Sn alloy at 120°C are: two solid phases ($\alpha + \beta$).

B. α refers to the solid solution of Sn in Pb (Sn is solute and Pb is solvent) whereas β refers to the solid solution of Pb in Sn.

C.   For the Pb-30wt% Sn alloy at 120°C, the composition of α solid solution is Pb-6 wt% Sn. For the Pb-30wt% Sn alloy at 120°C, the composition of β solid solution is Pb-99 wt% Sn.

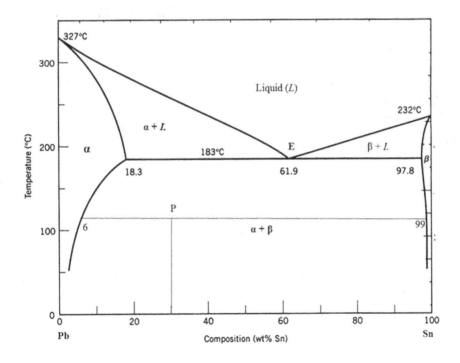

FIGURE E-6.8   The lead-tin phase diagram representing Pb-30 wt% Sn alloy at 120°C.

EXAMPLE 6.9 DETERMINING THE WEIGHT FRACTION OF EACH SOLID PHASE IN EUTECTIC SYSTEM

Refer to the lead-tin alloy in Example 6.8. Determine the relative amounts (weight fractions) of each phase present in the specified alloy composition at the specified temperature.

SOLUTION

In order to apply *lever rule*, we specify various compositions of Sn in Pb as follows: $C_0 = 30$ wt% Sn; $C_\alpha = 6$ wt% Sn; and $C_\beta = 99$ wt% Sn.

By using Equation 6.15 and Equation 6.16, we obtain:

$$W_\alpha = \frac{C_\beta - C_o}{C_\beta - C_\alpha} = \frac{99 - 30}{99 - 6} = 0.74$$

$$W_\beta = \frac{C_0 - C_\alpha}{C_\beta - C_\alpha} = \frac{30 - 6}{99 - 6} = \frac{24}{93} = 0.26$$

Weight fractions of solid phases α and β are 0.74 and 0.26, respectively.

EXAMPLE 6.10 CALCULATING THE VOLUME FRACTION OF EACH SOLID PHASE IN EUTECTIC SYSTEM

By using the data in Example 6.9, calculate the volume fraction of each solid phase in the eutectic system. Take densities of Pb and Sn at 120°C to be 11.27 and 7.29 g/cm³, respectively.

SOLUTION

$W_\alpha = 0.74$, $W_\beta = 0.26$, $\rho_{Pb} = 11.27$ g/cm³, $\rho_{Sn} = 7.29$ g/cm³; Wt% Sn in $\alpha = 6$, wt% Pb in $\alpha = 94$

Wt% Sn in $\beta = 99$; wt% Pb in $\beta = 1$

By using Equations 6.17 and 6.18,

$$\rho_\alpha = \frac{100}{\dfrac{wt\ \%\ Sn\ in\ \alpha}{\rho_{Sn}} + \dfrac{wt\ \%\ Pb\ in\ \alpha}{\rho_{Pb}}} = \frac{100}{\dfrac{6}{7.29} + \dfrac{94}{11.27}} = 10.92\ g/cm^3$$

$$\rho_\beta = \frac{100}{\dfrac{wt\ \%\ Sn\ in\ \beta}{\rho_{Sn}} + \dfrac{wt\ \%\ Pb\ in\ \beta}{\rho_{Pb}}} = \frac{100}{\dfrac{99}{7.29} + \dfrac{1}{11.27}} = 7.31\ g/cm^3$$

By using Equation 6.19,

$$V_\alpha = \frac{\dfrac{W_\alpha}{\rho_\alpha}}{\dfrac{W_\alpha}{\rho_\alpha} + \dfrac{W_\beta}{\rho_\beta}} = \frac{\dfrac{0.74}{10.92}}{\dfrac{0.74}{10.92} + \dfrac{0.26}{7.31}} = 0.63$$

By reference to Equation 6.12, we may write:

$$V_\alpha + V_\beta = 1$$
$$\text{or} \quad V_\beta = 1 - V_\alpha = 1 - 0.63 = 0.37$$

Hence, the volume fraction of each solid phase: $V_\alpha = 0.63$, $V_\beta = 0.37$.

EXAMPLE 6.11 DETERMINING COMPOSITIONS AT VARIOUS TEMPERATURES IN A BINARY PHASE DIAGRAM WITH PARTIAL SOLID SOLUBILITY

Refer to Figure 6.4. Consider the Pb-13wt%Sn alloy. (a) What is the phase at a temperature of 350°C? (b) The alloy is cooled to 170° C. What phase is present and what is the composition? (c) The alloy is cooled to 110°C. Identify the phase(s) and the composition(s). (d) The alloy is cooled to 60°C. What phases exist; and what are their compositions? (e) Based on your findings in (b–d), what is the relationship between the solid solubility (of Sn in Pb in α) and the temperature? (f) Which line represents the relationship; and which term is used for the line?

SOLUTION

A.   The stable phase in the Pb-13wt%Sn alloy at 350°C is *liquid*.
B.   The stable phase in the Pb-13wt%Sn alloy at 170°C is α (solid solution of Sn in Pb). The solid α contains 13 wt% Sn.
C.   The stable phases in the Pb-13wt%Sn alloy at 110°C are two solids ($\alpha + \beta$). The solid α contains 8 wt% Sn. The solid β contains 98 wt% Sn.
D.   The stable phases in the Pb-13wt%Sn alloy at 60°C are two solids ($\alpha + \beta$). The solid α contains 4 wt% Sn. The solid β contains 99 wt% Sn.
E.   The solubility of Sn in Pb in the solid α decreases with a decrease in temperature.
F.   The line BC represents the relationship: solid solubility decreases with a decrease in temperature; this line is called *solvus line*.

EXAMPLE 6.12 PHASE IDENTIFICATIONS AND APPLICATION OF LEVER RULE TO SN-BI SYSTEM

Figure E-6.12(A) shows a Sn-Bi phase diagram. (a) Which composition of tin-bismuth alloy is the most suitable for application as a solder? Justify. (b) What is the eutectic temperature of the tin-bismuth system? (c) What is the maximum solid solubility of Bi in Sn? (d) What is the maximum solid solubility of Sn in Bi? (e) Calculate the weight fractions of liquid and solid phases when a Sn-30wt%Bi ally is cooled from liquid phase to a temperature of 150°C. (f) What phase change will occur when a Sn-10wt%Bi alloy is slowly heated from room temperature to 150°C? (g) What phase changes will occur when a Sn-80wt%Bi alloy is slowly heated from room temperature to 250°C?

SOLUTION

In order to solve this problem, we may redraw Figure E-6.11(A) by labeling the tie line and cooling/heating lines as shown in Figure E-6.12(B).

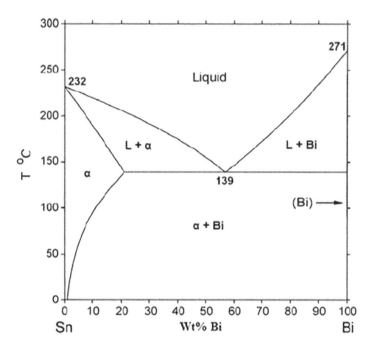

FIGURE E-6.12(A) Tin-bismuth eutectic phase diagram.

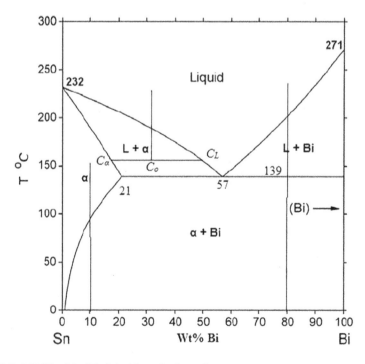

FIGURE E-6.12(B) Modified tin-bismuth phase diagram.

A. The composition of Sn-Bi alloy that is the most suitable for application as solder is Sn-57%Bi because this composition is the eutectic composition referring to the minimum temperature.
B. The eutectic temperature of the tin-bismuth system is 139°C.
C. The maximum solid solubility of Bi in Sn = 21 wt%
D. The maximum solid solubility of Sn in Bi = 0
E. By reference to Equations 6.13 and 6.14,

$$W_L = \frac{C_0 - C_\alpha}{C_L - C_\alpha} = \frac{30 - 15}{48 - 15} = 0.455$$

$$W_\alpha = \frac{C_L - C_0}{C_L - C_\alpha} = \frac{48 - 30}{48 - 15} = 0.545$$

F. At room temperature, the stable phases are the solids: α + Bi.
    At 150°C, the stable phase is the solid: α (solid solution of Bi in Sn).
G. At a temperature between room temperature and 139°C, stable phases: α + Bi.
    At a temperature between 139°C and 180°C, stable phase: L + Bi.
    At a temperature above 180°C, stable phase: L.

EXAMPLE 6.13 QUANTITATIVE ANALYSIS OF PERITECTIC PHASE DIAGRAM

Refer to Figure 6.5. Determine the composition and weight fraction of each phase present just above and below the *peritectic* isotherm for: (a) Fe-0.16 wt%C alloy, and (b) Fe-0.30%C alloy.

SOLUTIONS

(a) **Above the peritectic isotherm**, the compositions of each phase in Fe-0.16 wt%C alloy:
δ-phase composition: δ-Fe-0.11wt%C; Liquid composition: Fe-0.52 wt%C.
Below the peritectic isotherm, the composition of Fe-0.16 wt%C alloy: γ-phase (Fe-0.16%C).
Above the peritectic isotherm, the weight fractions of each phase in Fe-0.16 wt%C alloy can be calculated by rewriting Equations 6.13 and 6.14, as follows:

$$W_L = \frac{C_0 - C_\delta}{C_L - C_\delta} = \frac{0.16 - 0.11}{0.52 - 0.11} = \frac{0.05}{0.41} = 0.12$$

$$W_\delta = \frac{C_L - C_0}{C_L - C_\delta} = \frac{0.52 - 0.16}{0.52 - 0.11} = \frac{0.36}{0.41} = 0.88$$

Below the *peritectic* isotherm, the weight fractions of γ-phase in Fe-0.16 wt%C alloy: $W_\gamma = 1$

(b) Above the *peritectic* isotherm, the compositions of each phase in Fe-0.30 wt%C alloy:

δ-phase composition: δ-Fe-0.11wt%C; Liquid composition: Fe-0.52 wt%C

Below the peritectic isotherm, the compositions of each phase in Fe-0.30 wt%C alloy:

γ-phase composition: Fe-0.16 wt%C; Liquid composition: Fe-0.52 wt%C

Above the peritectic isotherm, the weight fraction of each phase in Fe-0.30 wt%C alloy:

$$W_L = \frac{C_0 - C_\delta}{C_L - C_\delta} = \frac{0.30 - 0.11}{0.52 - 0.11} = \frac{0.19}{0.41} = 0.46$$

$$W_\delta = \frac{C_L - C_0}{C_L - C_\delta} = \frac{0.52 - 0.30}{0.52 - 0.11} = \frac{0.22}{0.41} = 0.54$$

Below the *peritectic* isotherm, the weight fraction of each phase in Fe-0.30 wt%C alloy:

$$W_L = \frac{C_0 - C_\gamma}{C_L - C_\gamma} = \frac{0.30 - 0.16}{0.52 - 0.16} = \frac{0.14}{0.36} = 0.39$$

$$W_\gamma = \frac{C_L - C_0}{C_L - C_\gamma} = \frac{0.52 - 0.30}{0.52 - 0.16} = \frac{0.22}{0.36} = 0.61$$

EXAMPLE 6.14 IDENTIFYING PHASE TRANSFORMATION REACTIONS IN FE-FE₃C PHASE DIAGRAM

Refer to Figure 6.5. (a) Identify the various phase transformation reactions by specifying the compositions and temperatures in Figure 6.5. (b) Write down the equations for the various phase transformation reactions. (c) What is the maximum solubility of carbon in austenite (γ phase)?

SOLUTION

A. (i) A peritectic reaction occurs when the Fe-0.16%C (peritectic composition) alloy is cooled from a higher temperature to 1,499°C (peritectic temperature).

(ii) A eutectic reaction occurs when the Fe-4.27%C (eutectic composition) alloy is cooled from a higher temperature to 1,150°C (eutectic temperature).

(iii) A eutectoid reaction occurs when the Fe-0.76%C (eutectoid composition) alloy is cooled from a higher temperature to 727°C.

B. (i) The peritectic reaction is given as follows: Liq + δ → γ.

(ii) The eutectic reaction is given as follows: Liquid → γ + Fe₃C.

(iii) The eutectoid reaction is given as follows: γ → α + Fe₃C.

C. The maximum solid solubility of carbon in γ (austenite) = 2.04 wt%.

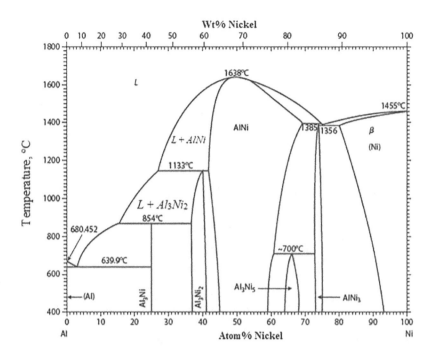

FIGURE E-6.15 Al-Ni phase diagram.

EXAMPLE 6.15 PHASE TRANSFORMATION REACTION POINTS IN A COMPLEX PHASE DIAGRAM

By reference to Figure E-6.15, identify the points and write equations for the following phase transformation reactions: (a) peritectic, and (b) eutectic.

SOLUTION

A. *Peritectic Points and the Reactions*
   1. 854°C, Al-25 at.% Ni: $L + Al_3Ni_2 \rightarrow Al_3Ni$
   2. 1,133°C, Al-41 at.% Ni: $L + AlNi \rightarrow Al_3Ni_2$
   3. 1,385°C, Al-73 at.% Ni: $L + AlNi \rightarrow AlNi_3$
B. *Eutectic Point and the Reaction*
   1,356°C, Al-75 at.% Ni: $L \rightarrow AlNi_3 + \beta(Ni)$

EXAMPLE 6.16 DETERMINING COMPOSITION IN A TERNARY PHASE DIAGRAM

Determine the composition of the point A in the hypothetical ternary (P-Q-R) phase diagram as shown in Figure E-6.16(A).

SOLUTION

In order to determine the composition of the point A in the hypothetical ternary phase diagram (Figure E-6.16(A)), two lines are drawn through the point A, parallel

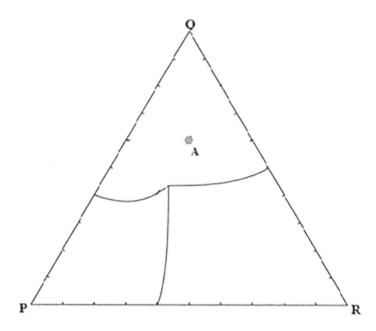

FIGURE E-6.16(A) Hypothetical P-Q-R (ternary) phase diagram.

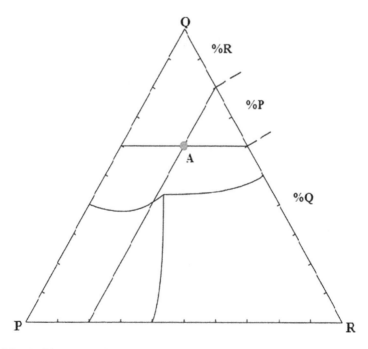

FIGURE E-6.16(B) Drawn lines to determine composition of the point A in Figure E-11.16(A).

to two sides (*PQ* and *PR*) of the triangle. The intersection of these two lines with the third side (*QR*) divides this side into three segments (see Figure E-6.16(B)). The lengths of the individual line segments (so obtained) are proportional to the relative amounts of the three components: *P*, *Q*, and *R*.

It is evident in Figure E-6.16(B) that the point *A* represents an alloy/material with the composition: P: 20%, Q: 60%, R: 20%.

QUESTIONS AND PROBLEMS

6.1. (a) Mathematically express Gibbs's phase rule, defining each symbol in the expression. (b) Write modified expressions for Gibbs's phase rule for: (i) invariant equilibrium, and (ii) univariant equilibrium.

6.2. Identify the following phase transformation reactions and give an example for each: (a) $Solid_1 + Solid_2 \rightarrow Solid_3$, (b) $Liquid \rightarrow Solid_1 + Solid_2$, and (c) $Solid_1 \rightarrow Solid_2 + Solid_3$.

6.3. (a) Draw Cu-Ni phase diagram. (b) What is *coring effect*? And why does this effect occur in copper-nickel alloys? (c) How can we overcome the *coring effect* problems?

6.4. Why are tin-lead alloys attractive for application as a solder? What limitation is imposed on the use of tin-lead solder? How is this limitation/problem overcome?

6.5. What phase transformation reactions occur in $Fe-Fe_3C$ diagram? Write expressions for the reactions indicating the composition and temperature for each phase transformation.

6.6. Classify phase diagrams on the basis of: (a) number of components, and (b) phase transformation reactions.

6.7. Identify the type of phase diagram (and its subcategory) for each of the following alloy/material systems: (a) stainless steel, (b) NiO-MgO, (c) aluminum-nickel, (d) tin-lead, (e) copper-nickel, (f) H_2O-NaCl, (g) tin-bismuth.

P6.8. Identify the peritectic reaction points in Cu-Zn phase diagram, as shown in Figure P10.8.

P6.9. Refer to Figure 6.2. A Cu-40%Ni alloy is cooled from 1,400°C to room temperature. Determine: (a) the temperature of the liquid at which first solid forms, (b) the temperature of the two phase mixture at which the last liquid solidifies, (c) the composition of the first solid to form, and (d) composition of the last liquid to solidify.

P6.10. A 60%Ab-40%An ceramic is cooled from 1,450°C to complete solidification. Calculate the weigh fractions of the liquid and solid phases at a temperature of 1,300°C.

P6.11. Determine the composition and weight fraction of each phase present just above and below the *eutectic* isotherm for Fe-4.27 wt%C alloy.

P6.12. Determine the composition of the point P in the hypothetical ternary (A-B-C) phase diagram, as shown in Figure P6.12.

P6.13. Refer to Figure 6.4. Consider a Pb-40 wt% Sn alloy at 150°C. (a) Determine the relative amounts (weight fractions) of each phase present in the specified alloy composition at the specified temperature. (b) Calculate the volume fraction of each solid phase in the eutectic system. Take densities of Pb and Sn at 120°C to be 11.27 and 7.29 g/cm^3, respectively.

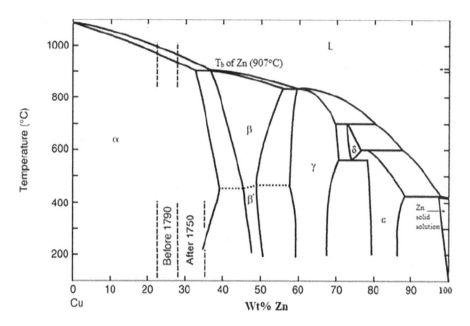

FIGURE P6.8 Cu-Zn phase diagram.

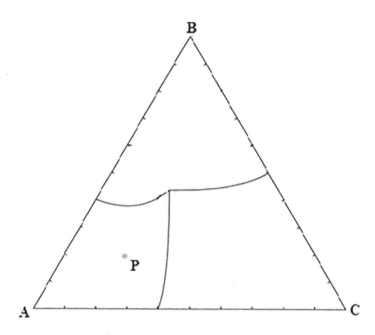

FIGURE P6.12 Hypothetical A-B-C phase diagram.

REFERENCES

Callister, W.D. (2007) *Materials Science and Engineering: An Introduction.* John Wiley & Sons Inc., New York.

Hillert, M. (2007) *Phase Equilibria, Phase Diagrams and Phase Transformations: Their Thermodynamic Basis.* 2nd ed. Cambridge University Press, Cambridge.

Huda, Z. & Bulpett, R. (2012) *Materials Science and Design for Engineers.* Trans Tech Publications, Switzerland.

Massalski, T.B., Baker, H., Bennett, L.H. & Murray, J.L. (1990) *Binary Alloy Phase Diagrams.* ASM International, Materials Park, OH.

McHale, A.E. (2010) *Phase Diagrams and Ceramic Processes.* 1st ed. Springer, New York.

Okamoto, H. (2002) *Phase Diagrams of Dilute Binary Alloys.* ASM International, Materials Park, OH.

Okamoto, H., Schlesinger, M.E. & Mueller, E.M. (eds) (2016) *ASM Handbook Volume 03: Alloy Phase Diagrams.* ASM International, Materials Park, OH.

Rhines, F. (1956) *Phase Diagrams in Metallurgy.* McGraw Hill, New York.

7 Phase Transformations and Kinetics

7.1 PHASE TRANSFORMATION AND ITS TYPES

The alteration in the microstructure of a metal, usually during its processing, is called *phase transformation*. The change in the microstructure may change the number and/or characters of phases; which in turn may result in either enhancement or loss in mechanical properties (Huda, 1996). There are four important three-point phase transformations in metals: (1) eutectic, (2) eutectoid, (3) peritectic, and (4) peritectoid reactions; these phase-transformation reactions have been explained in the preceding chapter (see section 6.2.1).

In general, phase transformations are divided into three classifications: (a) simple diffusion-dependent transformation (e.g. melting, solidification of a pure metal, allotropic transformation, recrystallizations, and grain growth) (see Chapter 14); (b) diffusion-dependent transformation involving change in phase composition and/or number of phases present (e.g. eutectoid reaction); and (c) diffusionless transformation (e.g. martensitic transformation).

7.2 THE KINETICS OF PHASE TRANSFORMATIONS

7.2.1 THE TWO STAGES OF PHASE TRANSFORMATION— NUCLEATION AND GROWTH

The time dependence of rate of phase transformation is referred to as its *kinetics*. *Phase transformations* involve change in structure and (for multi-phase systems) composition; the latter requires rearrangement and redistribution of atoms via diffusion. The process of phase transformation mainly involves two stages: (I) nucleation of new phase(s), and (II) growth of the new phase(s) at the expense of the original phase(s). In general, the kinetics of phase transformation may be graphically represented by an *S-shape curve*, as shown in Figure 7.1.

7.2.2 NUCLEATION

7.2.2.1 Nucleation and Its Types

Nucleation refers to the formation of stable small particles (nuclei) of the new phase(s). Nuclei are often formed at grain boundaries and other defects. There are two types of nucleation: (a) *homogeneous nucleation*, and (b) *heterogeneous nucleation* (Porter *et al.*, 2009). A distinction between the two types of nucleation can be made by considering solidification of a liquid phase that is supercooled below

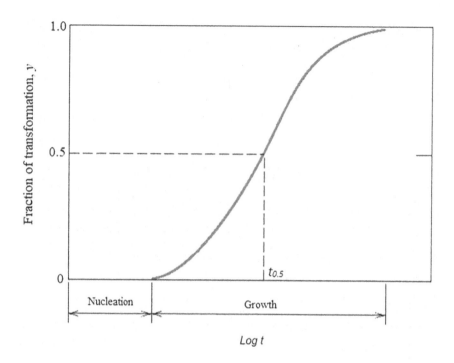

FIGURE 7.1 *S-shape curve* showing fraction of material transformed versus logarithm of heating time.

the melting temperature (see Figure 7.2). In *homogeneous nucleation*, the transition from supercooled liquid to a solid spherical particle in the liquid occurs spontaneously. In *heterogeneous nucleation*, the new phase appears on the walls of the container, at impurity particles, or the like. The two types of nucleation are discussed in the following subsections.

7.2.2.2 Homogeneous Nucleation

Homogeneous nucleation involves spontaneous appearance of solid nuclei within the supercooled phase (see Figure 7.2). The homogeneous nucleation stage of phase transformation can be explained in terms of the thermodynamic parameter: *Gibbs's free energy (G)*, defined by:

$$G = H - TS \tag{7.1}$$

where H is the enthalpy, and S is the entropy, and T is the temperature, K.

According to the thermodynamic principle, the equilibrium under conditions of constant temperature and pressure corresponds to the minimum of *G* and a phase transformation occurs spontaneously only when *G* value decreases during the transformation. In *homogeneous nucleation*, the formation of a solid nucleus leads to a *Gibbs's free energy change (ΔG)* of:

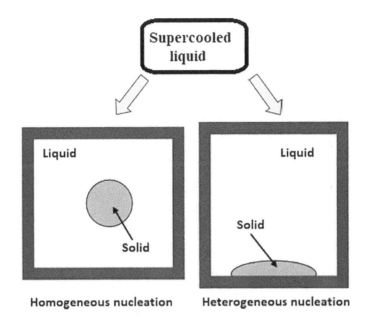

FIGURE 7.2   Difference between homogeneous nucleation and heterogeneous nucleation.

$$\Delta G = G_2 - G_1 = -V_s (G_v^L - G_v^S) + A_{SL}\gamma_{SL} \qquad (7.2)$$

where V_s is the volume of the solid spherical nucleus; and A_{SL} is the solid/liquid interfacial area; γ_{SL} is the solid/liquid interfacial energy; and G_v^L and G_v^S are the volume free energies of the liquid and solid, respectively. The simplification of Equation 7.2 yields:

$$\Delta G = -V_s (\Delta G_v) + A_{SL}\gamma_{SL} \qquad (7.3)$$

Since the nucleus is spherical, its volume and interfacial area can be expressed as $V_s = \frac{4}{3}\pi r^3$ and $A_{SL} = 4\pi r^2$, respectively. Accordingly, Equation 7.3 can be rewritten as:

$$\Delta G = -\frac{4}{3}\pi r^3 (\Delta G_v) + 4\pi r^2 \gamma_{SL} \qquad (7.4)$$

where ΔG is the *Gibbs's free energy change*, and r is the radius of the spherical nucleus. In Equation 7.4, the first term may be considered as the volume energy whereas the second term may be treated as interfacial energy; these two terms and the total free energy contributions are graphically represented in Figure 7.3.

It can be inferred from Figure 7.3 that firstly a solid particle begins to form as atoms in the liquid cluster together accompanying an increase in free energy. If the solid-particle radius reaches the critical value $r=r^*$, the corresponding free energy

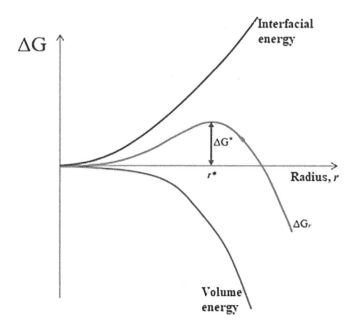

FIGURE 7.3 Plot of free energy versus nucleus radius (ΔG^* = critical free energy change, and r^*= critical nucleus radius).

change is called the critical free energy change, ΔG^*. If the nucleus grows i.e. $r >$ r^*, the *free energy* will decrease. By differentiating Equation 7.4 with respect to r, we obtain:

$$\frac{d(\Delta G)}{dr} = -\frac{4}{3}\pi(\Delta G_v)\frac{d}{dr}(r^3) + 4\pi\gamma_{SL}\frac{d}{dr}r^2 \tag{7.5}$$

At $r=r^*$, $\dfrac{d(\Delta G^*)}{dr}$ =0. Accordingly, Equation 7.5 can be rewritten as:

$$0 = -\frac{4}{3}\pi(\Delta G_v)(3r^{*2}) + 4\pi\gamma_{SL}(2r^*) \tag{7.6}$$

$$\text{or } r^* = -\frac{2\gamma_{SL}}{\Delta G_v} \tag{7.7}$$

The combination of Equations 7.4 and 7.7 yields:

$$\Delta G^* = \frac{16\pi\gamma_{SL}^3}{3(\Delta G_v)^2} \tag{7.8}$$

The volume free energy change (ΔG_v) (also called *"driving force"* for the phase transformation) is proportional to the supercooling below the melting temperature

($\Delta T = T_m - T$). The dependence of volume free energy change (ΔG_v) on temperature can be expressed as:

$$\Delta G_v \frac{(\Delta H_f)(\Delta T)}{T_m} = \frac{(\Delta H_f)(T_m - T)}{T_m} \tag{7.9}$$

where ΔH_f is the latent heat of fusion; T is the nucleation temperature, K; T_m is the melting temperature, K; and ΔT is the degree of supercooling, K.

By substituting the value of ΔG_v from Equation 7.9 into Equation 7.7, we obtain:

$$r^* = \left(-\frac{2\gamma_{SL} T_m}{\Delta H_f} \right) \left(\frac{1}{T_m - T} \right) \tag{7.10}$$

Equation 7.10 enables us to calculate the critical nucleus radius when the other physical quantities are known (see Example 7.1).

The combination of Equations 7.8 and 7.9 yields:

$$\Delta G^* = \left(\frac{16\pi\gamma_{SL}^3 T_m^2}{3(\Delta H_f)^2} \right) \frac{1}{(T_m - T)^2} \tag{7.11}$$

Since $(T_m - T) = \Delta T$, both the critical nucleus radius (r^*) and the critical free energy change (ΔG^*) decrease with increasing degree of supercooling, ΔT. The significance of Equation 7.11 is illustrated in Example 7.2.

The number of atoms in a nucleus of critical size can be determined by:

$$\text{No. of atoms/critical nucleus} = (\text{No. of atoms/unit cell})$$
$$\times \frac{critical\ nucleus\ volume}{unit\ cell\ volume} \tag{7.12}$$

The values of number of atoms/unit cell and the unit cell volume can be obtained from the crystallographic data (see Chapter 2) and the critical nucleus volume $= \frac{4}{3}\pi$ r^{*3}, respectively. The significance of Equation 7.12 is illustrated in Example 7.3.

In order to form a solid nucleus of critical size r^*, the energy barrier of ΔG^* must be reached. The rate of homogeneous nucleation is given by the following Arrhenius equation:

$$\dot{N} \sim v_d \exp(-\frac{\Delta G^*}{kT}) \tag{7.13}$$

where $\dot{N}$ is the rate of homogeneous nucleation, nuclei/m³/s; k is the Boltzmann's constant; and v_d is the frequency with which atoms from liquid attach to the solid nucleus $\left[v_d = \exp\left(-\frac{Q_d}{kT} \right) \right]$.

By substituting the value of v_d in Equation 7.13, we obtain:

$$\dot{N} \sim \exp\left(-\frac{Q_d}{kT} \right) \exp\left(-\frac{\Delta G^*}{kT} \right) \tag{7.14}$$

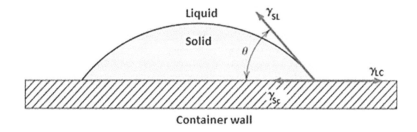

FIGURE 7.4 Heterogeneous nucleation of a solid from a liquid.

where Q_d is the activation energy for diffusion, J/mol; and T is the temperature, K.

The kinetics of nucleation plays an important role in the solidification of pure metals (see Chapter 5, section 5.1). If the rate of nucleation exceeds the rate of growth (e.g. in fast cooling), a fine-grained microstructure will result; and vice versa (see Figure 5.1).

7.2.2.3 Heterogeneous Nucleation

It has been introduced in the subsection 7.2.2.1 that nucleation is said to be *heterogeneous* if new phase appears on the walls of the container, at impurity particles, or the like. Let us consider a simple example of *heterogeneous nucleation* of a solid (spherical) particle from a liquid phase on a container's wall (Figure 7.4). Here, both (solid and liquid) phases spread out and cover the surface i.e. both phases "wet" the surface (container wall).

By balancing the interfacial surface tension forces acting on the wall of the container, we obtain:

$$\gamma_{LC} = \gamma_{SC} + \gamma_{LS} \cos \theta \tag{7.15}$$

where γ_{LC}, γ_{LS}, and γ_{SC} are the surface tensions between liquid-container interface, liquid-solid interface, and the solid-container interface, respectively; and θ is the wetting angle. It is evident in Figure 7.4 that for an effective wetting (spreading of the solid nucleus), the γ_{LC} term should be maximized, which is only possible when $\theta < 90°$. It means that for an effective heterogeneous nucleation of a solid from a liquid, the wetting angle must be acute (see Example 7.4).

7.3 GROWTH AND KINETICS

The growth step in a phase transformation starts once a stable nucleus of the new phase exceeding the critical size r* is formed. The new phase particles continue to grow until they meet i.e. growth process will cease in any region where the new-phase particles come in contact, since here the phase-transformation will have reached to completion.

Particle growth usually involve several steps; which may include: (I) diffusion through the parent phase, (II) diffusion across a phase boundary, and (III) diffusion

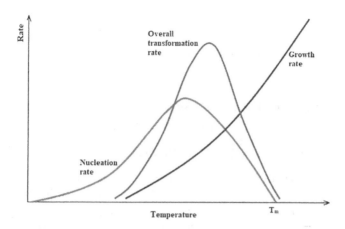

FIGURE 7.5 Schematic plot showing the kinetics of nucleation, growth, and overall transformation.

into the nucleus. For a diffusion-controlled thermally-activated transformation, the growth rate ($\dot{G}$) can be determined by:

$$\dot{G} = C \exp\left(-\frac{Q}{kT}\right) \tag{7.16}$$

where C is a pre-exponential constant, and Q is the activation energy for growth, J/mol.

The overall rate of a phase transformation induced by cooling is a product of the nucleation rate and the growth rate (see Figure 7.5). It is apparent in Figure 7.5 that (in a solidification process) a high transformation temperature (close to T_m) is associated with low nucleation rate and high growth rate; which results in a coarse-grained microstructure. On the other hand, a low transformation temperature (strong supercooling) involves a high nucleation rate and low growth rates; which results in a fine-grained structure.

7.4 KINETICS OF SOLID-STATE PHASE TRANSFORMATION

The rate of a phase transformation is defined as the reciprocal of time required for the transformation to proceed to half completion ($t_{0.5}$). Mathematically (Puri and Wadhawan, 2009),

$$\text{Rate} = \frac{1}{t_{0.5}} \tag{7.17}$$

In order to quantitatively analyze the kinetics of a solid-state phase transformation, data are plotted as the fraction of transformed metal versus the logarithm of time (the temperature is kept constant) (see Figure 7.1). By reference to Figure 7.1, the kinetics of a phase transformation can be expressed by *Avrami equation*, which relates the fraction of transformed metal (y) to time (t), as:

$$y = 1 - \exp(-k\,t^n) \tag{7.18}$$

where k and n are time-independent constants for the particular solid-state phase transformation (see Example 7.5). A notable example of solid-state phase transformation (involving the significance of Equation 7.18) is the recrystallization of metal (see Chapter 14, section 14.2).

7.5 MICROSTRUCTURAL CHANGES IN FE-C ALLOYS—EUTECTOID REACTION IN STEEL

7.5.1 S-shape Curves for Eutectoid Reaction

A number of phase transformation reactions, occurring in iron-carbons alloys, have been explained in Chapter 6 (section 6.2). Let us focus at the steel portion of iron-carbon phase diagram with particular reference to the formation of *pearlite*: a lamellar mixture of ferrite (α) and cementite (Fe_3C) (see Figure 7.6). The eutectoid reaction occurring at point p can be represented by:

$$\gamma \text{ (austenite)} \rightarrow \alpha \text{ (ferrite)} + Fe_3C \text{ (cementite)} \tag{7.19}$$

Let us consider an alloy of eutectoid composition (Fe-0.77 wt% C). Cooling from a temperature within the γ phase (say 820°C) and moving down through the point p (the eutectoid temperature, 727°C) will result in the transformation of *austenite* to *pearlite*. This eutectoid transformation occurs at the point p according to the reaction shown in Equation 7.19. The rate of austenite-to-pearlite transformation strongly depends on temperature; this temperature dependence is illustrated in Figure 7.7.

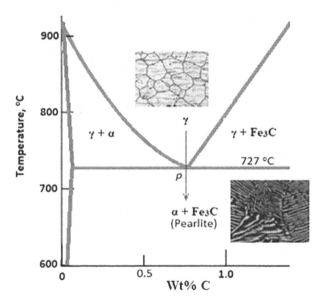

FIGURE 7.6 Steel portion of iron-carbon phase diagram showing the phase transformation: austenite-to-pearlite.

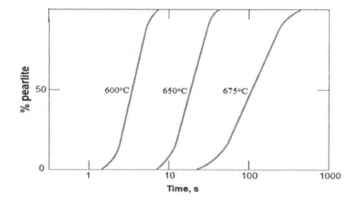

FIGURE 7.7 Isothermal fraction reacted versus the time (*log* scale) for the austenite-to-pearlite transformation for eutectoid composition (Fe-0.77 wt% C) alloy.

Figure 7.7 shows *S-shape* isothermal curves indicating plots of percent pearlite transformed versus the logarithm of time at three different temperatures in the range of 600–675°C (see also Figure 7.1). Each *S-shaped* curves was obtained by collecting data resulting from rapid cooling of specimen composed of 100% austenite to the temperature indicated at isothermal condition.

7.5.2 PEARLITE: *A QUANTITATIVE ANALYSIS*

Pearlite is a lamellar (plate-like) mixture of ferrite (α) and cementite (Fe_3C) (see Figure 7.8). By reference to the iron-carbon phase diagram (Figure 6.5), the maximum solid solubility of carbon in α, γ, and Fe_3C are 0.02 wt%, 0.77 wt%, and 6.67 wt%, respectively. The solid solubility data enable us to compute amounts of ferrite and cementite present in pearlite (see Example 7.6). The computed data (88.7 wt% ferrite and 11.3 wt% cementite) indicate that most of the pearlite is composed of ferrite.

A close observation of the pearilitic microstructure (Figure 7.8) reveals that cementite (a hard and brittle phase) lamella are surrounded by continuous ferrite phase (a soft and ductile phase); this microstructural design provides dispersion strengthening to the eutectoid steel (Askeland and Phule, 2008). The inter-lamellar spacing is strongly dependent on the heat treatment design (particularly the austenite transformation temperature) to generate *pearlite*. If the inter-lamellar spacing (λ) is larger ($\lambda > 4 \times 10^{-5}$ cm), the structure is called *coarse pearlite*, whereas smaller λ values refer to *fine pearlite*. The yield strength of pearlite varies from 200 MPa (for coarse pearlite) to 600 MPa (for fine pearlite) depending on the inter-lamellar spacing.

Figure 7.8 permits us to determine the inter-lamellar spacing (λ) by:

$$\lambda = \frac{Length\ shown\ above\ the\ micron\ bar\ in\ the\ photomicrograph}{Number\ of\ spacings\ over\ the\ length\ of\ the\ micron\ bar} \quad (7.20)$$

The significance of Equation 7.20 is illustrated in Example 7.7.

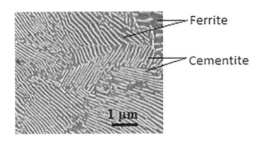

FIGURE 7.8 Photomicrograph of the lamellar pearlite.

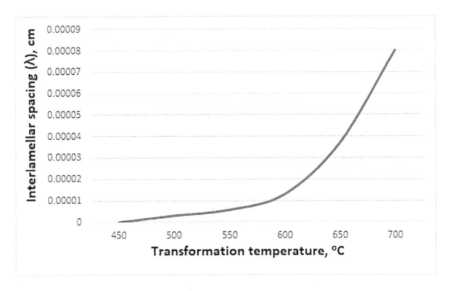

FIGURE 7.9 The effects of austenite transformation temperature on the pearlite inter-lamellar spacing.

The effect of austenite transformation temperature on the pearlite inter-lamellar spacing (λ) is graphically illustrated in Figure 7.9, and demonstrated in Example 7.8.

7.5.3 TIME-TEMPERATURE-TRANSFORMATION (TTT) DIAGRAM

Although the S-shape curves (Figure 7.7) provide some information about the kinetics of eutectoid phase transformation in steel, the standard way of representing both the time and temperature dependence of this transformation is the *time-temperature-transformation (TTT) diagram*, also called the *isothermal transformation* (IT) diagram (Figure 7.10). The TTT diagram (or IT diagram) is a plot that enables us to predict the structure, properties, and heat treatment required in steel (Vander Voort, 1991).

In order to develop the TTT diagram (Figure 7.10), transformation of austenite is plotted against temperature versus time on a logarithm scale. The TTT diagram indicates the beginning and end of a specific phase transformation; it also shows the

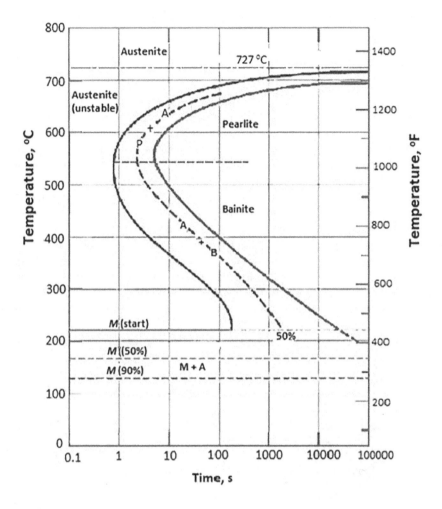

FIGURE 7.10 TTT diagram for Fe-0.77 wt% C alloy, including austenite-to-pearlite (A-P) and austenite-to-bainite (A-B) transformations.

percentage of transformation of austenite at a particular temperature. The shape of TTT diagram looks like either S or like C. In the TTT diagram, the area on the left of the transformation curve represents the austenite region. Austenite is stable above the lower critical temperature (LCT, 727°C) but unstable below LCT. The left curve and the right curve indicate the start and finish of a transformation, respectively. It is evident in the TTT diagram (Figure 7.10) that when austenite is rapidly cooled to a temperature below the LCT (over the temperature range of 540–727°C), and held at the temperature for a duration long enough to complete the transformation followed by quenching to room temperature, the resulting structure is *pearlite*. By heating and cooling a series of samples with various cooling rates, the history of the austenite transformation (to other structures) may be recorded. Cooling rates in the order of increasing severity are achieved by quenching from austenitic temperatures as follows: furnace cooling, air cooling, oil quenching, quenching in liquid salts, water quenching, and brine quenching. The application of TTT diagram is illustrated in Examples 7.9–7.11.

7.5.4 BAINITE AND MARTENSITE

Transformation Temperatures. The kinetics of the eutectoid reaction controls the arrangement of ferrite and cementite in the reaction product. It is evident in the TTT diagram that two types of micro-constituents are generated as a result of the austenite transformation. Above 550°C, *pearlite* is formed, whereas below the temperature, *bainite* is formed. It means that rapid cooling from austenitic temperature to a lower temperature (215–540°C), holding for a long enough time to complete the transformation followed by quenching to room temperature, results in *bainite*. If austenite is cooled at very high cooling rate (e.g. quenched in cold water) to a temperature in the vicinity of the ambient, the resulting microstructure is *martensite* (see Figure 7.10).

Bainite may be either upper or lower. The microstructure of upper bainite shows gray, feathery plates (Figure 7.11a) whereas the lower bainite's structure shows dark needles (Figure 7.11b). Bainite microstructure ensures excellent combinations of hardness, strength, and toughness in steel. This is why control (kinetics) of eutectoid

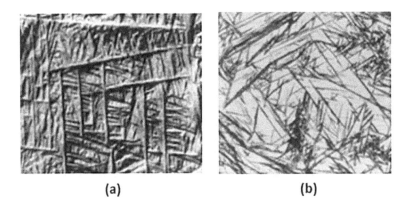

(a)   (b)

FIGURE 7.11   The microstructures of bainite; (a) upper bainite, (b) lower bainite.

FIGURE 7.12   Plate martensite microstructure.

reaction plays an important role in the formation of bainite. In particular, when quenching steel from austenitic temperature to a temperature in the range of 215–540°C, it is important to hold the steel for a long enough time to ensure complete transformation to bainite (see Example 7.12).

Martensite. *Martensite* is a non-equilibrium phase that forms when steel is quenched from austenitic temperature to a temperature in the vicinity of ambient. *Martensite* formation is the result of diffusionless solid-state transformation. *Martensite* in steels is very hard and brittle, provided the carbon content in the steel is higher than 1.0 wt%; such steels have plate martensite structure (Figure 7.12). *Martensite* formation by water quenching of steel results in distortion; this problem can be overcome by the addition of alloying elements in steel (see section 7.6).

7.6 CONTINUOUS COOLING TRANSFORMATION (CCT) DIAGRAMS

The development of a TTT diagram requires transferring steel specimen to a separate furnace for an isothermal treatment; which is a time-consuming and expensive process. This difficulty is overcome by continuous cooling transformation (CCT) diagrams. A *CCT diagram* measures the extent of transformation as a function of time for a continuously decreasing temperature. It means that (for developing a CCT diagram) a specimen is austenitized and then cooled at a predetermined rate and the degree of transformation is measured (Atking, 1980). Figure 7.13 illustrates the

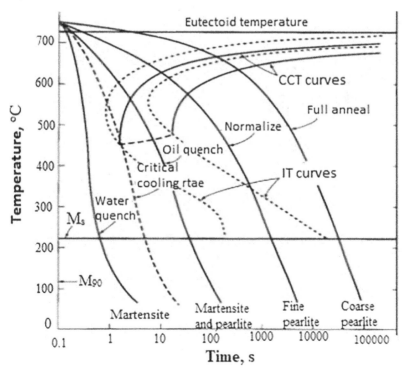

FIGURE 7.13   Continuous cooling transformation (CCT) curves for eutectoid steel.

various phase transformations when eutectoid steel is cooled with a continuously decreasing temperature from the austenitic range to form pearlite, bainite, and martensite, respectively.

A close observation of the *CCT* diagram (Figure 7.13) indicates that there exists a *critical cooling rate* to obtain martensite structure. The *critical cooling rate (CCR)* is the cooling rate associated with the cooling curve that is tangent to the nose of TTT curve (see Figure 7.13). Any cooling rate equal to or faster than *CCR* will form only martensite. The *CCR* for eutectoid steel is 140°C/s; the high CCR value of carbon steel (water quenching to obtain martensite) causes distortion. This problem is overcome by the addition of alloying elements in steel (see Chapter 12, section 12.3).

7.7 CALCULATIONS—EXAMPLES ON PHASE TRANSFORMATIONS AND KINETICS

EXAMPLE 7.1 CALCULATING THE CRITICAL NUCLEUS RADIUS FOR NICKEL

If nickel homogeneously nucleates at 1,136°C, calculate the critical nucleus radius given values of −2.53 x 10^9 J/m³ and 0.255 J/m², respectively, for the latent heat of fusion and the interfacial free energy. The melting temperature of nickel is 1,455°C.

SOLUTION

$T = 1136$ °C $= 1409$K, $T_m = 1455$ °C $= 1728$K, $H_f = -2.53$ x 10^9 J/m³,
   $\gamma_{SL} = 0.255$ J/m²

By using Equation 7.10,

$$r^* = \left(-\frac{2\gamma_{SL}T_m}{\Delta H_f}\right)\left(\frac{1}{T_m - T}\right) = \left(-\frac{2 \times 0.255 \times 1728}{-2.53 \times 10^9}\right)\left(\frac{1}{1728 - 1409}\right)$$
$$= 1.09 \ x \ 10^{-9} m$$

The critical nucleus radius for nickel = 1.09 nm.

EXAMPLE 7.2 COMPUTING THE CRITICAL FREE ENERGY CHANGE FOR A METAL

By using the data in Example 7.1, calculate the critical free energy change for nickel.

SOLUTION

$T = 1409$K, $T_m = 1728$K, $\Delta H_f = -2.53$ x 10^9 J/m³, $\gamma_{SL} = 0.255$ J/m², $\Delta G^* = ?$

By using Equation 7.2,

$$\Delta G^* = \left(\frac{16\pi\gamma_{SL}^3 T_m^2}{3\left(\Delta H_f\right)^2}\right)\frac{1}{\left(T_m - T\right)^2}$$

$$= \left(\frac{16\pi \times 0.255^3 \times 1728^2}{3\left(-2.53\times 10^9\right)^2}\right)\frac{1}{\left(1728 - 1409\right)^2}$$

$$= 1.27 \times 10^{-18}\,J$$

The critical free energy change for nickel = 1.27 x 10^{-18} J.

EXAMPLE 7.3 COMPUTING THE NUMBER OF ATOMS IN A NUCLEUS OF CRITICAL SIZE

Nickel crystallizes as FCC, and its atomic radius is 0.1246 nm. By using the data in Example 7.1, calculate the number of nickel atoms in a nucleus of critical size.

SOLUTION

$r^* = 1.09$ x 10^{-9} m; No. of atoms/unit cell in FCC nickel = 4,

The critical nucleus volume $= \frac{4}{3}\pi r^{*3} = \frac{4}{3}\pi \, (1.09 \text{ x } 10^{-9})^3 = 5.45 \times 10^{-27} m^3$

The lattice parameter of Ni $= a = 2\sqrt{2} = 2\times 1.414 \times 0.1246 \times 10^{-9} = 0.352 \times 10^{-9}\, m$

The unit cell volume of Ni $= a^3 = (0.352)^3$ x $10^{-27} = 0.0437$ x 10^{-27} m^3
By using Equation 7.12,

No. of atoms/critical nucleus $= \left(\text{No. of atoms/unit cell}\right) \text{ x } \dfrac{critical\ nucleus\ volume}{unit\ cell\ volume}$

$$= 4\times\frac{5.45 \text{ x}10^{-27}}{0.0437 \text{ x}10^{-27}} \text{ No. of atoms/critical nucleus}$$

$$= 499.2$$

There are around 500 nickel atoms in a nucleus of critical size.

EXAMPLE 7.4 SELECTING THE BEST WETTING ANGLE FOR EFFECTIVE HETEROGENEOUS NUCLEATION

Determine the extent of heterogeneous nucleation for the following values of wetting angles, and hence select the best wetting angle for extensive heterogeneous nucleation:

(a) $\theta = 110°$, (b) $\theta = 70°$, (c) $\theta = 90°$.

SOLUTION

(a) By using Equation 7.15,

$$\gamma_{LC} = \gamma_{SC} + \gamma_{SL} \cos \theta = \gamma_{SC} + \gamma_{SL} \cos 110° = \gamma_{SC} - 0.34 \, \gamma_{SL}$$

For $\theta = 110°$, the surface tension γ_{LC} is low. The heterogeneous nucleation is not extensive.

(b) For $\theta = 70°$, $\gamma_{LC} = \gamma_{SC} + \gamma_{SL} \cos \theta = \gamma_{SC} + \gamma_{SL} \cos 70° = \gamma_{SC} + 0.34 \, \gamma_{SL}$

For $\theta = 70°$, the surface tension γ_{LC} is high. The heterogeneous nucleation is extensive.

(c) For $\theta = 90°$, $\gamma_{LC} = \gamma_{SC} + \gamma_{SL} \cos \theta = \gamma_{SC} + \gamma_{SL} \cos 90° = \gamma_{SC} + 0 = \gamma_{SC}$

For $\theta = 90°$, the surface tension γ_{LC} is low. The heterogeneous nucleation is not extensive.

Since $\theta = 70°$ yields the high γ_{LC}, case (b) ($\theta = 70°$) is the best wetting angle for an extensive heterogeneous nucleation. This confirms that an acute angle is favorable for heterogeneous nucleation.

EXAMPLE 7.5 CALCULATING THE RATE OF PHASE TRANSFORMATION

A solid-state phase transformation obeys Avrami kinetics with k and n having values of 5 x 10^{-4} and 2, respectively. Calculate the rate of the transformation in s^{-1}.

SOLUTION

For $y = 50\% = 0.5$, $t = t_{0.5}$; $k = 5$ x 10^{-4}, $n = 2$
By using Equation 7.18,

$$y = 1 - \exp(-kt^n)$$
$$0.5 = 1 - \exp[-5 \text{ x } 10^{-4} \, t_{0.5}^2]$$
$$0.5 = \exp[-5 \text{ x } 10^{-4} \, t_{0.5}^2]$$
$$\ln 0.5 = -5 \text{ x } 10^{-4} \, t_{0.5}^2]$$
$$t_{0.5}^2 = 0.138 \text{ x } 10^4$$
$$t_{0.5} = 37.1 \text{ s}$$

By using Equation 7.17,

$$\text{Rate} = \frac{1}{t_{0.5}} = \frac{1}{37.1} = 2.69 \sim 2.7 \, \text{s}^{-1}$$

The rate of transformation = 2.7 per second.

EXAMPLE 7.6 COMPUTING THE WEIGHT PERCENTAGES OF FERRITE AND CEMENTITE IN PEARLITE

The maximum solid solubility of carbon in ferrite (α), austenite (γ), and cementite (Fe$_3$C) are 0.02 wt%, 0.77 wt%, and 6.67 wt%, respectively. Compute the wt. percentages of α and Fe$_3$C in pearlite.

SOLUTION

Wt% C in α = 0.02, wt% C in γ = 0.77, wt% C in Fe_3C = 6.67
By using *lever rule* (Equation 6.15 and Equation 6.16),

$$W_\alpha = \frac{C_{Fe3C} - C_{pearlite}}{C_{Fe3C} - C_\alpha} = \frac{6.67 - 0.77}{6.67 - 0.02} = 0.887 = 88.7 \text{ wt\%}$$

$$W_{Fe3C} = \frac{C_{pearlite} - C_\alpha}{C_{Fe3C} - C_\alpha} = \frac{0.77 - 0.02}{6.67 - 0.02} = 0.113 = 11.3 \text{ wt\%}$$

Hence, the amounts of ferrite and cementite in pearlite are 88.7 wt% and 11.3 wt%, respectively.

EXAMPLE 7.7 CALCULATING THE PEARLITE INTER-LAMELLAR SPACING FROM A PHOTOMICROGRAPH

By reference to Figure 7.8, calculate the inter-lamellar spacing in the pearlite, and hence identify the type of pearlite (coarse or fine).

SOLUTION

The inter-lamellar spacing λ is measured from one α plate to the next α plate in the photomicrograph. By counting the number of lamellar spacing in Figure 7.8, we find six spacings over the 0.7-cm distance (= length of the micron bar). By referring to the micron bar in the micrograph, and by using Equation 7.20, we obtain:

$$\lambda = \frac{Length\ shown\ above\ the\ micron\ bar\ in\ the\ photomicrograph}{Number\ of\ spacings\ over\ the\ length\ of\ the\ micron\ bar}$$
$$= \frac{1\,\mu m}{6\,spacings} = 1.67 \times 10^{-5} \text{cm}$$

Pearlite inter-lamellar spacing = 1.67×10^{-5} cm.
Since $\lambda < 4 \times 10^{-5}$ cm, the microstructure in Figure 7.8 shows *fine pearlite*.

EXAMPLE 7.8 DETERMINING THE AUSTENITE TRANSFORMATION TEMPERATURE WHEN Λ IS KNOWN

By reference to the data in Example 7.7 and Figure 7.9, determine the austenite transformation temperature to generate the pearlite.

SOLUTION

From the plot, $\lambda = 1.67 \times 10^{-5}$ cm corresponds to austenite transformation temperature of 605°C (see Figure E-7.8).

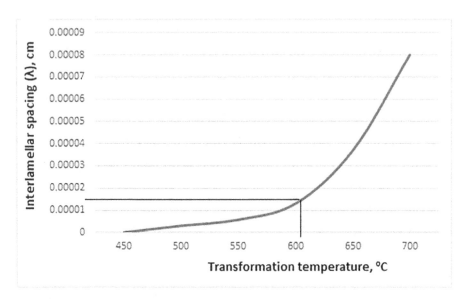

FIGURE E-7.8   Graphical plot with drawn lines to determine the austenite transformation temperature to generate pearlite with specified inter-lamellar spacing.

EXAMPLE 7.9 USING TTT DIAGRAM TO SPECIFY THE MICROSTRUCTURE AFTER 300°C/8,000 S

A specimen of iron-carbon alloy of eutectoid composition is heated to 780°C and held at this temperature long enough to obtain austenite. The specimen is rapidly cooled to 300°C, held for 8,000 s, and quench to room temperature. By using the TTT diagram, specify the nature of the resulting final microstructure.

SOLUTION

The time-temperature path for this the treatment is shown as (E-7.9) in Figure E-7.9-E-7.11.

On cooling from 780°C to 300°C, the stable austenite becomes unstable. Then holding at 300°C for enough time results in the transformation of unstable austenite to bainite; this transformation begins after about 80 s and reaches completion at about 1050 s. Thus, by 8,000 s (as stipulated in this Example) 100% of the specimen is *bainite*. Final quenching to room temperature will not change the microstructure because the phase transformation has been completed.

EXAMPLE 7.10 USING TTT DIAGRAM TO SPECIFY THE MICROSTRUCTURE AFTER 650°C/10,000 S

A specimen of iron-carbon alloy of eutectoid composition is heated to 780°C and held at this temperature long enough to obtain austenite. The specimen is rapidly cooled to 650°C, held for 10,000 s, and quenched to room temperature. By using the TTT diagram, specify the nature of the resulting final microstructure.

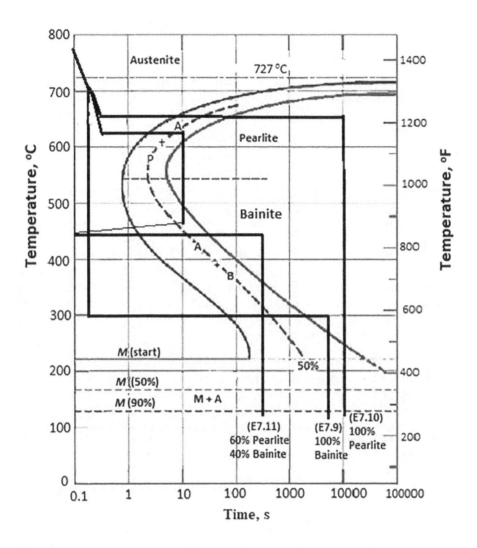

FIGURE E-7.9   TTT diagram showing treatments in Examples 7.9.

SOLUTION

The time-temperature path for this the treatment is shown as (E7.10) in Figure E-7.9.

Cooling the steel from 780°C to 650°C destabilizes the austenite. Then holding at 650°C for enough time results in the transformation of unstable austenite to *coarse pearlite*; this transformation begins after about 8 s and reaches completion at about 90 s. Thus, by 10,000 s (as stipulated in this Example) 100% of the specimen is *coarse pearlite*. Final quenching to room temperature will not change the micro-structure due to completion of the phase transformation.

EXAMPLE 7.11 USING TTT DIAGRAM TO SPECIFY THE MICROSTRUCTURE AFTER MULTI-STAGE TREATMENT

A specimen of iron-carbon alloy of eutectoid composition is heated to 780°C and held at this temperature long enough to obtain austenite. The specimen is rapidly cooled to 620°C, held for 10 s, rapidly cooled to 440°C, held for 500 s, and quenched to room temperature. By using the TTT diagram, specify the nature of the resulting final microstructure.

SOLUTION

The time-temperature path for this the treatment is shown as (E7.11) in Figure E-7.9.

Cooling from 780 to 620°C destabilized the austenite. For the isothermal line at 620°C, pearlite starts to form after about 5 s; by the time 10 s has passed, about 60% of the specimen has transformed to pearlite. Then rapid cooling to 440°C will initiate the transformation of remaining austenite to *bainite*. At 440°C, we start timing at zero time; hence, by the time 500 s have elapsed, all of the remaining (40%) austenite will have completely transformed to bainite. Final quenching to room temperature will not result in any change in the microstructure. The final microstructure at room temperature comprises of 60%pearlite-40%bainite.

EXAMPLE 7.12 USING TTT DIAGRAM TO SPECIFY THE MICROSTRUCTURE AFTER 670°C/100 S

A specimen of iron-carbon alloy of eutectoid composition is heated to 740°C and held at this temperature for a long time. The steel is quenched to 670°C, held at the temperature for about 1.5 min, and then quenched to room temperature. What microstructures will result?

SOLUTION

By using TTT diagram, cooling from 780 to 620°C destabilized the austenite. For the isothermal line at 670°C, pearlite starts to form after about 20 s; by the time 100 s (1.5 min) has passed, about 50% of the specimen has transformed to pearlite. The remaining austenite transforms to martensite by quenching to room temperature (below $M_{90\%}$).

EXAMPLE 7.13 PREDICTING SUCCESS OF A HEAT TREATMENT BY USING TTT DIAGRAM

A eutectoid steel was heat treated to obtain a microstructure suitable for an application that involves moderately high stresses with impact loading. The steel's heat treatment involved austenizing at 770°C; then quenching and holding at 260°C for three hours. Finally, the steel was quenched to room temperature. Was the heat treatment successful?

SOLUTION

3 h = 10,800 s; this application requires a strong and tough material with bainite microstructure.

Quenching austenite to 260°C will initiate transformation to bainite at 200 s; by 10,000 s all the unstable austenite transforms completely to 100% bainite. Thus the heat treatment is successful.

This transformation will finish at about 10,000 s.

EXAMPLE 7.14 PREDICTING MICROSTRUCTURE FOR A STRESSED APPLICATION BY USING TTT DIAGRAM

A eutectoid steel was austenized at 760°C; then quenched and held at 260°C for 30 min. Finally, the steel was quenched to room temperature. Is the resulting micro-structure of the steel is suitable for an application that involves moderately high stresses with impact loading?

SOLUTION

30 min = 1,800 s; this application requires a strong and tough material with bainite microstructure.

Quenching the steel from 760°C to 260°C destabilizes the austenite. Holding at 260°C will initiate transformation to bainite at 200 s. After 30 minutes (1,800 s), about 70% lower bainite has formed and the remainder of the steel still contains unstable austenite; the latter transforms to martensite. Thus the final microstructure is a mixture of lower bainite and hard, brittle martensite. Owing to the presence of the brittle phase, the resulting structure is unsuitable for the specified application that involves moderately high stresses with impact loading.

QUESTIONS AND PROBLEMS

7.1. Differentiate between homogeneous nucleation and heterogeneous nucleation with the aid of a sketch.

7.2. Define free energy and derive the following relationship:

$$\Delta G^* = \frac{16\pi \gamma_{SL}^3}{3(\Delta G_v)^2}$$

7.3. What is the role of wetting angle in the heterogeneous nucleation of a solid from a liquid?

7.4. State the meaning of the terms in the following relationship:

$$\dot{N} \sim \exp\left(-\frac{Q_d}{kT}\right)\exp\left(-\frac{\Delta G^*}{kT}\right)$$

7.5. What are the effects of rates of nucleation and growth (during solidification of a pure metal) on the grain size?

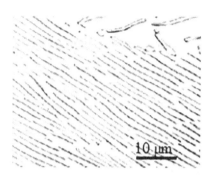

FIGURE P7.11

7.6. Draw the schematic plot showing the kinetics of nucleation, growth, and overall transformation

7.7. Draw the steel portion of iron-carbon phase diagram showing the phase transformation: austenite-to-pearlite; and briefly explain the phase transformation.

P7.8. The degree of supercooling for homogeneous nucleation of aluminum is 190°C. The interfacial energy and melting temperature of aluminum are 0.125 J/m² and 660°C, respectively. The heat of fusion of aluminum is 8.7 x 10⁸ J/m³. Calculate the following parameters for aluminum: (a) nucleation temperature, (b) the critical nucleus radius, (c) the critical free energy change, and (d) the number of nickel atoms in a nucleus of critical size.

P7.9. Determine the extent of heterogeneous nucleation for the following values of wetting angles: (a) $\theta = 90°$, (b) $\theta = 60°$, (c) $\theta = 120°$.

P7.10. A solid-state phase transformation obeys Avrami kinetics with k and n having values of 4 x 10⁻⁴ and 1.8, respectively. Calculate the rate of the transformation in s⁻¹.

P7.11. By reference to Figure P7.11, calculate the inter-lamellar spacing in the pearlite, and hence identify the type of pearlite (coarse or fine).

P7.12. By reference to the data in Figure P7.11 and Figure 7.9, determine the austenite transformation temperature to generate the pearlite.

P7.13. Design a heat treatment to obtain 100% bainite in a eutectoid steel.

P7.14. A specimen of iron-carbon alloy of eutectoid composition is heated to 760°C and held at this temperature long enough to obtain austenite. The specimen is rapidly cooled to 610°C, held for 7,000 s, and quenched to room temperature. By using the TTT diagram, specify the nature of the resulting final microstructure.

REFERENCES

Askeland, D.R. & Phule, P.P. (2008) *The Science and Engineering of Materials.* Carnage Learning, Mason, OH.

Atking, M. (1980) *Atlas of Continuous Cooling Transformation Diagrams for Engineering Steels.* British Steel Corporation, Sheffield.

Huda, Z. (1996) Influence of phase transformation on the kinetics of grain growth in P/M IN-792 superalloy. In: Haq, A.U., Tauqir, A. & Khan, A.Q. (eds) *Phase Transformations-96.* A.Q. Khan Research Laboratories, Kahuta, pp. 24–30.

Porter, D.A., Easterling, K.E. & Sherif, M.Y. (2009) *Phase Transformations in Metals and Alloys.* 3rd ed. CRC Press, Boca Raton.

Puri, S. & Wadhawan, V. (2009) *Kinetics of Phase Transitions.* CRC Press, Boca Raton.

Vander Voort, G. (ed) (1991) *Atlas of Time Temperature Diagrams for Irons and Steels.* ASM International, Materials Park, OH.

Part III

Engineering/Mechanical Metallurgy and Design

8 Mechanical Properties of Metals

8.1 MATERIAL PROCESSING AND MECHANICAL PROPERTIES

Relationship between Processing and Properties. The mechanical properties of a material strongly depend on its structure; which in turn depends on the processing given to the material. Hence, there exists a strong relationship between processing technique and mechanical properties. For example, the tensile and fatigue strengths of forged metallic components are significantly superior to those of cast components. Another example is the effect of rate of cooling in metal casting/welding. Rapid cooling results in fine-grain-sized microstructure associated with higher strength whereas slow cooling rate produces coarse-grain-sized microstructure indicating a lower-strength material.

Mechanical Properties. Mechanical properties refer to the response of a material to an applied load or stress. They determine the amount of deformation which a material, under stress, can withstand without failure. Mechanical properties and behaviors include: elasticity, plasticity, strength, hardness, ductility, toughness, and the like (see Figure 8.1).

It is evident in Figure 8.1 that the mechanical properties/behaviors of metals can be divided into four groups: (a) elastic properties, (b) plastic properties, (c) fatigue behavior, and (d) creep behavior. *Elasticity* refers to the ability of a material to be deformed under load and then return to its original shape and dimensions when the load is removed. *Plasticity* is the ability of a material to be permanently deformed by applying a load. *Strength* determines the ability of a material to withstand load or stress without failure. *Tensile strength* of a material may be defined as the maximum force required to fracture in tension a bar of unit cross-sectional area. *Hardness* is the ability of a material to resist indentation, scratching, or wear. *Ductility* refers to the ability of a material to undergo deformation under tension without rupture. *Toughness* is the ability of a material to absorb energy when receiving a blow or a shock. Additionally, resistance to fatigue and creep are also important mechanical behavior characteristics of engineering materials. These mechanical properties and behaviors are discussed in subsequent sections.

8.2 STRESS AND STRAIN

Tensile Loading. Consider a tensile test specimen that is subject to a tensile force (F); which is increased by suitable increments (see Figure 8.2a).

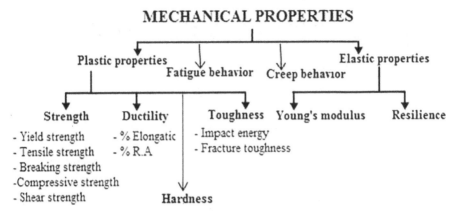

FIGURE 8.1 Classification chart of mechanical properties of metals.

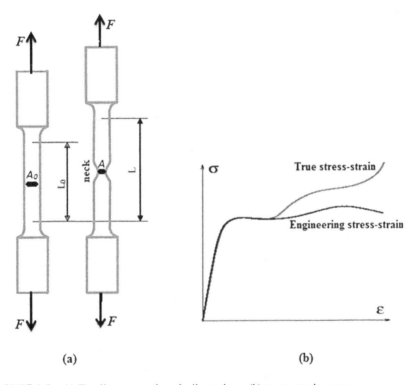

(a)   (b)

FIGURE 8.2 (a) Tensile test specimen's dimensions, (b) stress-strain curves.

The force and elongation recorded during the tensile test enables us to calculate stress and strain. For engineering stress (σ_{eng}) and engineering strain (ε_{eng}), the original (gauge) dimensions of specimen are used, as follows:

$$\sigma_{eng} = \frac{F}{A_0}$$
(8.1)

$$\varepsilon_{eng} = \frac{\Delta L}{L_0} = \frac{L-L_0}{L_0} \tag{8.2}$$

where σ_{eng} is the engineering stress, MPa; F is the applied force, N; A_o is the original cross-sectional area of the specimen (before the test begins), mm^2; ε_{eng} is the engineering strain; L_o is the original length of the specimen (or distance between gauge marks), mm; and L is the final gauge length or the length at any instant during tensile testing, mm (see Figure 8.2b; Example 8.1).

True stress (σ_{true}) and true strain (ε_{true}) are used for accurate specification of plastic behavior of ductile materials by considering the actual (instantaneous) dimensions, as follows:

$$\sigma_{true} = \frac{F}{A} \tag{8.3}$$

$$\varepsilon_{true} = \int_{L_0}^{L} \frac{dL}{L} = \ln\left(\frac{L}{L_0}\right) \tag{8.4}$$

where A and L are the cross-sectional area and (gauge) length at any instant during the tensile loading, mm^2 and mm, respectively.

Since volume remains constant during tensile testing, the following relation applies:

$$V = A \cdot L = A_0 \cdot L_0 \tag{8.5}$$

By combining Equation 8.3 and Equation 8.5,

$$\sigma_{true} = ..\frac{F}{A}.. = ..\frac{F*L}{A_0*L_0} \tag{8.6}$$

By combining Equations 8.1, 8.2, and 8.6, we obtain:

$$\sigma_{true} = \sigma_{eng}(1+\varepsilon_{eng}) \tag{8.7}$$

By combining Equation 8.2 and 8.4, we obtain:

$$\varepsilon_{true} = \ln(1+\varepsilon_{eng}) \tag{8.8}$$

Equations 8.7 and 8.8 can be used to derive the true stress-strain curve from the engineering curve, as shown in Figure 8.2b (see also Example 8.2).

By combining Equation 8.2 and 8.5, we obtain:

$$\varepsilon_{true} = \ln\left(\frac{A_0}{A}\right) \tag{8.9}$$

The technological significance of Equations 8.3–8.9 is explained in Chapter 9 (section 9.4)

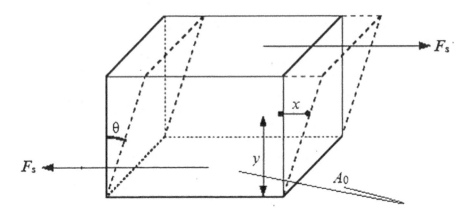

FIGURE 8.3 Shear deformation in a solid ($\tau = F_s/A_0$; $\gamma = \tan \theta = x/y$).

TABLE 8.1
Young's and Shear Moduli of Various Metals at Ambient Temperature

Material	Lead	Magnesium	Aluminum	Copper	Nickel	Steel	Tungsten
Young's modulus, GPa*	15	45	70	110	207	207	407
Shear modulus, GPa	6	17	25	46	76	83	160

* 1 GPa = 10^9 N/m^2

Shear Loading. When a shear force (F_s) is applied to a solid, a shear deformation (angle of twist, θ) is produced (see Figure 8.3). It means a shear stress (τ) produces a shear strain (γ).

It has been shown that shear stress is directly related to shear strain through the expression:

$$\tau = G\,\gamma \tag{8.10}$$

where G is the shear modulus or modulus of rigidity, MPa (see Example 8.3; see Table 8.1). Besides Equation 8.10, the shear modulus can also be determined by measuring the slope of the linear elastic region in the shear stress-strain curve (Callister, 2007).

8.3 TENSILE TESTING AND TENSILE PROPERTIES

8.3.1 TENSILE TESTING

A tensile test enables us to determine the tensile mechanical properties of a metal. In order to conduct tensile testing, first it is necessary to produce a tensile test specimen, as shown in Figure 8.2a. The dimensions of the test specimen have been standardized by the American Society of Testing of Materials (ASTM). Before the test, the original gauge length (l_0) and the original diameter (d_0) are measured so as to enable calculations of original cross-sectional area.

The tensile test is conducted by use of a tensile testing machine. In tensile testing, the specimen is gripped in place by holding jaws (see Figure 8.4). One end of the specimen is held firm, while a hydraulic piston forces the other grip away from it thereby increasing the tensile load within the specimen. In modern machines, an electronic device for measuring the specimen extension (extensometer) is mounted on the specimen; which has to be removed once the specimen approaches its proportional limit or it will be damaged when the specimen breaks. Following the failure of the material, the specimen is reassembled; and the gauge length at failure (l_f), and the final cross-sectional area (A_f) are measured. The force and elongation data recorded during the tensile test are used to calculate stress, strain, strength, and ductility (see Equations 8.1–8.9).

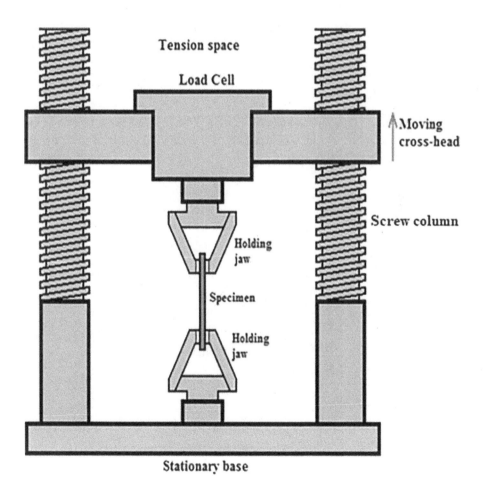

FIGURE 8.4 Tensile testing principle.

8.3.2 Tensile Mechanical Properties

Stress-Strain Curve. In the preceding subsection, we learned that tensile testing involves subjecting a test specimen to a pulling force. If the material is continually pulled until it breaks, a complete tensile profile is obtained in the form of a stress-strain curve, as shown in Figure 8.5.

Strength Properties. The stress-strain curve (Figure 8.5) indicates that in the first stage of elastic behavior, the rate of straining is very small and such strain is proportional to the stress up to the proportional limit i.e. *Hooke's law* is obeyed from *O* to *A*. This linear elastic behavior is given by:

$$\sigma = E \epsilon \tag{8.11}$$

where σ is the *elastic stress*; ε is the *elastic strain*; and the constant of proportionality *E* is called *Young's modulus* or modulus of elasticity. Young's modulus is a measure of the stiffness of the material in tension.

The stress at the elastic point (*B*) is called the *elastic limit* (see Figure 8.5). By taking the stress and strain up to the yield point (*C*), Equation 8.11 can be written as:

$$\sigma_{ys} = E\epsilon_y \tag{8.12}$$

where σ_{ys} is the yield stress or yield strength, *E* is Young's modulus, and ϵ_y is the elastic strain taken to yielding. The modulus of elasticity (*E*) values of some metals are listed in Table 8.1.

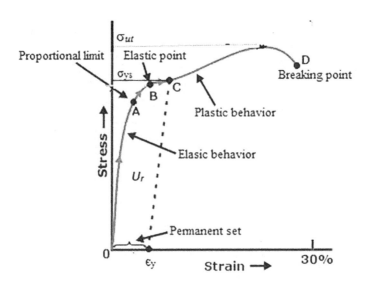

FIGURE 8.5   Stress-strain curve for a ductile metal (σ_{ys} = yield strength, σ_{ut} = tensile strength, U_r = resilience).

Young's modulus may be computed by rewriting Equation 8.11 as follows:

$$E = \frac{Elastic\ stres}{Elastic\ strain} = \frac{\dfrac{F}{A_0}}{\dfrac{\delta l}{l_0}} = \frac{F l_0}{A_0 (\delta l)} \tag{8.13}$$

where F is the elastic force, N; and δl is the elastic elongation, mm; and E is the Young's modulus, MPa (see Examples 8.4–8.6).

It is evident in Figure 8.5 that the metal behaves elastic deformation up to the elastic point (B); after which the metal yields up to C. The stress at the yield point C is called the yield stress, σ_{ys}.

Beyond point C, there is *strain hardening* i.e. there is an increase in stress with increasing strain (see Figure 8.5; see Chapter 9, Sect. 9.4). The maximum stress during the tensile test, is called the tensile strength (σ_{ut}) (see Figure 8.5). The tensile strength is computed by:

$$\sigma_{ut} = \frac{F_{max}}{A_0} \tag{8.14}$$

where σ_{ut} is the tensile strength, MPa; F_{max} is the maximum force that is applied during the tensile test, N; and A_0 is the original cross-sectional area, mm^2. It is evident in Figure 8.5 that on reaching the maximum stress (σ_{ut}), necking starts and fracture finally occurs at the point D. The stress at the fracture point is called the *breaking strength*. The computations for yield strength (σ_{ys}), tensile strength (σ_{ut}), and breaking strength are illustrated in Example 8.7.

Ductility Properties. The stress-strain curve enables us to determine not only strength properties but also ductility properties; the latter include percent elongation and % reduction in area (% RA).

$$\%\ Elongation = \frac{\Delta l}{l_0} \times 100 = \frac{l_f - l_0}{l_0} \times 100 \tag{8.15}$$

$$\%\ RA = \frac{A_0 - A_f}{A_0} \times 100 \tag{8.16}$$

The significance of Equations 8.15 and 8.16 is illustrated in Example 8.8. Table 8.2 lists the tensile mechanical properties of some important metals/alloys. It may be

TABLE 8.2
Tensile Mechanical Properties of Some Metals

Material	Aluminum	Copper	Nickel	Mg-alloy	Steels	Cast iron	Titanium
Yield strength (σ_{ys}), MPa	35	70	140	170–200	220–1,000	100–150	450
Tensile strength (σ_{ts}), MPa	90	200	480	300–320	350–2,000	300–1,000	520
% Elongation (in l_0=50mm)	40	45	40	10–13	12–30	0–7	25

noted that the tensile strength of carbon steels strongly depend on their carbon contents (see Chapter 12, Figure 12.2).

8.4 ELASTIC PROPERTIES: *YOUNG'S MODULUS, POISSON'S RATIO, AND RESILIENCE*

Anisotropic and Isotropic Materials. In the preceding section, we have learned that Young's modulus (E) of a material is a measure of its stiffness in tension. Besides Young's modulus, there are two other important elastic properties: Poisson's ratio and the resilience. It is worth mentioning that E values of many materials strongly depend on their crystallographic orientations; such materials are called *anisotropic materials*. When the mechanical behavior of a material is the same for all crystallographic orientations, the material is called *isotropic*. The elastic behavior of *isotropic materials* is expressed in terms of *Poisson's ratio* (v).

Poisson's Ratio. It has been experimentally shown that an axial elastic elongation of a metal under tension is accompanied by a corresponding constriction in the lateral direction (see Figure 8.6). It means the axial strain is accompanied by a corresponding lateral strain.

Poisson's ratio (v) is the ratio of the lateral strain to the axial strain. Mathematically,

$$v = -\frac{Lateral\,strain}{Axial\,strain} \tag{8.17}$$

In general, *Poisson's ratio* values for metals lie in the range of 0.28–0.34 (see Example 8.9). For isotropic materials, there exists a mathematical relation between Young's modulus (E), the shear modulus (G), and Poisson's ratio (v) according to:

$$E = 2G\,(1 + v) \tag{8.18}$$

The significance of Equation 8.18 is illustrated in Example 8.10.

Resilience. Resilience of a material is the amount of energy absorbed per unit volume when it is deformed elastically and then, upon unloading, to have this energy

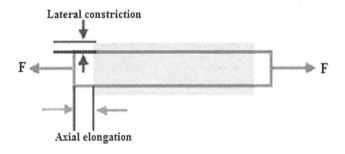

FIGURE 8.6 Axial elongation (positive strain) and lateral constriction (negative strain).

recovered. Mathematically, *resilience* (U_r) is the area under the stress-strain curve taken to yielding (see Figure 8.5), or

$$U_r = \int_0^{\epsilon_y} \sigma \, d\epsilon = \frac{1}{2} \sigma_{ys} \epsilon_y \tag{8.19}$$

where U_r is the resilience, J/m³; σ is the elastic stress, Pa; ϵ_y is the elastic strain up to yielding, and σ_{ys} is the yield strength, Pa. By combining Equation 8.12 and Equation 8.19, we obtain:

$$U_r = \frac{1}{2} \sigma_{ys} \epsilon_y = \frac{1}{2} \sigma_{ys} \left(\frac{\sigma_{ys}}{E} \right) = \frac{\sigma_{ys}^2}{2E} \tag{8.20}$$

Equation 8.20 indicates that a resilient material is one having a high yield strength but a low modulus of elasticity (see Example 8.11). Highly resilient alloys find applications in mechanical springs.

8.5 HARDNESS

8.5.1 Hardness and Its Testing

The term *hardness* refers to the ability of a material to resist plastic deformation by indentation (penetration), scratching, or abrasion. *Hardness* gives a material the ability to resist being permanently deformed (bent, broken, or have its shape changed), when a load is applied on it. In general, hardness is measured as the indentation hardness; which is defined as the resistance of a metal to indentation. In hardness testing, a pointed or rounded indenter is pressed into a surface under a substantially static load (Bhaduri, 2018). Hardness tests are routinely conducted in manufacturing industries for specification purposes, for checking heat-treatment procedures or effectiveness of surface-hardening methods, and as a substitute for tensile tests on small parts. There are five methods of hardness testing: (1) Brinell hardness test, (2) Rockwell hardness test, (3) Vickers test, (4) Knoop hardness test, and (5) microhardness test.

8.5.2 Brinell Hardness Test

The Brinell hardness test involves indenting the test material with a 10-mm-diameter hardened- steel/carbide ball that is subjected to a specified load, F (see Figure 8.7). For softer materials, the applied load is in the range of 500–1,500 kgf whereas for hard metals, the full load of 3,000 kgf is used. The full load is normally applied for 10 to 15 seconds in the case of ferrous metals; this duration is about 30 s in the case of nonferrous metals. The diameter of the indentation produced in the test material is measured with the aid of a low-power microscope.

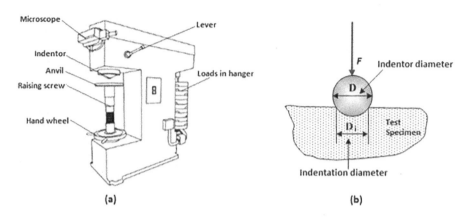

FIGURE 8.7   Brinell hardness testing; (a) equipment, (b) principle.

TABLE 8.3
Brinell Hardness Numbers of Some Metals at 25°C

Material	Annealed brass	Carbon steels	Cast irons	Hardened steels	Al-alloys
BHN	60	130–235	350–415	800–900	30–160

Brinell harness number (BHN) is calculated by dividing the load applied by the surface area of the indentation, according to:

$$BHN = \frac{F}{\frac{\pi}{2}\left(D - \sqrt{D^2 - D_i^2}\right)} \tag{8.21}$$

where the symbols F, D, and D_i are illustrated in Figure 8.7b (see Example 8.12).

A well-defined BHN reveals the test conditions. For example, 80 HB 10/1,000/30 means that Brinell Hardness of 80 was obtained using a 10-mm-diameter hardened steel ball with a 1,000 kgf load applied for 30 seconds. On tests of extremely hard metals, a tungsten carbide (WC) ball is substituted for the steel ball. Brinell hardness test method is the best for achieving the bulk or macro-hardness of a material, particularly those with heterogeneous structures. Table 8.3 lists BHN values of some metals and alloys.

For most steels, the Brinell hardness is related to the tensile strength according to:

$$\sigma_{ut} = 4.35 \ HB \tag{8.22}$$

where σ_{ut} is the tensile strength, MPa; and HB is the Brinell hardness number.

8.5.3 ROCKWELL HARDNESS TEST

Rockwell hardness testing involves indenting the test material with a diamond cone or hardened steel-ball indenter. The indenter is forced into the test material under a minor load $F0$, usually 10 kgf (see Figure 8.8). On reaching equilibrium, an indicating device is set to a datum position. While the minor load is still applied, an additional major load, $F1$, is applied with resulting increase in penetration. When equilibrium has again been reached, the additional major load is removed but the minor load is still maintained. This reduces the depth of penetration (Figure. 8.8).

Rockwell hardness number (HR) is calculated by (Huda and Bulpett, 2012):

$$HR = R - h \tag{8.23}$$

where h is the permanent increase in the depth of penetration due to partial elastic recovery, measured in units of 0.002 mm (see Figure 8.8), and R is a constant depending on the form of indenter: 100 units for diamond indenter, and 130 units for steel ball indenter.

Rockwell hardness testing method is employed for rapid routine testing of finished materials and products since it indicates the final test results directly on a dial which is calibrated with a series of scales. In general, scales A–C are used for metals and alloys (see Table 8.4). Brasses and aluminum alloys have Rockwell hardness numbers in the range of 20 HRB–60 HRB whereas steels have hardness from 20 HRC (for easily machined steels) to 80 HRC for nitrided steels. This author has successfully assured the specified hardness of around 50 HRC in a forged-hardened-tempered 1050 steel for application in axle-hubs of motor cars (Huda, 2012).

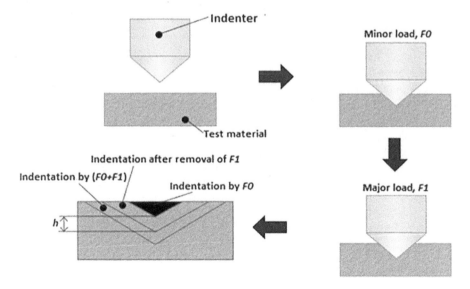

FIGURE 8.8 Rockewell hardness testing principle (The penetration h is measured by the Rockwell hardness tester).

TABLE 8.4

Rockwell Hardness Scales for Typical Applications

Scale	Indenter	Minor load F0, kgf	Major load F1, kgf	Total load F, kgf	R	Typical applications
A	Diamond cone	10	50	60	100	Cemented carbides, thin steel, and shallow case-hardened steel
B	$\frac{1}{16}$ steel ball	10	90	100	130	Copper alloys, soft steels, aluminum alloys, malleable irons, etc.
C	Diamond cone	10	140	150	100	Steel, hard cast irons, case-hardened steel, and other materials harder than 100 HRB

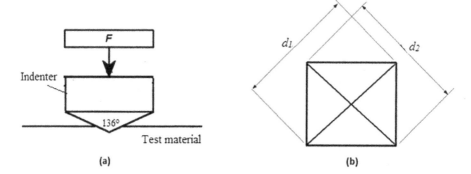

FIGURE 8.9   Vickers hardness testing; (a) indenter is subjected to load, F, (b) indentation.

8.5.4   VICKERS HARDNESS TEST

The Vickers hardness test involves indenting the test material with a diamond indenter, in the form of a right pyramid with a square base and an angle of 136° between opposite faces. The indenter is subjected to a load in the range of 1–100 kgf (see Figure 8.9a). The full load is normally applied for 10 to 15 seconds. The two diagonals of the indentation left in the surface of the material after removal of the load are measured using a microscope (see Figure 8.9b).

The two diagonals d_1 and d_2 (in mm) enable us to calculate d; which is the arithmetic mean of the two diagonals (Figure 8.9b). Vickers hardness number (VHN) can be calculated using the formula:

$$\text{VHN} = \frac{2F\sin\dfrac{136}{2}}{d^2} = \frac{1.854\,F}{d^2} \tag{8.24}$$

The significance of Equation 8.24 is illustrated in Example 8.13.

The Vickers hardness should be reported in the form: 700 HV/20, which means a Vickers hardness of 700, was obtained using a 20 kgf force. The Vickers hardness test has several advantages; which include: (1) very accurate readings can be taken, (2) just one type of indenter is used for all types of metals and surface treatments, and the like.

8.5.5 Knoop Hardness Test

Knoop hardness test, also called Knoop microhardness test, involves the use of a rhombic-based pyramidal diamond indenter that forms an elongated diamond-shaped indent. The indenter is pressed into the sample by an accurately controlled test force, P, in the range of 1–1,000 g. The load is maintained for a specific dwell time (10–15 s). Then, the indenter is removed leaving an elongated diamond-shaped indent in the sample. An optical microscope is used to measure the longest diagonal (L) of the diamond-shaped indent (see Figure 8.10).

Knoop hardness number (*KHN*) is determined by formula:

$$KHN = \frac{14.23\,P}{L^2} \tag{8.25}$$

where P is the test force, kgf; and L is the longest diagonal of indent, mm (see Example 8.14).

The Knoop hardness number normally ranges from HK 60 to HK 1,000 for metals.

8.5.6 Microhardness Test

Microhardness testing method is used when test samples are very small or thin, or when small regions in a multi-phase metal sample or plated metal are to be measured. A *microhardness tester* can measure surface to core (case depth) hardness on carburized or case-hardened parts, as well as surface conditions (e.g. grinding burns, de-carburization, etc.).

Microhardness testing is based on either Knoop hardness test or Vickers hardness test (see the preceding paragraphs). For each test, a small diamond indenter with pyramidal geometry is forced (1–1,000 gf) into the surface of the test material. The resulting indentation is observed under an optical microscope; and measured. Equation 8.24 or Equation 8.25 may be used to calculate KHN or VHN. Modern microhardness equipment have been automated by linking the indenter apparatus to a computerized image analyzer; this system is capable of determining indent location, indent spacing, hardness value (HV/KH), and the like.

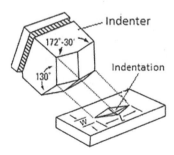

FIGURE 8.10 Knoop microhardness testing principle.

8.5.7 Hardness Conversion

It is often desirable to convert hardness on one scale to another. The hardness conversion data/tables are contained elsewhere (ASTM, 2017; Kinney, 1957). For example, the hardness of a free-machining steel may be around: 250 HB = 100 HRB = 30 HRC = 300 HK.

8.6 IMPACT TOUGHNESS—*IMPACT ENERGY*

The impact toughness is the amount of energy absorbed before fracturing of the metal when loaded under a high strain rate (impact condition). The impact testing of a material involves the use of a pendulum with a known mass at the end of its arm; which swings down and strikes a notched specimen while it is held securely in position (see Figure 8.11a). Two methods are generally used for impact testing: (a) Charpy test, and (b) Izod test. In the *Charpy test*, the impact test specimen is securely held in a horizontal position; and the notch is positioned facing away from the striker (see Figure 8.11b). In the *Izod test*, the notched specimen is held in a vertical position with the notch positioned facing the striker. Impact tests are used as an economical quality control method to assess notch sensitivity and impact toughness of engineering materials.

In the impact test a heavy pendulum, released from a known height, strikes the specimen on its downward swing thereby fracturing it (see Figure 8.11). By knowing the mass of the pendulum and the difference between its initial and final heights $(h–h')$, the energy absorbed by the fracturing specimen can be measured. This energy is called the *impact energy*. Mathematically,

$$U_i = m\,g\,(h - h') \tag{8.26}$$

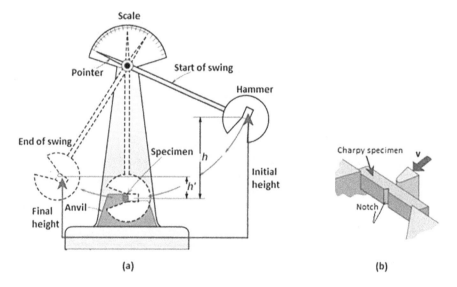

FIGURE 8.11 (a) Impact testing principle; (b) Charpy test specimen.

TABLE 8.5
Impact Energy of Some Alloys at 25°C

Alloy	Cast irons	Magnesium alloy	Aluminum alloy	Steels	Titanium	70-30 brass
Izod impact energy (J)	1–3	8–10	22–27	44–75	60	90

where U_i is the impact energy of the specimen's material, J; m is the mass of the pendulum, kg; and g is the gravitational acceleration ($g = 9.81$ m/s^2) (see Example 8.15).

The impact toughness of some metals and alloys is presented in Table 8.5.

Materials with high impact energies are referred to as *ductile materials* (e.g. lead, brass, mild steel). On the other hand, brittle materials are characterized by their low impact energy values; notable examples include: glass, cast iron, magnesium alloys, and the like (see Table 8.5). The impact energy of a metal is strongly dependent on temperature. In particular, there is a sharp reduction in impact energy when the temperature falls below the ductile-brittle transition (DBT) temperature; which is generally below 0°C. The DBT material behavior has great technological significance on the fracture behavior of materials (see Chapter 10).

8.7 FATIGUE AND CREEP PROPERTIES

The core mechanical properties of materials have been discussed in the previous sections. In addition to the tensile properties, hardness, and impact toughness, there are special loading condition (e.g. cyclic stresses, high temperature, etc.) to which a material may be subjected during service. Fatigue properties relate to the failure of a material under cyclic stresses whereas creep properties are associated with the failure of a material due to high temperature or slow straining. Since fatigue and creep reflect the failure mechanical behaviors, they are discussed in Chapter 10.

8.8 CALCULATIONS—EXAMPLES ON MECHANICAL PROPERTIES OF METALS

EXAMPLE 8.1 COMPUTING ENG. STRESS, ENG. STRAIN, AND FINAL C.S.A

A 140-mm-long metal rod having a diameter of 1.7 mm is subject to a pulling force of 1,600 N. As a result, the length increases to 160 mm. Calculate the engineering stress, engineering strain, and the final cross-sectional area.

SOLUTION

$L_o = 140$ mm, $L = 160$ mm, $d_0 = 1.7$ mm, $F = 1600$ N, $\sigma_{eng} = ?$, $\varepsilon_{eng} = ?$, $A = ?$

$$A_0 = \frac{\pi}{4}d_0^2 == \frac{\pi}{4}(1.7)^2 = 2.27 \text{ mm}^2$$

By using Equations 8.1 and 8.2

$$\sigma_{eng} = \frac{F}{A_0} = \frac{1600}{2.27} = 705 \text{ MPa}$$

$$\varepsilon_{eng} = \frac{\Delta L}{L_0} = \frac{L - L_0}{L_0} = \frac{160 - 140}{140} = 0.143$$

By using Equation 8.5,

$$A \cdot L = A_0 \cdot L_0$$
$$A = (A_0 \cdot L_0) / L = (2.27 \times 140) / 160 = 1.98 \text{ mm}^2$$

Engineering stress = 705 MPa, Engineering strain = 0.143, Final cross-sectional area = 1.98 mm²

EXAMPLE 8.2 COMPUTING TRUE STRESS AND TRUE STRAIN WHEN ENG. STRESS AND STRAIN ARE KNOWN

By using the data in Example 8.1, calculate the true stress and true strain.

SOLUTION

σ_{eng} = 705 MPa, ε_{eng} = 0.143, σ_{true} = ?, ε_{true} = ?

By using Equations 8.7 and 8.8,

$$\sigma_{true} = \sigma_{eng}(1 + \varepsilon_{eng}) = 705(1 + 0.143) = 805.8 \text{ MPa}$$

$$\varepsilon_{true} = ln(1 + \varepsilon_{eng}) = ln(1 + 0.143) = 0.134$$

True stress = 805.8 MPa, True strain = 0.134

EXAMPLE 8.3 DETERMINING THE SHEAR MODULUS OF A MATERIAL

A shear force of 90 kN acts along the base area of 10 mm² of a metallic solid. As a result, the angle of twist is 5°. Calculate the shear modulus of the metal.

SOLUTION

F_s = 90 kN = 90,000 N, A_0 = 10 mm², θ = 5°, G = ?

By reference to Figure 8.3,

$$\tau = F_s / A_0 = \frac{90,000}{10} = 9,000 \, \text{N/mm}^2$$
$$\gamma = \tan \theta = \tan 5° = 0.087$$

By rewriting Equation 8.10, we obtain:

$$G = \frac{\tau}{\gamma} = \frac{9000}{0.087} = 103,450 \, \text{MPa}$$

The shear modulus = 103,450 MPa = 103.45 GPa.

EXAMPLE 8.4 COMPUTING THE YOUNG'S MODULUS OF A METAL; AND IDENTIFYING THE METAL

A 140-mm-long rod, having a diameter of 2 mm, is subject to an elastic force of 300 N. The length at the proportional limit is 140.3 mm. Calculate the Young's modulus of the rod material; and hence identify the metal.

SOLUTION

$l_0 = 140$ mm, $d_0 = 2$ mm, $l = 140.3$ mm, $F = 300$ N, $E = ?$

$$A_0 = \frac{\pi}{4} d_0^2 == \frac{\pi}{4} 2^2 = \pi = 3.142 \, \text{mm}^2$$
$$\delta l = l - l_0 = 140.3 - 140 = 0.3 \, \text{mm}$$

By using Equation 8.13,

$$E = \frac{F l_0}{A_0 (\delta l)} = \frac{300 \times 140}{3.142 \times 0.3} = 44,563 \, \text{MPa}$$

Young's modulus = E = 44,563 MPa = 44.56 GPa
By reference to Table 8.1, we identify the metal as magnesium.

EXAMPLE 8.5 COMPUTING THE ELASTIC FORCE WHEN THE MATERIAL IS KNOWN

An aluminum rod of length 150 mm and 2 mm diameter is subject to an elastic force. The length of the rod at the proportional limit is 150.4 mm. What elastic force was applied to the rod?

SOLUTION

$l_0 = 150$ mm, $d_0 = 2$ mm, $l = 150.4$ mm, $F = ?$

By reference to Table 8.1, we obtain $E = 69$ GPa = 69,000 x 10^6 Pa = 69,000 MPa

$$A_0 = \frac{\pi}{4} d_0^2 == \frac{\pi}{4} 2^2 = \pi = 3.142 \text{ mm}^2$$

$$\delta l = l - l_0 = 150.4 - 150 = 0.4 \text{ mm}$$

By rewriting Equation 8.13, we obtain:

$$F = \frac{E A_0 (\delta l)}{l_0} = \frac{69,000 \times 3.142 \times 0.4}{150} = 578 \text{ N}$$

The elastic force = 578 N.

EXAMPLE 8.6 DRAWING A LOAD-ELONGATION CURVE; AND COMPUTING YOUNG'S MODULUS

An 80-mm-gage-length brass specimen is subject to tensile testing. The original diameter of the specimen was 16 mm. The test data obtained for the cold-worked brass is given in Table E-8.6.

Plot the load-elongation curve for the brass; and hence compute the Young's modulus.

SOLUTION

By using the MS Excel software package, we obtain the load-elongation curve (see Figure E-8.6).

It can be deduced from the curve in Figure E-8.6 that the material obeys Hooke's law for elastic strain up to $\delta l = 0.4$ mm; this elongation corresponds to an elastic force, F = 80 kN.

$l_0 = 80$ mm, $\delta l = 0.4$ mm, F = 80,000 N, $A_0 = \frac{\pi}{4} d_0^2 == \frac{\pi}{4} (16)^2 = 201 \text{ mm}^2$

By using Equation 8.13,

$$E = \frac{F l_0}{A_0 (\delta l)} = \frac{80,000 \times 80}{201 \times 0.4} = 79,601 \text{ MPa}$$

The modulus of elasticity or Young's modulus = E = 79,601 MPa = 79.6 GPa.

TABLE E-8.6
Tensile Test Data for Cold-Worked Brass

Force, kN	23	69	89	94	102	110	123	131	139	132	118 (break)
Elongation (Δl), mm	0.1	0.3	0.5	0.6	0.8	1.0	1.5	2.0	3.0	4.0	4.3

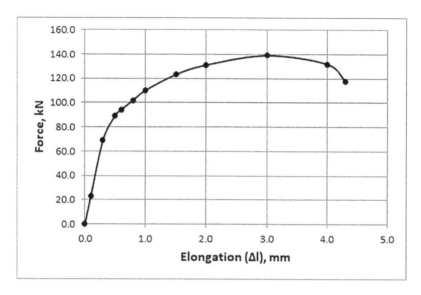

FIGURE E-8.6   Load-elongation curve for cold-worked brass.

EXAMPLE 8.7 COMPUTING THE STRENGTH PROPERTIES FROM LOAD-ELONGATION CURVE

By using the data in Example 8.6, compute the following strength properties of the cold-worked brass: (a) yield strength, (b) tensile strength, and (c) breaking strength.

SOLUTION

By reference to Figure E-8.6, the force at yield point is 89 kN; the maximum force is 139 kN; and the force at break = 118 kN.

By using Equation 8.14,

$$\sigma_{ut} = \frac{F_{max}}{A_0} = \frac{139,000}{201} = 691 \, \text{MPa}$$

$$\text{Yield strength} = \sigma_{ys} = \frac{F_{yield}}{A_0} = \frac{89,000}{201} = 442 \, \text{MPa}$$

$$\text{Breaking strength} = \frac{F_{break}}{A_0} = \frac{118,000}{201} = 587 \, \text{MPa}$$

Yield strength = 442 MPa, Tensile strength = 691 MPa, Breaking strength = 587 MPa

EXAMPLE 8.8 CALCULATING THE DUCTILITY PROPERTIES FROM TENSILE TEST DATA

By reference to the data in Table E-8.6, calculate the ductility properties of the material.

SOLUTION

$l_0 = L_0 = 80$ mm, $\Delta l = 4.3$ mm, $A_0 = 201$ mm²; $L - L_0 = 4.3$ mm, $L = 4.3$ $+ 80 = 84.3$ mm

By using Equation 8.15,

$$\% \text{ Elongation} = \frac{\Delta l}{l_0} \times 100 = \frac{4.3}{80} \times 100 = 5.37$$

By using Equation 8.5,

$$A \cdot L = A_0 \cdot L_0$$

$$A \cdot (84.3) \cdot = \cdot 201 \cdot \times \cdot 80$$

$$A \cdot = \cdot 190.7 \cdot \text{mm}^2$$

By using Equation 8.16,

$$\% \text{ RA} = \frac{A_0 - A_f}{A_0} \times 100 = \frac{201 - 190.7}{201} \times 100 = 5.12$$

% Elongation = 5.37; % Reduction in area = 5.12

EXAMPLE 8.9 CALCULATING POISSON'S RATIO FROM A TENSILE TEST DATA

A 100-mm-long rod, made of an isotropic metal, is subject to tensile stress that resulted in an elastic axial elongation of 0.0657 mm. The original diameter of the rod is 10 mm.; and the change in diameter is 0.002 mm. Calculate Poisson's ratio for the metal.

SOLUTION

$L_0 = 100$ mm; $\delta L = 0.0657$ mm; $\delta d = 0.002$ m; $d = 10$ mm

By using Equation 8.2,

$$\text{Axial strain} = \frac{\delta L}{L_0} = \frac{0.0657}{100} = 6.57 \times 10^{-4}$$

$$\text{Lateral strain} = \frac{\delta d}{d_0} = -\frac{0.002}{10} = -0.0002$$

By using Equation 8.17,

$$v = -\frac{Lateral\,strain}{Axial\,strain} = \frac{0.00020}{0.000657} = 0.3$$

Poisson's ratio for the metal = 0.3

EXAMPLE 8.10 COMPUTING POISSON'S RATIO USING E AND G DATA TABLE

By reference to Table 8.1, calculate Poisson's ratio for lead.

SOLUTION

$E = 15$ GPa (see Table 8.1), $G = 6$ GPa (see Table 8.1), $v = ?$
By using Equation 8.18,

$$E = 2G (1 + v)$$

$$\text{or } v = \frac{E - 2G}{2G} = \frac{15 - (2 \times 6)}{2 \times 6} = 0.25$$

Poisson's ratio for lead = 0.25

EXAMPLE 8.11 SELECTING A MATERIAL FOR SPRING APPLICATION

By reference to the data in Tables 8.1 and 8.2, compute the resilience of aluminum and steel; and select one of them for spring application.

For aluminum: $\sigma_{ys} = 35$ MPa, $E = 70$, GPa = 70,000 MPa
By using Equation 8.10,

$$U_r = \frac{\sigma_{ys}^2}{2E} = \frac{35^2}{2 \times 70,000} = 0.00875 \text{ MPa} = 0.00875 \times 10^6 \text{Pa} = 8.75 \text{ kJ/m}^3$$

For steel: $\sigma_{ys} = 700$ MPa, $E = 207$ GPa = 207,000 MPa

$$U_r = \frac{\sigma_{ys}^2}{2E} = \frac{700^2}{2 \times 207,000} = 1.18 \text{ MPa} = 1.18 \text{ MJ/m}^3$$

The resilience of steel is in mega Joules whereas that of aluminum is in kilo Joules. Hence, steel must be selected for spring application.

EXAMPLE 8.12 CALCULATING THE BRINELL HARDNESS NUMBER WHEN THE LOAD IS KNOWN

A load of 800 kgf is applied by using a 10-mm-diameter steel ball that results in the depression diameter of 7 mm in a test material. Calculate the BHN for the test material.

SOLUTION

$F = 800$ kgf, $D = 10$ mm, $D_i = 7$ mm, BHN = ?

By using Equation 8.21,

$$\text{BHN} = \frac{F}{\frac{\pi}{2}\left(D - \sqrt{D^2 - D_i^2}\right)} = \frac{800}{\frac{\pi}{2}\left(10 - \sqrt{10^2 - 7^2}\right)} = \frac{800}{\frac{\pi}{2}(10 - 7.14)} = 178$$

EXAMPLE 8.13 COMPUTING THE VICKERS HARDNESS NUMBER OF A TEST MATERIAL

The indenter of a Vickers hardness tester is subjected to a load of 100 kgf for 10 s. The two diagonals of the indentation left in the surface of the material are 0.7 mm and 1 mm, respectively. Compute the Vickers hardness number of the material.

SOLUTION

$$F = 100 \text{ kgf}, d = \frac{d_1 + d_2}{2} = \frac{0.7 + 1}{2} = 0.85 \text{ mm}$$

By using Equation 8.24,

$$VHN = \frac{1.854 F}{d^2} = \frac{1.854 \times 100}{0.85^2} = 257$$

The Vickers hardness number = 257.

EXAMPLE 8.14 COMPUTING THE KNOOP HARDNESS NUMBER

A Knoop indenter presses a test material with a load of 500 gf. The longest diagonal of the indent is measured to be 30 μm. Calculate the Knoop hardness number for the test material.

SOLUTION

P = 50 gf = 0.05 kgf, L = 30 μm = 0.03 mm, KHN = ?

By using Equation 8.25,

$$KHN = \frac{14.23P}{L^2} = \frac{14.23 \times 0.05}{0.03^2} = 790$$

The Knoop hardness number = 790.

EXAMPLE 8.15 COMPUTING THE IMPACT ENERGY AND ASSESSING ITS DUCTILITY/BRITTLENESS

In a Charpy impact test, a heavy pendulum of 2 kgf, released from a height of 15 cm, strikes the specimen on its downward swing thereby fracturing it. The height at the end of swing is 4 cm. Calculate the impact energy. Is the material ductile or brittle?

SOLUTION

m = 2 kg, g = 9.81, $h - h'$ = 15 − 4 = 11 cm = 0.11 m, U_i = ?

By using Equation 8.26,

$$U_i = m\,g\left(h-h'\right) = 2 \times 9.81 \times 0.11 = 2.16\ J$$

The impact energy of the material is low; hence the material is brittle.

QUESTIONS AND PROBLEMS

8.1. Encircle the most appropriate answer for the following multiple choice questions (*MCQs*).
   (a) Which mechanical test involves a high strain rate? (i) hardness test, (ii) impact test, (iii) tensile test, (iv) shear test.
   (b) Which mechanical test involves indentation in the test material? (i) hardness test, (ii) impact test, (iii) tensile test, (iv) shear test.
   (c) Which mechanical test is the best for determining the yield strength? (i) hardness test, (ii) impact test, (iii) tensile test, (iv) shear test.
   (d) Which mechanical property involves elastic energy absorption? (i) impact toughness, (ii) Young's modulus, (iii) resilience, (iv) ductility.
   (e) Which hardness test involves the use of rhombic-based pyramidal diamond indenter? (i) Brinell test, (ii) Knoop test, (iii) Rockwell test, (iv) Vickers test.
   (f) The ratio of the lateral strain to the axial strain is called: (i) true strain, (ii) strain ratio, (iii) engineering strain, (iv) Poisson's ratio.
   (g) Which mechanical test involves the angle of twist? (i) hardness test, (ii) impact test, (iii) tensile test, (iv) shear test.
   (h) Which hardness test allows us to use hardened steel ball indenter? (i) Brinell test, (ii) Knoop test, (iii) Rockwell test, (iv) Vickers test.
8.2. Does there exist a strong relationship between a material's processing technique and its mechanical properties? If yes, support your answer by giving at least three examples.
8.3. (a) Differentiate between the following terms: (i) elasticity and plasticity, (ii) engineering stress and true stress, and (iii) engineering strain and true strain. (b) Classify the various mechanical properties by drawing a classification chart.
8.4. (a) Describe the tensile testing of metals with the aid of a diagram. (b) List the tensile properties under: (i) strength properties and (ii) ductility properties.
8.5. Describe the Vickers hardness testing of metals with the aid of sketch.
8.6. What mechanical properties are required in mechanical springs? Justify your answer with the aid of a mathematical expression.
P8.7. A 155-mm-long metal rod having a diameter of 1.8 mm is subject to a pulling force of 1,800 N. As a result, the length increases to 175 mm. Calculate the (a) engineering stress, (b) engineering strain, (c) true stress, and (d) true strain.
P8.8. A 13-mm-wide and 7-mm-thick steel specimen is subject to tensile testing. The original gauge length of the specimen was 50 mm. The test data so obtained is given in Table P8.8. Plot the stress-strain curve for the steel, taking the last stress as the breaking point; and hence compute the Young's modulus.

TABLE P8.8

Tensile Test Data for Carbon Steel

Stress, MPa	100	200	300	420	500	520	550	580
Strain	0.0005	0.0010	0.0015	0.002	0.0035	0.005	0.01	0.013

P8.9. By using the data in Table P8.8, compute the tensile strength and the breaking strength for the steel.

P8.10. By using the data in Table P8.8, compute the ductility properties of the steel.

P8.11. A 80-mm-long rod, made of an isotropic metal, is subject to tensile stress that resulted in an elastic axial elongation of 0.058 mm. The original diameter of the rod is 8 mm; and the final diameter is 8.002 mm. Calculate *Poisson's ratio* for the metal.

P8.12. Based on their resilience values, which of the following two metals is more suitable for application in mechanical springs? Nickel or copper?

P8.13. By using the data in Table 8.1, calculate Poisson's ratio for copper.

P8.14. A load of 900 kgf is applied by using a 10-mm-diameter steel ball that results in the depression diameter of 8 mm in a test material. Calculate the BHN for the test material.

P8.15. The indenter of a Vickers hardness tester is subjected to a load of 80 kgf for 12 s. The two diagonals of the indentation left in the surface of the test material are 0.8 mm and 0.9 mm, respectively. Compute the Vickers hardness number of the material.

REFERENCES

ASTM (2017) *Standard Hardness Conversion Tables for Metals Relationship Among Brinell Hardness, Vickers Hardness, Rockwell Hardness, Superficial Hardness, Knoop Hardness, Scleroscope Hardness, and Leeb Hardness*. ASTM International Inc, West Conshohocken, PA.

Bhaduri, A. (2018) *Mechanical Properties and Working with Metals and Alloys*. Springer, Singapore.

Callister, W.D. (2007) *Materials Science and Engineering: An Introduction*. John Wiley & Sons, Inc, Hoboken, NJ.

Huda, Z. (2012) Reengineering of manufacturing process design for quality assurance in axle-hubs of a modern motor-car—A case study. *International Journal of Automotive Engineers*, 13(7), 1113–1118.

Huda, Z. & Bulpett, R. (2012) *Materials Science and Design for Engineers*. Trans Tech Publications, Switzerland.

Kinney, G.F. (1957) *Engineering Properties and Applications of Plastics*. John Wiley & Sons Inc, Hoboken, NJ.

9 Strengthening Mechanisms in Metals

9.1 DISLOCATION MOVEMENT AND STRENGTHENING MECHANISMS

Movement of Dislocation. The movement of dislocations in crystalline solids is of fundamental importance in understanding strengthening mechanisms in metals. We learned in Chapter 3 that when a crystal is subjected to shear stress, there are atomic displacements due to dislocation motion (see section 3.3). Figure 9.1(a–c) shows a crystal under shear stress (τ). When the resolved shear stress reaches the critical value τ_{crss}, the movement of edge dislocation occurs; which results in plastic deformation. It is evident in Figure 9.1 that the edge dislocation moves in the direction of applied stress (from left to right). The dislocation in the top half of the crystal slips one plane as it moves to the right from its position (Figure 9.1[I]) to its position in (II) and finally in (III). The movement of the dislocation across the plane eventually causes the top half of the crystal to move with respect to the bottom half by a *unit slip step* (see Figure 9.1[III]).

Strengthening Mechanisms. The discussion in the preceding paragraph leads us to conclude that a metal can be strengthened by making it more difficult for dislocations to move. This may be accomplished by introducing obstacles, such as interstitial atoms or grain boundaries, to "pin" the dislocations. In the presence of an obstacle, greater stress is required to move dislocations; which results in strengthening of the metal. For example, in cold working: as a material plastically deforms, more dislocations are produced and intermingle with each other thereby impeding dislocation movement. This is why strain or work hardening occurs in a cold-worked metal. The principal strengthening mechanisms in metals/alloys include: solid-solution strengthening, grain-boundary strengthening, strain hardening, precipitation strengthening, and dispersion strengthening (Argon, 2007; Geddes *et al.*, 2010). These strengthening mechanisms are discussed in the following sections.

9.2 SOLID-SOLUTION STRENGTHENING

Solid-solution strengthening refers to the strengthening caused by the presence of solute atoms in a crystal lattice. It occurs when the atoms of the new element (solute) form a solid solution with the base metal (solvent), but there is still only one phase. The presence of solute atoms in a crystal lattice causes strain and distortion in the lattice; hence greater shear stress is required to move dislocations through the lattice. This is why an alloy is generally stronger than pure metal. For example, the tensile

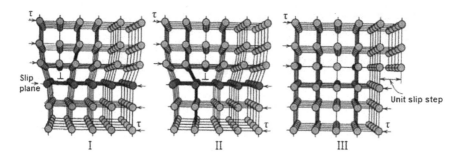

FIGURE 9.1 Movement of edge dislocation (τ = shear stress; $\perp$ = edge dislocation line).

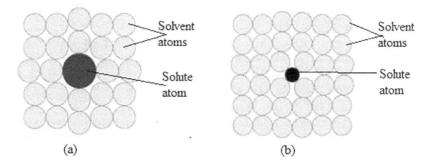

FIGURE 9.2 Solid-solution strengthening; (a) lattice distortion in substitutional solid solution, (b) stress-field creation in interstitial solid solution.

strength of an aluminum alloy containing 2% Mg is around 75 MPa as compared to 30 MPa for pure aluminum. In general, higher the concentration of an alloying element, greater will be the strengthening contribution of the alloying element to the alloy.

We have learned in Chapter 3 that there are two types of solid solutions: (a) substitutional solid solutions and (b) interstitial solid solutions. In the case of substitutional solid solutions, if atom(s) of solvent element were to be substituted by solute atom(s); the host lattice would experience a misfit stress associated with the difference in size between the solute and solvent atoms (Figure 9.2a). An example of substitutional solid solution is Fe-Cr alloy; where atomic radii difference is small. An interstitial solid solution is formed when small-sized solute atoms are dissolved into interstices of the solvent lattice thereby creating a stress field (Figure 9.2b).

The degree of solid-solution strengthening strongly depends on whether the solute atom possesses a symmetrical or nonsymmetrical stress field. In particular, interstitial solid solutions play an important role in strengthening. It can be shown that strengthening by dissolving carbon in BCC iron is more pronounced as compared to FCC iron. When carbon atoms are dissolved into small non-symmetrical interstices of the BCC lattice, lattice distortion is great and strengthening contribution is enhanced. This is because the stress field surrounding the solute atom (in BCC iron) is non-symmetrical in character so the solute atom interacts strongly with both edge and screw dislocations. On the other hand, the strengthening potential for carbon in FCC iron is much less since the strain field surrounding the interstitial atom site is symmetrical.

The strengthening contribution is generally measured as the lattice misfit strain ($\varepsilon_{lattice}$), expressed as (Mohri and Suzuki, 1999):

$$\varepsilon_{lattice} = k \frac{da}{dc} \tag{9.1}$$

where k, the constant of proportionality, is the reciprocal of lattice parameter ($k=1/a$), a is the lattice parameter, and c is the solute concentration.

9.3 GRAIN-BOUNDARY STRENGTHENING— HALL-PETCH RELATIONSHIP

Grain-boundary strengthening refers to the strengthening of a polycrystalline material owing to the presence of grain boundaries in its microstructure. We learned in Chapter 2 that solidification of a polycrystalline material results in a grained microstructure with grain boundaries (see Figure 2.13). Strictly speaking, a grain boundary is an array of dislocations (Huda, 1993). When a solid is under a shear stress, dislocations tend to move through the lattice (see section 9.1). However, a dislocation approaching a grain boundary will not be able to easily cross it into the adjacent grain (see Figure 9.3). In order for the dislocation to easily cross the grain boundary, greater stress is needed to be applied. This mechanism is called *grain-boundary strengthening*.

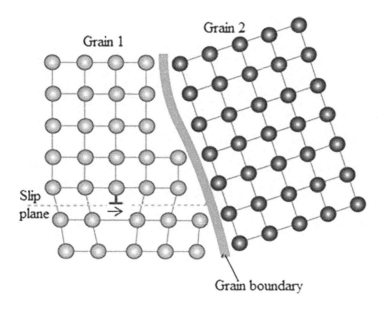

FIGURE 9.3 A grain boundary as an obstacle in the movement of an edge dislocation.

Grain boundaries may be regarded as barriers to slip. Since a fine grain-sized material has a greater grain-boundary area as compared to a coarse grain-sized material, a higher stress is required to move a dislocation in the former as compared to the latter. This is why, finer the grain size, greater is the strength of the material (see Figure 9.4); however this relationship is applicable at low and moderate temperatures (Huda, 2007).

Hall-Petch Relationship. It has been experimentally shown from the work of Petch and Hall that the yield strength of a polycrystalline material is related to the grain size by (Hansen, 2004):

$$\sigma_{ys} = \sigma_0 + \frac{K}{\sqrt{d}}$$ (9.2)

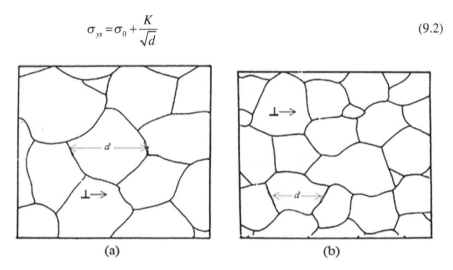

FIGURE 9.4   Dislocations to cross grain boundaries in: (a) coarse grained material, (b) fine grained material (d = grain diameter).

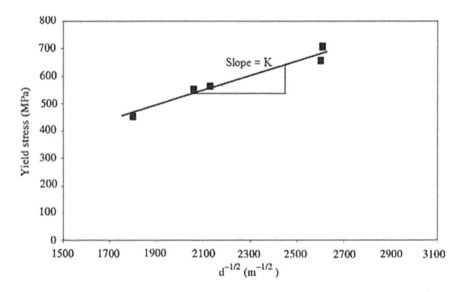

FIGURE 9.5   Hall-Petch relationship's plot for cryo-milled aluminum.

Where σ_{ys} is the yield strength of the polycrystalline sample, MPa; σ_0 is the yield strength of the single crystal (material with no grain boundaries), MPa; K is the constant (= relative hardening contribution of grain boundaries); and d is the average grain diameter, mm (see Examples 9.1–9.4). Equation 9.2 is known as Hall-Petch relationship; this equation is a linear equation in the slope-intercept form. If a graph of yield stress σ_{ys} is plotted versus $d^{-\frac{1}{2}}$, a straight line is obtained having a slope = K, and intercept = σ_0 (see Figure 9.5).

9.4   STRAIN HARDENING

Strain hardening, also referred to as work hardening, involves an increase in hardness and strength due to cold working. The strain-hardening strengthening mechanism can be explained in terms of obstructions in the movement of dislocations as follows (see Chapter 3). As a metal is deformed during the cold-working, dislocations moving through the metal entangle with one another thereby dramatically increasing the number of dislocation-dislocation interactions. It means a greater stress is required to move a dislocation through the entangled-dislocation regions. This mechanism is called *strain hardening*.

Strain hardening results in an increase in the dislocation density. The dislocation density may be defined as the number of dislocations in a unit area of the metal sample (see Figure 9.6). Mathematically, dislocation density is expressed as (Amelinckx, 1964):

$$\rho_D = \frac{n}{A} \tag{9.3}$$

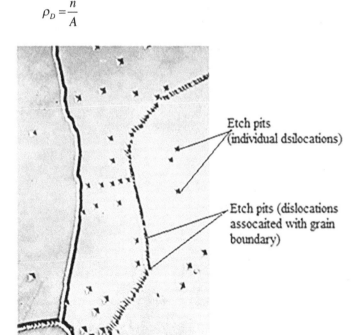

FIGURE 9.6   Etched metal sample showing etch pits that represent dislocations.

where ρ_D is the dislocation density, dislocations cm^{-2}; n is the number of dislocations (or etch pits); and A is the area of the etched sample, cm^2 (see Example 9.5).

Strain hardening of a metal results in a sudden rise in the dislocation density; which in turn results in an increase in hardness and the flow stress of the metal. It has been shown that for many metals, the flow stress may be related to the dislocation density by (Schaffer *et al.*, 1999):

$$\sigma_{flow} = \sigma_0 + k\sqrt{\rho_D} \qquad (9.4)$$

where σ_{flow} is the flow stress, MPa; ρ_D is the dislocation density, dislocations cm^{-2}; and σ_0 and k are the constants for the material (see Examples 9.6 and 9.7).

The strength of a cold-worked metal strongly depends on the amount of deformation; which is generally expressed as percent cold work (% CW), as follows:

$$\% \text{ CW} = \left(\frac{A_0 - A_{Cw}}{A_0} \right) \times 100 \qquad (9.5)$$

where A_0 is the original cross-sectional area, and A_{CW} is the cross-sectional area after cold work. Figure 9.7 illustrates the variation of tensile mechanical properties with the percent cold work for grade-310 austenitic stainless steel (see Example 9.8). It is evident in Figure 9.7 that the yield strength and the ultimate tensile strength of grade-310 stainless steel increases with % CW whereas ductility (% elongation) decreases with percent cold work.

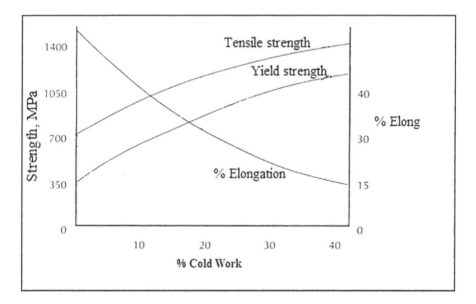

FIGURE 9.7   Variation of tensile properties with % CW for grade-310 stainless steel.

TABLE 9.1

Values of *n* and *K* (see Equation 9.7) for Some Alloys

Alloy	2024-T3 Al alloy	Naval brass (annealed)	Low-carbon steel (annealed)	304 stainless steel	4340 alloy steel	Copper (annealed)
n	0.17	0.21	0.21	0.44	0.12	0.44
K (MPa)	780	585	600	1,400	2,650	530

Many work-hardened metals and alloys have stress-strain curves which can be represented by:

$$\sigma = K\,\varepsilon_{true}^{n} \qquad (9.6)$$

where σ is the true stress or the yield stress, MPa; ε_{true} is the true strain, K is the strength coefficient, MPa; and n is the work-hardening exponent (see Examples 9.9–9.11). For a perfectly plastic solid, n=0 whereas n=1 represents a 100% elastic solid. In order to strengthen a metal by strain hardening, an n value in the range of 0.10–0.50 should be selected. Table 9.1 presents the values of n and K for various alloys.

9.5 PRECIPITATION STRENGTHENING

Precipitation strengthening, also called age hardening, is a heat-treatment process that produces uniformly dispersed particles as coherent precipitates within a metal's microstructure. These coherent precipitates hinder dislocation motion thereby strengthening the metal. A precipitate (second-phase particle) is said to be *coherent* when solute atoms in a lattice are concentrated in certain regions but are still occupying sites in the parent metal lattice, and there is no distinct boundary between the main lattice and the *precipitate* (see Figure 9.8).

Notable examples of alloys which are strengthened by precipitation strengthening include: age-hardenable aluminum alloys, Ni-base superalloys, and the like (Huda, 2007, 2009). For example, the formation of coherent θ' precipitates in a copper-containing age-hardenable aluminum alloy strengthens the alloy according to the following phase-transformation reaction:

$$Cu + 2Al \longrightarrow \theta' \qquad (9.7)$$

[Al lattice] [Al lattice] [coherent precipitate]

where the coherent precipitate θ' is the intermetallic compound: $CuAl_2$ (see Figure 9.8). Precipitation strengthening of aluminum-copper alloy is further discussed in Chapters 13 and 15.

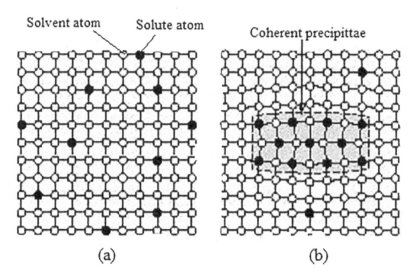

FIGURE 9.8   Precipitation strengthening; before heat treatment (a); after heat treatment (b).

It is evident in Figure 9.8 that there is a lattice mismatch due to precipitation strengthening. In superalloys, it is desirable to determine lattice misfit as a %, as follows (Lahrman, 1982):

$$\% \text{ Lattice mismatch} = \frac{a_{\gamma'} - a_{\gamma}}{a_{\gamma}} \times 100 \tag{9.8}$$

where a_{γ} is the lattice parameter of the matrix (FCC, γ) phase, nm; and $a_{\gamma'}$ is the lattice parameter of the (second phase) gamma prime (γ') phase, nm (see Examples 9.12 and 9.13).

9.6   DISPERSION STRENGTHENING—MECHANICAL ALLOYING

Dispersion strengthening, refers to the increase in strength of an alloy owing to the presence of nano-sized, widely-dispersed, non-coherent particles in the alloy's microstructure (see Figure 9.9b). There must be at least two phases in the microstructure of any dispersion-strengthened alloy. The continuous phase is called the matrix whereas the second phase is the dispersed precipitate.

For effective dispersion strengthening, the matrix should be soft and ductile whereas the precipitate should be strong (see Figure 9.10). The dispersed precipitate hinders dislocation motion, while the matrix provides some ductility to the overall alloy. Notable examples of dispersion-strengthened alloys include oxide-dispassion-strengthened (ODS) alloys (e.g. TD-(thoria-dispersed) nickel, sintered alumina powder [SAP], etc.), mechanically alloyed dispersion strengthened superalloys, and the like.

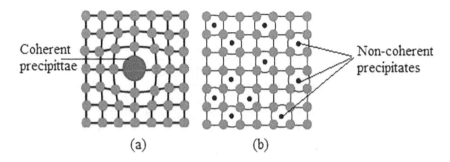

FIGURE 9.9   Difference between precipitation strengthening (a), and dispersion strengthening (b).

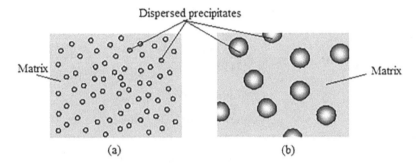

FIGURE 9.10   The schematic of the microstructure of dispersion strengthened alloy; (a) microstructure showing widely dispersed precipitates, (b) spherical dispersed precipitates.

9.7   CALCULATIONS—EXAMPLES ON STRENGTHENING MECHANISMS IN METALS

EXAMPLE 9.1   COMPUTING THE AVERAGE GRAIN DIAMETER FROM HALL-PETCH RELATION'S PLOT

Refer to Figure 9.5. What is the average grain diameter (in nanometer) of cryo-milled aluminum that corresponds to the highest yield strength in the graphical plot?

SOLUTION

From Figure 9.5, the highest yield strength is 700 MPa; which corresponds to $d^{-\frac{1}{2}} = 2670$ m$^{-\frac{1}{2}}$

$$d^{-\frac{1}{2}} = 2670 \text{ m}^{-\frac{1}{2}}$$
$$\frac{1}{\sqrt{d}} = 2670 \frac{1}{\sqrt{m}}$$
$$\frac{1}{d} = (2670)^2 \frac{1}{m}$$

$$d = \frac{1}{2670^2} m = \frac{1}{7128900} m = 1.4 \times 10^{-7} \text{ m} = 140 \times 10^{-9} \text{ m} = 140 \text{ nm}$$

The average grain diameter corresponding to the maximum yield strength = 140 nanometer.

EXAMPLE 9.2 DETERMINATION OF THE CONSTANT IN HALL-PETCH RELATION BY GRAPHICAL PLOT

By reference to the graphical plot in Figure 6.5, determine the constants K and σ_0 in the Hall-Petch relationship for cryo-milled aluminum.

SOLUTION

In order to determine the values of the slope K and the intercept σ_0, the graphical plot is reproduced as shown in Figure E-9.2.

By reference to Figure E-9.2,

$$\text{Slope} = K = \frac{\Delta\sigma_{ys}}{\Delta d^{-\frac{1}{2}}} = \frac{630 - 530}{2450 - 2050} = \frac{100\,MPa}{400\,m^{-\frac{1}{2}}} = \frac{100\,MPa}{400\left(10^3\,mm\right)^{-\frac{1}{2}}}$$

$$= 7.9\text{MPa-}\sqrt{mm}$$

$$\text{Intercept} = \sigma_0 = 380\,\text{Mpa}$$

The Hall-Petch relation's constants are: $K = 7.9\,\text{MPa-}\sqrt{mm}$, $\sigma_0 = 380$ MPa.

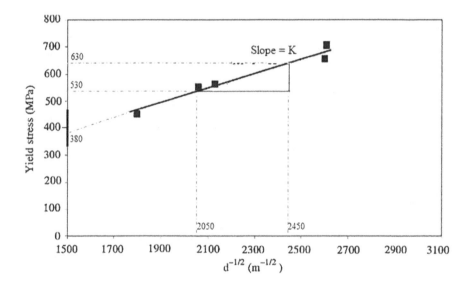

FIGURE E-9.2 Determination of the constant in Hall-Petch relation by graphical plot.

EXAMPLE 9.3 DETERMINING THE CONSTANTS IN HALL-PETCH EQUATION BY CALCULATION

A steel sample with an average grain size of 0.022 mm has a yield strength of 318 MPa. The same steel was heat treated to produce another sample with an average grain size of 0.011 mm having a yield strength of 375 MPa. Calculate the Hall-Petch relation's constants: K and σ_0.

SOLUTION

Sample 1: $d_1 = 0.022$ mm, $\sigma_{ys1} = 318$ MPa; Sample 2: $d_2 = 0.011$ mm, $\sigma_{ys2} = 375$ MPa

By using Equation 9.2 for sample 1,

$$\sigma_{ys} = \sigma_0 + \frac{K}{\sqrt{d}}$$
$$318 = \sigma_0 + \frac{K}{\sqrt{0.002}} \tag{E-9.4a}$$

By using Equation 9.2 for sample 2,

$$375 = \sigma_0 + \frac{K}{\sqrt{0.011}} \tag{E-9.4b}$$

By performing the subtraction of Equations: (E-9.4b–E-9.4a),

$$375 - 318 = \frac{K}{\sqrt{0.011}} - \frac{K}{\sqrt{0.022}}$$

$$K\left(\frac{1}{\sqrt{0.011}} - \frac{1}{\sqrt{0.022}}\right) = 57$$

$$K = 20.4 \, \text{MPa-}\sqrt{mm}$$

By substituting the value of K in Equation E-9.4a,

$$318 = \sigma_0 + \frac{20.4}{\sqrt{0.022}}$$
$$\sigma_0 = 318 - \frac{20.4}{\sqrt{0.022}} = 180 \, \text{MPa}$$

The constants are: $\sigma_0 = 180$ MPa, $K = 20.4 \, \text{MPa-}\sqrt{mm}$

EXAMPLE 9.4 CALCULATING THE YIELD STRENGTH WHEN THE AVERAGE GRAIN DIAMETER IS KNOWN

By using the data in Example 9.3, calculate the yield strength of the steel sample having an average grain diameter of 0.016 mm.

SOLUTION

$K = 20.4\,\text{MPa-}\sqrt{mm}$, $\sigma_0 = 180$ MPa, $d = 0.016$ mm, $S_y = ?$

By using Equation 9.2,

$$\sigma_{ys} = \sigma_0 + \frac{K}{\sqrt{d}} = 180 + \frac{20.4}{\sqrt{0.016}} = 180 + 161.3 = 341.3$$

The yield strength of the steel sample = 341.3 MPa.

EXAMPLE 9.5 CALCULATING THE DISLOCATION DENSITY BY ETCH-PIT COUNTS

5×10^7 etch pits (dislocations) are counted in a 5 mm² area of an etched sample of a deformed metal. Calculate the dislocation density.

SOLUTION

$n = 5 \times 10^7$ dislocations, $A = 5$ mm² $= 5 \times (10^{-1}$ cm$)^2 = 5 \times 10^{-2}$ cm², $\rho_D = ?$

By using Equation 9.3,

$$\rho_D = \frac{n}{A} = \frac{5 \times 10^7}{5 \times 10^{-2}} = 10^9 \text{ cm}^{-2}$$

The dislocation density $= 10^9$ cm⁻².

EXAMPLE 9.6 CALCULATING THE CONSTANTS FOR THE FLOW STRESS- DISLOCATION DENSITY RELATION

Cold working of a metal results in an increase in the dislocation density from 10^6 to 10^9 cm⁻²; which increases the flow stress from 2 to 35 MPa. Calculate the constants σ_0 and k for the metal.

SOLUTION

$\rho_{D1} = 10^6$ cm⁻², $\sigma_{flow1} = 2$ MPa, $\rho_{D1} = 10^9$ cm⁻², $\sigma_{flow2} = 35$ MPa, $\sigma_0 = ?$, k $= ?$

By using Equation 9.4 for pre-cold working,

$$\sigma_{flow1} = \sigma_0 + k\sqrt{\rho_{D1}}$$
$$2\,\text{MPa} = \sigma_0 + k\sqrt{10^6 \text{ cm}^{-2}} \qquad\qquad \text{(E-9.7a)}$$

By using Equation 9.4 for post-cold working,

$$\sigma_{flow2} = \sigma_0 + k\sqrt{\rho_{D2}}$$
$$33\,\text{MPa} = \sigma_0 + k\sqrt{10^6 \text{ cm}^{-2}} \qquad\qquad \text{(E-9.7b)}$$

By performing the subtraction of Equations: E-9.7b-E-9.7a,

$$33 \text{ MPa} = k\left(\sqrt{10^9 \ cm^{-2}} - \sqrt{10^6 \ cm^{-2}}\right)$$

$$k = \frac{33\,MPa}{\sqrt{10^9 \ cm^{-2}} - \sqrt{10^6 \ cm^{-2}}} = \frac{33\,MPa}{\sqrt{10}\,10^4 \ cm - 10^3 \ cm} = 1.96 \times 10^{-3}\,\text{MPa-cm}$$

By substituting the value of k in Equation E-9.7a,

$$2 \text{ MPa} = \sigma_0 + 1.96 \times 10^{-3} \text{MPa-cm} \sqrt{10^6 \ cm^{-2}}$$
$$2 \text{ MPa} = \sigma_0 + (1.96 \times 10^{-3} \text{MPa-cm} \times 10^3 \ cm^{-1})$$
$$\sigma_0 = 2 \text{ MPa} - 1.96 \text{ MPa} = 0.04 \text{ MPa}$$

The constants are: $k = 1.96 \times 10^{-3}$ MPa-cm and $\sigma_0 = 0.04$ MPa.

EXAMPLE 9.7 CALCULATING THE FLOW STRESS WHEN THE DISLOCATION DENSITY IS KNOWN

By using the data in Example 9.6, calculate the flow stress of the cold-worked metal after further deformation when its dislocation density increases to 10^{10} cm^{-2}.

SOLUTION

$k = 1.96 \times 10^{-3}$ MPa-cm, $\sigma_0 = 0.04$ MPa, $\rho_D = 10^{10}$ cm^{-2}, $\sigma_{flow} = ?$

By using Equation 9.4,

$$\sigma_{flow} = \sigma_0 + k\sqrt{\rho_D} = 0.04 + (1.96 \times 10^{-3} \times \sqrt{10_{10}}) = 0.04 + (1.96 \times 100)$$
$$= 196.04 \text{ MPa}$$

The flow stress = 196.04 MPa.

EXAMPLE 9.8 COMPUTING THE TENSILE PROPERTIES BY CALCULATING % CW AND BY PLOT

A 2-mm-thick and 10-mm-wide sheet made of grade-310 stainless steel is cold rolled to 1.4-mm thickness with no change in width. Compute the tensile mechanical properties of the steel.

SOLUTION

$$A_0 = t_0 \times w_0 = 2 \times 10 = 20 \text{mm}^2$$
$$A_{cw} = t_f \times w_f = 1.4 \times 10 = 14 \text{mm}^2$$

By using Equation 9.5,

$$\% \text{ CW} = \left(\frac{A_0 - ACw}{A_0}\right) \times 100 = \left(\frac{20-14}{20}\right) \times 100 = 30$$

By reference to Figure 9.7, the 30% cold work corresponds to the following mechanical properties:

Yield strength = 900 MPa, Tensile strength = 1,300 MPa, % Elongation = 20

EXAMPLE 9.9 CALCULATING THE TRUE STRAIN WHEN THE YIELD STRESS IS KNOWN

The following relationship holds good for stress-strain behavior of annealed 304 stainless steel:

$$\sigma_{true} = 1275 (\varepsilon_{CW} + 0.002)^{0.45}$$

Where σ_{true} is the true stress, and ε_{CW} is the true strain. (a) What is the value of strength coefficient? (b) Calculate the true strain when the yield stress is 745 MPa.

SOLUTION

(a) By comparing the given equation with Equation 9.5, we obtain: $K = 1,275$ MPa.

(b) By substituting the value of the yield stress in the given equation,

$$745 = 1275 (\varepsilon_{CW} + 0.002)^{0.45}$$

$$(\varepsilon_{CW} + 0.002)^{0.45} = 0.584$$

$$(\varepsilon_{CW} + 0.002)^{\frac{0.45}{0.45}} = 0.584^{\frac{1}{0.45}}$$

$$\varepsilon_{CW} + 0.002 = 0.584^{2.22} = 0.303$$

$$\varepsilon_{CW} = 0.303 - 0.002 = 0.3$$

True strain = $\varepsilon = 0.3$

EXAMPLE 9.10 CALCULATING THE PERCENT COLD WORK WHEN THE TRUE STRAIN IS KNOWN

By using the data in Example 9.9, calculate the percent cold working that has to be inflicted to induce the required true strain in the steel.

SOLUTION

$\varepsilon_{true} = 0.3$, % CW = ?

By using Equation 8.9,

$$\varepsilon_{true} = ln\left(\frac{A_0}{A}\right)$$

$$0.3 = ln\left(\frac{A_0}{A_{CW}}\right)$$

$$\frac{A_0}{A_{CW}} = e^{0.3} = 1.35$$

By using Equation 3.16,

$$\% CW = \frac{A_0 - A_{cw}}{A_0} \times 100$$

$$or \quad \% CW = \left[1 - \left(A_{CW} / A_o\right)\right] \times 100 = [1 - (\frac{1}{1.35})] \times 100 = 26$$

The degree of cold work = 26%.

EXAMPLE 9.11 CALCULATING THE TRUE STRAIN WHEN N AND K VALUES ARE KNOWN

A 304 stainless steel (annealed) is subject to a true stress of 450 MPa. Calculate the true strain that will be produced in the steel. (Hint: refer to Table 9.1).

SOLUTION

$n = 0.44$ (see Table 9.1), $K = 1,400$ MPa (see Table 9.1), $\sigma_{true} = 450$ MPa, ε_{true} ?

By rewriting Equation 9.6 in logarithmic form,

$$\log \sigma_{true} = \log K + n \log \varepsilon_{true}$$

$$\log 450 = \log 1400 + 0.44 \log \varepsilon_{true}$$

$$\log \varepsilon_{true} = -1.1199$$

$$\varepsilon_{true} = 10^{-1.1199} = 0.076$$

True strain = 0.076

EXAMPLE 9.12 CALCULATING THE PERCENT LATTICE MISMATCH IN A SUPERALLOY

A superalloy's microstructure comprises of FCC γ phase having a lattice parameter of 0.358 nm and the γ' phase having a lattice parameter of 0.357 nm. Calculate the percent lattice mismatch.

SOLUTION

$a_\gamma = 0.358$ nm, $a_{\gamma'} = 0.357$ nm, % Lattice mismatch = ?

By using Equation 9.8,

$$\% \text{ Lattice mismatch} = \frac{a_{\gamma'} - a_\gamma}{a_\gamma} \times 100 = \frac{0.357 - 0.358}{0.358} \times 100 = -0.279$$

Lattice mismatch = −0.279%

EXAMPLE 9.13 CALCULATING THE PERCENT LATTICE MISMATCH IN AN *AL-CU* ALLOY

Aluminum (FCC lattice) has the atomic radius of 0.1431 nm. A precipitation-strengthened aluminum-copper alloy's microstructure comprises of FCC γ phase (matrix) and the precipitated θ' phase having a lattice parameter of 0.29 nm. Calculate the percent lattice mismatch.

SOLUTION

$R_{Al} = 0.1431$ nm, $a_{\theta'} = 0.290$ nm, % Lattice mismatch = ?

By rewriting Equation 2.4, we obtain:

$$a_{Al} = 2\sqrt{2}\, R_{Al} = 2 \times 1.414 \times 0.1431 = 0.404 \text{ nm}$$

By modifying Equation 9.8 for the aluminum-copper alloy, we obtain:

$$\% \text{ Lattice mismatch} = \frac{a_{\theta'} - a_{Al}}{a_{Al}} \times 100 = \frac{0.290 - 0.404}{0.404} \times 100 = -28.2$$

Lattice mismatch = −28.2%

QUESTIONS AND PROBLEMS

9.1. Explain how the movement of a dislocation results in the deformation of a solid.

9.2. (a) What is the role of dislocation in a strengthening mechanism?
(b) List the various strengthening mechanisms.

9.3. (a) Why is an alloy stronger than pure metal? (b) Differentiate between interstitial and substitutional solid solutions with the aid of diagrams. (c) Why is strengthening by dissolving carbon in BCC iron more pronounced as compared to FCC iron?

9.4. Why is a fine-grained material stronger than a coarse-grained material? Explain with the aid of diagrams.

9.5. What is the role of dislocation density in strain hardening?

9.6. Explain precipitation strengthening with the aid of sketches.

9.7. Differentiate between precipitation strengthening and dispersion strengthening.

P9.8. Refer to Figure 9.5. What is the average grain diameter (in nanometer) of cryo-milled aluminum that corresponds to the lowest yield strength in the graphical plot?

P9.9. A steel sample with an average grain size of 0.022 mm has a yield strength of 318 MPa. The same steel was heat treated to produce another sample with an average grain size of 0.011 mm having a yield strength of 375 MPa. Calculate the yield strength of the steel sample with an average grain diameter of 0.019 mm.

P9.10. By using the data in P9.9, draw a graphical plot for $d^{-\frac{1}{2}}$ versus S_y; and hence compute the constants K and S_0 in the Hall-Petch relationship for the steel.

P9.11. 8×10^6 etch pits (dislocations) are counted in a 4 mm² area of an etched sample of a deformed metal. Calculate the dislocation density.

P9.12. Nickel (FCC lattice) has the atomic radius of 0.1246 nm. A precipitation-strengthened Ni-base alloy's microstructure comprises of FCC γ phase (matrix) and the precipitated γ' phase having a lattice parameter of 0.357 nm. Calculate the percent lattice mismatch.

P9.13. Calculate the flow stress of a metal when its dislocation density is 9×10^8 cm⁻². The values of the constants are: k = 1.7 x 10⁻³ MPa-cm and $\sigma_0 = 0.05$ MPa.

P9.14. The stress-strain behavior of annealed 304 stainless steel can be expressed as:

$$\sigma = 1275 \, (\varepsilon_{CW} + 0.002)^{0.45}$$

where σ is the true stress, and ε_{CW} is the true strain. Calculate the percent cold work that has to be inflicted for obtaining a yield strength of 755 MPa for the steel.

P9.15. A 4-mm-thick and 15-mm-wide sheet made of grade-310 stainless steel is cold rolled to 3-mm thickness with no change in width. Compute the tensile properties of the steel.

REFERENCES

Amelinckx, S. (1964) *The Direct Observation of Dislocations*. Academic Press, New York.

Argon, A.S. (2007) *Strengthening Mechanisms in Crystal Plasticity*. Oxford University Press, Oxford.

Geddes, B., Leon, H. & Huang, X. (2010) *Strengthening Mechanisms*. ASM International, Russell Township, OH.

Hansen, N. (2004) Hall-Petch relation and boundary strengthening. *Scripta Materialia*, 51(8), 801–806.

Huda, Z. (1993) Grain-boundary engineering applied to grain growth in a high temperature material. In: Haq, A.U., Habiby, F. & Khan, A.Q. (eds) *Advanced Materials-93*. A.Q. Khan Research Labs, Islamabad, pp. 391–400.

Huda, Z. (2007) Development of heat treatment process for a P/M superalloy for turbine blades. *Materials & Design*, 28(5), 1664–1667.

Huda, Z. (2009) Precipitation strengthening and age hardening in 2017 aluminum alloy for aerospace application. *European Journal of Scientific Research*, 26(4), 558–564.

Lahrman, D.F. (1982) *Investigation of Lattice Mismatch in Rhenium Containing Nickel-base Superalloy*. M.S. Thesis, Department of Metallurgy and Mining Engineering, University of Illinois at Urbana-Champaign.

Mohri, T. & Suzuki, T. (1999) Solid solution hardening by impurities. In: Briant, C.L. (ed) *Impurities in Engineering Materials*. CRC Press, Boca Raton, FL, pp. 259–270.

Schaffer, J.P., Saxena, A., Antolovich, S.D., Sanders, Jr., T.H. & Warner, S.B. (1999) *The Science and Design of Engineering Materials*. WCB-McGraw Hill Inc, New York.

10 Failure and Design

10.1 METALLURGICAL FAILURES— CLASSIFICATION AND DISASTERS

10.1.1 FRACTURE AND FAILURE—HISTORICAL DISASTERS

A component or a structure is said to have fractured if it develops a crack leading to its breakage. However, the term failure is used in a broader sense. A component or a structure is said to have failed if one or more of the following conditions exist: (a) it becomes completely inoperative, (b) it ceases to perform the function for which it was made, (c) it fractures, and (d) it becomes unsafe or unreliable due to serious material deterioration. Metals can fracture during loading in one of a number of ways: cleavage/intergranular brittle fracture, transgranular ductile fracture, fatigue, creep, hydrogen embrittlement, and stress-corrosion cracking. In systems with moving parts, excessive friction and/or contact between two components often leads to *wear* or more extensive material damage. Since fractures and failures impose a threat to society, it is important to prevent them through a systematic failure analysis (Tawancy *et al.*, 2004).

Ductile failures are accompanied by gross plastic deformation prior to failure; and hence they do not offer much threat to society. On the other hand, brittle failures are often catastrophic, and usually occur without warning at stresses far below the yield strength of the material. Thus brittle fractures have a high impact threat to society. Many reported engineering failures have been associated with brittle fractures. In particular, the failure of the *Liberty* ship that cracked in half during World War II is attributed to brittle (fast) fracture. The *Liberty* ship was fabricated by welding steel plates together to form a continuous body. Initially, cracks started at sharp corners or arc-weld spots and subsequently propagated right round the hull; so that finally the ship broke in half. A possible design improvement against this failure could be to incorporate some riveted joints in the ship structures so as to act as crack arresters. Other examples of historical failures include: the failure of the *Titanic* ship (1912), bridge failures (e.g. the bridge that failed in Belgium in 1951 only three years after its fabrication), and the molasses tank failure in Boston, USA (1919); all these failures are attributed to brittle fractures (Huda *et al.*, 2010; Hertzberg, 1996).

10.1.2 CLASSIFICATION OF METALLURGICAL FAILURES

We just learned in the preceding subsection that metals can fail during monotonic loading in one of the following ways: cleavage/intergranular brittle fracture (fracture along grain boundaries), trans-granular ductile fracture (fracture through grains), fatigue, creep, hydrogen embrittlement, and stress-corrosion cracking. These failure

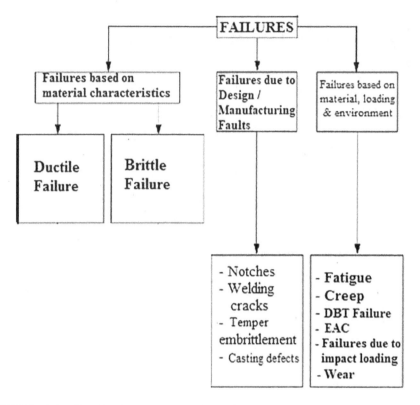

FIGURE 10.1 Classification of metallurgical failures (DBT = ductile-brittle transition; EAC = environmental assisted cracking).

mechanisms are presented in the classification chart of metallurgical failures, as shown in Figure 10.1.

It is evident in Figure 10.1 that all metallurgical failures can be classified into three groups: (a) failures based on material characteristics, (b) failures due to design/manufacturing faults, and (c) failures based on material, loading, or environmental conditions. Most of the various types of failure (shown in Figure 10.1) are explained in the subsequent sections.

10.2 DUCTILE AND BRITTLE FRACTURES

Based on the material characteristic, a failure may be either ductile or brittle. We have learned in the preceding section that ductile fractures are accompanied by gross plastic deformation prior to failure whereas brittle fractures usually occur without warning at stresses far below the yield strength of the material (see Figure 10.2).

It is evident in Figure 10.2 that a ductile fracture usually results in either 100% reduction in area (e.g. lead, thermoplastics, etc.) (see Figure 10.2a), or the fracture surface showing a cup-and-cone feature (e.g. mild steel, brass, etc.) (see Figure 10.2b). Both of these types of ductile fracture involve gross plastic deformation prior to

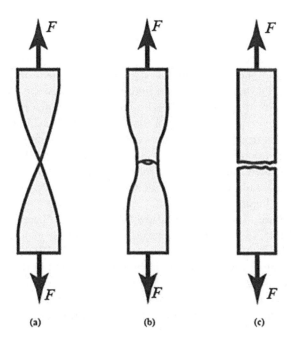

FIGURE 10.2 Three types of fracture; (a) very ductile fracture; (b) cup-and-cone ductile fracture, and (c) brittle fracture.

TABLE 10.1
Ductile and Brittle Fractures

#	Ductile fracture	Brittle fracture
1.	There is a gross plastic deformation of the	There is no or very little plastic deformation of
2.	material prior to failure.	the material.
3.	The material has high impact energy.	The material has a low impact energy.
4.	The fracture surface appears dull and fibrous showing a cup-and-cone. Scanning electron microscopy (*SEM*) fractographs usually show spherical dimples due to micro-void coalescence (MVC) in metals.	The fracture surface is usually flat and shining. SEM fractographs of metals show transgranular cleavage and intergranular fracture as important fracture mechanisms.

failure. On the other hand, there is no or very little plastic deformation in brittle fractures (e.g. glass, cast iron, etc.) (see Figure 10.2c). Table 10.1 presents a comprehensive distinction between ductile and brittle fractures.

Many reported engineering failures have been brittle fractures (see section 10.1). Brittle fractures generally occur in brittle materials; however they can also occur in ductile materials under certain loading conditions or at very low temperature (see section 10.3). For example, an impact loading condition or triaxial stress can cause a ductile material to behave in a brittle manner.

10.3   DUCTILE-BRITTLE TRANSITION FAILURE

We learned in Chapter 8 that materials with high impact energies exhibit *ductile failure* whereas brittle materials are characterized by their low impact energy values. The impact energy of a metal is strongly dependent on temperature. In particular, there is a sharp reduction in impact energy when the temperature falls below the ductile-brittle transition (DBT) temperature; which is generally below 0°C. It has been experimentally shown that metals which exhibit normal ductile fracture at ambient (room) temperature, can fail at low temperatures by a sudden cleavage (brittle) fracture at comparatively low stresses (see Figure 10.3). This phenomenon is called *ductile-brittle transition (DBT) failure*; and the temperature at which DBT failure occurs is called the ductile-brittle transition temperature (DBTT) (see Figure 10.3).

DBT failures are more common at low temperatures. There have been many instances of failures of metals by unexpected brittleness at low temperatures. For instance, the failure of the ***Titanic*** is attributed to DBT failure. In 1912, the *Titanic* (during voyage) struck an iceberg that resulted in the fracture of the ship structure causing the death of 1,500 passengers. The failure investigations revealed the DBT failure that occurred at −2°C was due to a relatively higher DBTT of the material. It means the lower the DBTT, the safer is the design against the DBT failure.

Metals, with FCC crystal structures maintain ductility at low temperatures, whilst some metals with structures other than FCC tend to exhibit brittleness. Steels (with BCC ferritic structure) fail with brittle fracture along the (100) plane. For designing such steels against DBT failure, it is important to significantly reduce the DBT temperature (DBTT) of the steel by minimizing the carbon and phosphorous contents in the steels for applications at low temperatures.

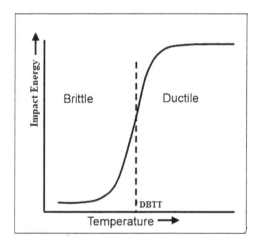

FIGURE 10.3   Drop in impact energy at low temperatures (below DBTT) (DBTT = ductile-brittle transition temperature).

10.4 GRIFFITH'S CRACK THEORY

A.A. Griffith, a British engineer, initiated the development of *Fracture Mechanics* during World War I. He reported that the stress becomes infinity large at the tip of a sharp crack in a brittle material. Griffith assumed that growth of a crack requires some surface energy; which is supplied by the loss of strain energy accompanying the relaxation of local stresses as the crack grows. According to Griffith's crack theory, failure occurs when the surface energy increases significantly due to the loss of strain energy (Anderson, 2017).

Mathematical Modeling of Griffith's Crack Theory. Let us consider a stressed large plate, made of a brittle material, having a through-thickness central crack of length $2a$ as shown in Figure 10.4. The stress applied σ is magnified as one moves from center of the crack to its tip; the stress is the maximum at the tip of crack. Let the thickness of the plate be t.

The total crack surface area, A, is the sum of the upper and lower cracks' surface areas, i.e.

$$A = \text{upper crack's area} + \text{lower crack's area}$$
$$= 2a \times t + 2a \times t = 4a \times t \tag{10.1}$$

The specific energy of a crack (γ_s) is defined as the crack surface energy per unit surface area. Thus the increase in surface energy due to the introduction of the crack (ΔU_1) is given by:

$$\Delta U_1 = 4a \times t \gamma_s \tag{10.2}$$

By using stress analysis for an infinitely large plate containing an elliptical crack, Griffith expressed the decrease in potential energy of the cracked plate (ΔU_2), as follows:

$$\Delta U_2 = -\frac{\pi \sigma^2 a^2 t}{E} \tag{10.3}$$

where E is the Young's modulus of the plate material. By combining Equations 10.2 and 10.3, the net change in potential energy of the plate due to the introduction of the crack is given by:

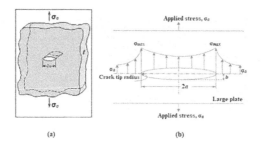

(a) (b)

FIGURE 10.4 The stressed plate with a through-thickness central crack (the stress magnifies at the crack tip).

$$U - U_0 = \Delta U_1 + \Delta U_2 = 4a \times t \times \gamma_s - \frac{\pi \sigma^2 a^2 t}{E} \tag{10.4}$$

where U_0 is the potential energy of the plate without a crack, and U is the potential energy of the plate after the introduction of the crack.

By rearranging the terms in Equation 10.4, we obtain:

$$U = 4a \times t \times \gamma_s - \frac{\pi \sigma^2 a^2 t}{E} + U_0 \tag{10.5}$$

At equilibrium, there is no change in the potential energy (constant U) i.e.

$$\text{At equilibrium, } \frac{\partial U}{\partial a} = 0 \tag{10.6}$$

By taking partial derivatives (with respect to a) of both sides of Equation 10.5, we obtain:

$$\frac{\partial U}{\partial a} = 4t\gamma_s - \frac{2\pi\sigma^2 at}{E} \tag{10.7}$$

By combining Equations 10.6 and 10.7, we get:

$$\frac{\pi\sigma^2 a}{E} = 2\gamma_s \tag{10.8}$$

$$\text{or } \sigma = \frac{2E\gamma_s}{\pi a} \tag{10.9}$$

where σ is stress applied normal to the major axis of the crack, a is the half crack length, t is the plate thickness, E is the modulus of elasticity, and γ_s is the specific surface energy. Equation 10.9 represents the mathematical form of Griffith's theory. It is evident in Equation 10.9 that the growth of a crack requires the creation of surface energy, which is supplied by the loss of strain energy. Failure occurs when the loss of strain energy is sufficient to provide the increase in surface energy. The use of Griffith's theory (Equation 10.9) enables us to compute the maximum length of the crack that is allowable without causing fracture (see Example 10.1).

10.5 STRESS CONCENTRATION FACTOR

It was established in the preceding section that the stress (σ) applied normally to a central crack is magnified as one moves from the center of the crack to its tip; the stress is the maximum at the tip of the crack (see Figure 10.4). The stress concentration factor is a simple measure of the degree to which an external stress is magnified at the tip of a small crack. Numerically, the *stress concentration factor* is the ratio of the maximum stress (σ_{max}) to the nominal applied tensile stress (σ_0).

The stress concentration factor (for static loading) can be expressed in terms of crack geometry as follows:

$$K_s = \frac{\sigma_m}{\sigma_0} = 2\sqrt{\frac{a}{\rho_t}} \tag{10.10}$$

where K_s is the stress concentration factor for static loading; σ_m is the maximum stress at the crack tip; σ_0 is the stress applied normally to the long axis of the central crack, a is one-half the crack length; and ρ_t is the radius of curvature at the crack tip (see Examples 10.2). It is evident in Equation 10.10 that the *stress concentration factor* (K_s) varies directly as the square root of the crack length, and inversely as the square root of the crack-tip radius of curvature. For a good design of a structure/component with a crack and/or notch, it is important to keep K_s to a minimum. This is why engineering components are provided with a reasonably large radius of curvature (ρ_t) in their design. As a case study, we may consider the side window in an aircraft's fuselage; the window geometry requires an adequate corner radius. Unfortunately, the de Havilland Comet—the world's first pressured commercial jet airliner (which first flew in 1949)—crashed owing to the failure due to a small corner radius (ρ_t) in its window (Ashby *et al.*, 2010). Thus, today's aircrafts' windows are provided with large corner radii.

By rearranging the terms in Equation 10.10, we obtain:

$$\sigma_m = 2\sigma_0 \sqrt{\frac{a}{\rho_t}} \tag{10.11}$$

Equation 10.11 is useful in computing the radius of curvature (ρ_t) when the stresses (σ_m and σ_0) are known (see Example 10.3).

10.6 SAFETY AND DESIGN

10.6.1 THE FACTOR OF SAFETY IN DESIGN

We learned in Chapter 8 that the stress at the fracture point during tensile testing, is called the *breaking strength*. It means that a component will operate safely as long as the stress on the component during service (or the design stress) is less than the allowable stress (or the breaking strength). This design aspect is expressed by the term *factor of safety* (*FoS*), as follows:

$$\text{FoS} = \frac{Breaking\ strength\ of\ component}{Stress\ on\ the\ component} = \frac{\sigma_{all}}{\sigma_d} \tag{10.12}$$

where σ_{all} is the allowable stress or the breaking strength of the component, MPa; and σ_d is the design stress or the stress on the component during service, MPa (see Example 10.4). The *FoS* is actually a margin of safety in design. It is obvious from Equation 10.12 that a design is adequate or safe when $FoS > 1$. On the other hand, if $FoS \leq 1$, the design is inadequate. The *FoS* usually range from 1.3 (when material properties and operating conditions are known in detail) to 4 (for untried materials under average operating conditions) (Huda and Jie, 2016). For elevator applications, FoS may be as high as 7.

10.6.2   Fracture Mechanics—A Design Approach

10.6.2.1 K, K_c, and the *Plain Strain Fracture Toughness* (K_{IC})

In fracture mechanics, three terms are used to express stress and strength conditions in a flawed component: (1) stress intensity factor, K, (2) critical stress intensity factor, K_c, and (3) plane strain fracture toughness, K_{IC}. In order to distinguish between the three terms, we consider a large plate of infinite width containing a central flaw or crack, as shown in Figure 10.5.

Stress Intensity Factor, K. The stress applied to a component or structure, containing a crack, is intensified at the tip of the crack. The stress level at the crack tip is called the *stress intensity factor, K*; which depends on the local stress as well as the crack size. For a large plate containing a through-thickness central crack (Figure 10.5a), the stress intensity factor, K, can be expressed as:

$$K = \sigma\sqrt{\pi a} \qquad (10.13)$$

where σ is the stress applied in a direction normal to the major axis of the crack in the plate, MPa; and a is half the crack length, m; and K is the stress intensity factor, Mpa-$\sqrt{m}$. For various specimen sizes and crack geometries, Equation 10.13 can be generalized as follows:

$$K = f(\sigma, a) = Y\sigma\sqrt{\pi a} \qquad (10.14)$$

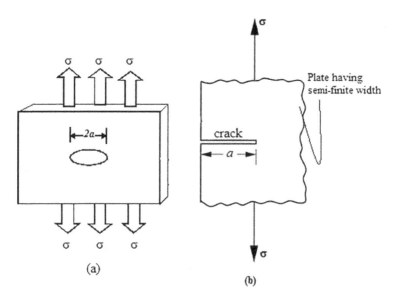

FIGURE 10.5   Crack geometries in stressed large plates containing: (a) a central crack, (b) an edge crack.

where Y is the geometric correction factor—a dimensionless parameter. By using fracture mechanics principles, various different values of Y have been developed for other crack-specimen geometries (Blake, 1996). For example, for a plate of infinite width containing a central crack of length $2a$ (Figure 10.5a), $Y=1.0$. For a plate of semi-finite width with an edge crack of length a (Figure 10.5b), $Y=1.1$. For a plate of semi-finite width containing embedded circular flaw of length $2a$, $Y=\dfrac{2}{\pi}$ (see Example 10.5).

Critical Stress Intensity Factor, K_c. It is evident in Equation 10.14 that the stress intensity factor, K, varies with the crack length and the magnitude of applied load. The critical level of stress intensity factor that causes the crack to grow and result in failure, is called the *critical stress intensity factor, K_c.* Hence, K_c can be expressed as

$$K_c = K \text{ required to grow a crack} = K \text{ for fast fracture} \qquad (10.15)$$

Plain Strain Fracture Toughness, K_{IC}. *Fracture toughness* is a measure of a material's resistance to brittle fracture when there exists a crack. For thick specimens with the thickness much greater than the crack size, the fracture toughness is equal to K_c; which can be computed by using Equation 10.15. However, for thin specimens, the value of K_c will strongly depend on the thickness of the specimen. As the specimen-thickness increases, K_c decreases to a constant value; this constant is called the *plane strain fracture toughness, K_{IC}* (pronounced as kay-one-see). For example, the K_c of a 207-MPa yield-strength steel (thin specimen) decreases from $K_c=230$ MPa√m to a constant value around 93MPa√m as the thickness of the specimen increases; this constant value of K_c is reported as the *plane strain fracture toughness, K_{IC}* of the steel. In general, it is the K_{IC} that is reported as the fracture toughness property of the material since the value of K_{IC} does not depend on the thickness of the component. The fracture toughness value ranges for some engineering materials are presented in Table 10.2.

TABLE 10.2
Fracture Toughness (K_{IC}) of Some Materials at 25°C

Material	Fracture toughness (MPa√m)	Material	Fracture toughness (MPa√m)
Stainless steels	62–280	Polystyrene (PS)	0.7–1.1
Low-alloy steels	14–200	Polypropylene (PP)	3–4.5
Carbon steels	12–92	PVC	1.5–4.1
Cast irons	22–54	Polyethylene (PE)	1.4–1.7
Aluminum alloys	22–35	Silica glass	0.6–0.8
Copper alloys	30–90	Concrete	0.3–0.4
Nickel alloys	80–110	Glass-fiber polymer	7–23
Titanium alloys	14–120	Carbon-fiber polymer	6.1–88

10.6.2.2 Design Philosophy of Fracture Mechanics

The design philosophy of fracture mechanics enables us to design components against fracture by using the relationship involving the critical crack size, the critical stress, and K_{IC}, as follows:

$$K \leftrightarrow K_{IC} = Y \sigma_c \sqrt{\pi a_c} \qquad (10.16)$$

where σ_c is the (critical) stress level to cause fracture; and a_c is the critical (maximum allowable) crack size. It is obvious from Equation 10.16 that there are three variables a_c, σ_c, and K_{IC}; if two variables are known, the third can be determined. The design philosophy (Equation 10.16) enables us to decide whether or not the design of a component is safe for an application. For example, if the crack geometry and the stress level are known, the stress intensity factor (K) can be computed by using Equation 10.14; this K value is then compared with K_{IC} of the material. If $K < K_{IC}$, the crack would not propagate i.e. the **design is safe.** On the other hand, if $K \geq K_{IC}$, the **design is unsafe** (see Example 10.6).

Equation 10.16 is also helpful in selecting an appropriate material for an engineering application. For example, if the a_c and σ_c values for an application are known, K_{IC} can be computed by using Equation 10.16; the K_{IC} value so obtained enables a designer to select the appropriate material having the right fracture toughness (K_{IC}) for the application (see Example 10.7). Another usefulness of Equation 10.16 lies in design of a testing/NDT method. If we have selected a material (on the basis of its K_{IC} value) and if we have also specified the design/critical stresses (σ_c), we can compute the maximum (critical) size of the flaw (a_c) that can be allowed by using Equation 10.16. An NDT technique that detects any flaw greater than the computed a_c can help assure safety against fracture (see Example 10.8). The design philosophy (Equation 10.16) also enables us to compute the design stress for a stressed application (see Example 10.9).

10.7 FATIGUE FAILURE

10.7.1 What Is Fatigue?

In the preceding section, we considered failures occurring at uniform (static) loading; however in many engineering situations it is important to know how metals behave under cyclic (dynamic) loading. When a material is subjected to alternating or cyclic stresses over a long period of time, it may fail after a number of cycles even though the maximum stress in any cycle is considerably less than the breaking strength of the material. This failure is called *fatigue*. Typical examples of fatigue failures include: rotating shafts, springs, turbine blades, airplane wings, gears, bones, and the like. There are three main stages in a fatigue failure: (I) *crack initiation*, the crack originates at a point of stress concentration (such as a notch) or a metallurgical flaw (e.g. inclusion); (II) *crack propagation*, the crack propagates across the part under cyclic or repeating stresses, and (III) *crack termination*, here final fracture occurs. These stages are illustrated in Figure 10.6; which shows the fracture surface of a steel car-wheel that failed by fatigue.

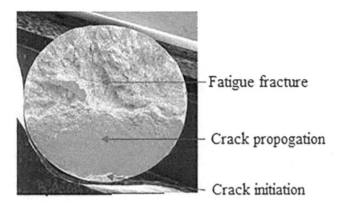

FIGURE 10.6 Fatigue failure in a steel wheel.

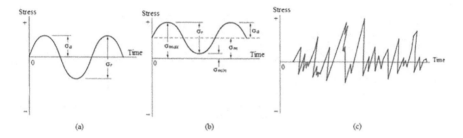

FIGURE 10.7 Types of stress cycles: (a) alternating, (b) repeating, (c) fluctuating.

10.7.2 STRESS CYCLES

There are three types of stress cycles with which loads may be applied to a component: (a) reversed or alternating stress cycle, (b) repeated stress cycle, and (c) fluctuating stress cycle (see Figure 10.7). The simplest type of stress cycle is the *alternating stress cycle*; which has a symmetrical amplitude about the *x*-axis (Figure 10.7a). Here, the maximum and minimum stresses are equal, but opposite in sign. An example of alternating stress cycle is the rotating axle; where after every half turn or half period (as in the case of the sine wave), the stress on a point would be reversed.

The most common type of cycle found in engineering applications is the *repeated stress cycle*; where the maximum stress (σ_{max}) and minimum stress (σ_{min}) are asymmetric, unequal, and opposite (Figure 10.7b). In *fluctuating stress cycle*, stress and frequency vary randomly (Figure 10.7c). An example of *fluctuating stress cycle* is the automobile shocks; here the frequency magnitude of imperfections in the road produces varying minimum and maximum stresses.

The *mean stress* is the average of maximum and minimum stresses in a cycle. Mathematically,

$$\sigma_m = \frac{\sigma_{max} + \sigma_{min}}{2} \tag{10.17}$$

where σ_m is the mean stress; σ_{max} is the maximum stress; and σ_{min} is the minimum stress (see Example 10.10). In case of alternating stress cycle, the mean stress is zero, 0 (see Figure 10.7a).

The stress range (σ_r) is the difference of the maximum and minimum stresses. Numerically,

$$\sigma_r = \sigma_{max} - \sigma_{min} \tag{10.18}$$

The stress amplitude (σ_a) of the alternating stress cycle is one-half of the stress range i.e.

$$\sigma_a = \frac{1}{2}\sigma_r \tag{10.19}$$

The stress ratio (R_s) is the ratio of the minimum to the maximum stress amplitude. Numerically,

$$R_s = \frac{\sigma_{min}}{\sigma_{max}} \tag{10.20}$$

The *amplitude ratio* (R_a) is the ratio of the stress amplitude to the mean stress i.e.

$$R_a = \frac{\sigma_a}{\sigma_m} \tag{10.21}$$

where σ_a is the stress amplitude, MPa; and σ_m is the mean stress, MPa. The significance of Equations 10.18–10.21 is illustrated in Example 10.11.

10.7.3 FATIGUE TESTING—*FATIGUE STRENGTH AND FATIGUE LIFE*

In fatigue testing, a cylindrical test piece is subjected to alternating stress cycles with a mean stress of zero; the results so obtained are plotted in the form of an S-N curve (stress versus cycles to failure curve). The S-N curve is plotted with the stress amplitude (S) on the vertical axis and the logarithm of "N" on the horizontal axis (see Figure 10.8).

It is evident in Figure 10.8 that steel exhibits endurance limit (S_e) or fatigue limit whereas a nonferrous metal does not. *Endurance limit* (S_e) or the fatigue limit is the stress level below which a specimen can withstand cyclic stress indefinitely without exhibiting *fatigue* failure. It means that if a material has a *fatigue limit*, then on loading the metal below this stress, it will not fail, no matter how many times it is loaded. Most steels (particularly plain carbon and low-alloy steels) exhibit a definite fatigue limit (S_e); which is generally about half of the tensile strength (σ_{ut}):

$$S_e = \frac{1}{2}\sigma_{ut} \tag{10.22}$$

It has been experimentally shown that the 316L stainless steel (manufactured by selective laser melting, SLM) has a good fatigue performance for cyclically loaded

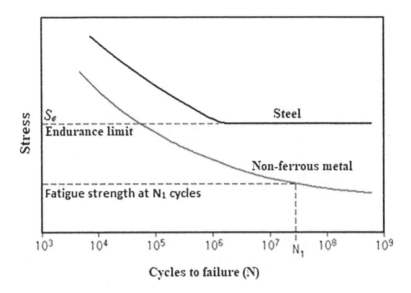

FIGURE 10.8   S-N plots for steel and a nonferrous alloy (S_e = endurance limit).

parts (Riemer *et al.*, 2014). On the other hand, nonferrous alloys do not generally show a fatigue limit i.e. they will fail at any stress and number of cycles (see
Figure 10.8).

It is important to differentiate between the terms: *fatigue strength* and *fatigue life*.
The stress at which failure occurs for a given number of cycles is called the *fatigue
strength*. The number of cycles required for a material to fail at a certain stress is
known as its *fatigue life* (N_f). *Fatigue strength* refers to the stress amplitude from
an S-N curve at a particular life of interest. For example, the fatigue strength at 10^6
cycles is the stress corresponding to $N_f = 10^6$. *High cycle fatigue* refers to the situation of long fatigue life; which is typically in the range of 10^3–10^7cycles. Examples
of machine components that experience high-cycle fatigue include: aircraft wings,
turbine blades, rotating shafts, and the like.

10.7.4   THE MODIFIED GOODMAN LAW

The modified Goodman equation is an empirical relationship between stress amplitude, mean stresses, tensile strength, and fatigue strength, expressed as:

$$\frac{\sigma_a}{\sigma_{FS}} + \frac{\sigma_m}{\sigma_{ut}} = 1 \tag{10.23}$$

where σ_a is the stress amplitude, σ_{FS} is the fatigue strength, σ_m is the mean stress,
and σ_{ut} is the tensile strength. Equation 10.23 enables us to compute the fatigue
strength of a metal when the values of stress amplitude, mean stress, and tensile
strength are known (see Example 10.12).

10.7.5 AVOIDING FATIGUE FAILURE IN COMPONENT DESIGN

Fatigue cracks generally initiate at a surface; and then propagate through the bulk material. Hence, it is important to minimize surface defects (such as scratches, machining marks, fillets, dents, etc.) since the surface defects are regions of stress concentration. Additionally, any notch (hole, groove, etc.) in the component design must be avoided since the former will result in stress concentration sites leading to fatigue failure. The fatigue stress concentration factor (*s.c.f*) is a function of flaw geometry as well as the material and the type of loading. The material aspect of the fatigue *s.c.f* is often expressed as a notch sensitivity factor, q_n, defined as:

$$q_n = \frac{K_f - 1}{K_s - 1} \tag{10.24}$$

where K_s is the stress concentration factor for static loading; and K_f is the stress concentration factor for fatigue loading; which is defined as:

$$K_f = \frac{Fatigue\,limit\,for\,notch-free\,specimen}{Fatigue\,limit\,for\,notched\,specimen} \tag{10.25}$$

Equation 10.25 indicates that the fatigue strength of a component is reduced by a factor K_f by the effect of notch root surface area (see Example 10.13). We can also note from Equation 10.24 that the range of q_n is between 0 (when $K_f = 1$) and 1 (when $K_f = K_s$).

   The major techniques for avoiding fatigue failure include: (a) avoiding notches/stress concentration sites in the component design, (b) control of microstructural defects and grain size, (c) improvement of surface conditions, (d) corrosion control, (e) minimizing manufacturing defects (e.g. initial flaws, machining marks, etc.), (f) maximize the critical crack size, and (g) introducing compressive residual stresses (Huda and Bulpett, 2012).

10.8 CREEP FAILURE

10.8.1 WHAT IS CREEP?

Creep refers to the progressive deformation of a material at constant stress, usually at a high temperature (that exceeds 0.4 to 0.5 of the melting temperature, expressed in Kelvin). Some notable examples of systems that have components experiencing creep include: gas turbine engines, nuclear reactors, boilers, and ovens. In nuclear reactors, the metal tubes carrying the fuel may undergo creep failure in response to the pressures and forces exerted on them at high temperatures. Creep also occurs in structures at ambient temperatures. For example, pre-stressed concrete beams, which are held in compression by steel rods that extend through them, may creep. The creep and stress relaxation in the steel rods eventually leads to a reduction of the compression force acting in the beam; which can result in failure.

10.8.2 CREEP TEST AND CREEP CURVE

A creep test involves the application of a constant load to a tensile specimen at a high temperature; thus the loaded material undergoes progressive rate of deformation.

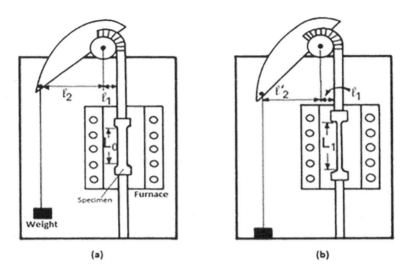

FIGURE 10.9 Creep testing principle; (a) initial position, (b) the specimen length has increased from L_0 to L_1.

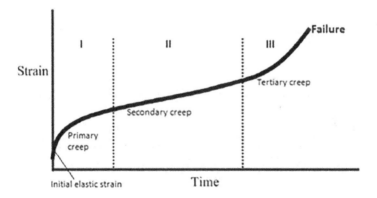

FIGURE 10.10 A typical creep curve.

Figure 10.9 illustrates a creep testing machine with variable lever arms to ensure constant stress on the specimen; note that l_2 decreases as the length of the specimen increases. The corresponding amount of strain is recorded over the period of time, as shown by a typical creep curve in Figure 10.10.

It is evident in the creep curve (Figure 10.10) that there are three stages in creep: (I) primary creep, (II) secondary creep, and (III) tertiary creep. The stage II shows a steady-state creep behavior; here, the rate of creep can be determined by the slope of the curve, as follows:

$$\text{Creep strain rate} = \dot{\varepsilon} = \frac{\Delta \varepsilon}{\Delta t} \tag{10.26}$$

where $\Delta \varepsilon$ is the increment in the strain over the period of time Δt, h^{-1} (see Example 10.14).

10.8.3 EFFECTS OF STRESS AND TEMPERATURE ON CREEP RATE

The rate of creep of a material is strongly dependent on stress as well as temperature. An increase in the applied stress would result in an increase in the creep rate. The steady-state creep strain rate $\left(\dot{\varepsilon}\right)$ is related to the stress (σ) at constant temperature by the following expression:

$$\dot{\varepsilon} = C_1 \sigma^n \tag{10.27}$$

where C_1 and n are the material's constants (see Example 10.15).

An increase in temperature has a pronounced effect on creep strain rate. The steady-state creep rates vary exponentially with temperature according to the following Arrhenius-type expression:

$$\dot{\varepsilon} = C_2 \sigma^n \exp\left(-\frac{Q_c}{RT}\right) \tag{10.28}$$

where $\dot{\varepsilon}$ is the creep rate, /s; T is the temperature, K; C_2, n and the stress σ are constants; R = 8.3145 J/mol·K; and Q_c is the activation energy for creep, J/mol (see Example 10.16). By equating the term $C_2 \sigma^n$ to another constant A, Equation 10.28 can be simplified as:

$$\dot{\varepsilon} = A \exp\left(-\frac{Q_c}{RT}\right) \tag{10.29}$$

By taking the natural logarithms of both sides, Equation 10.29 can be rewritten as:

$$ln\,\dot{\varepsilon} = ln\,A - \frac{Q_c}{RT} \tag{10.30}$$

Thus, the activation energy for creep, Q_c, can be determined experimentally by plotting the natural log of creep rate against the reciprocal of temperature.

10.8.4 LARSON-MILLER (LM) PARAMETER

The Larson-Miller parameter (LMP) is a useful tool for determining the stress-rupture time or creep life for any stress-temperature combination for a given material. It is defined as:

$$LMP = \frac{T}{1000}(C + log\,t) \tag{10.31}$$

where LMP is the Larson-Miller parameter; T is the temperature in Rankin, R; t is the creep-life time, h; and C is a material's constant (usually C=20). Figure 10.11 shows the LMPs for Ti alloys. Creep lives for most materials lie in the range of 100–100,000 h (see Example 10.18).

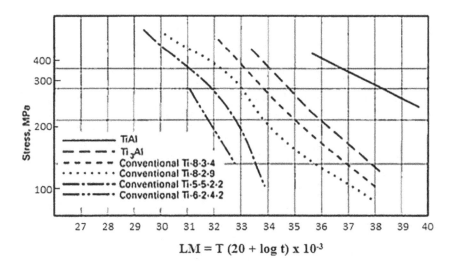

$$LM = T\,(20 + \log t) \times 10^{-3}$$

FIGURE 10.11 The graphical plots showing Larson-Miller parameters for titanium alloys.

10.9 CALCULATIONS—EXAMPLES ON FAILURE AND DESIGN

EXAMPLE 10.1 COMPUTING THE MAXIMUM CRACK LENGTH THAT IS ALLOWABLE TO PREVENT FAILURE

A tensile stress of 52 MPa is applied normal to the major axis of a central flaw in a relatively large plate of glass ceramic (Pyroceram). The Young's modulus and the specific surface energy for Pyroceram are 69 GPa and 0.3 Jm⁻², respectively. Calculate the maximum length of the crack that is allowable without causing failure.

SOLUTION

σ = 52 MPa = 52 x 10^6 Pa, E = 69 GPa = 69 x 10^9 Pa, γ_s = 0.3 J-m⁻² = 0.3 N-m⁻¹

By rewriting Equation 10.8 for a, we obtain:

$$a = \frac{2E\gamma_s}{\pi\sigma^2} = \frac{2\,x\,69\,x10^9\,N\,m^{-2}\,x0.3\,N\,m^{-1}}{\pi\,x\,52^2\,x10^{12}\,N^2\,m^{-4}} = \frac{41.4\,x10^{-3}}{2704\,m^{-1}} 0.0153 \times 10^{-3}$$
$$= 4.87\,\mu m$$

The maximum length of the central flaw that is allowable without causing failure = $2a$ = 9.74 μm.

EXAMPLE 10.2 CALCULATING THE RADIUS OF CURVATURE AT THE CRACK TIP

Silicon-aluminum-oxynitride (SIALON), an advanced flaw-free ceramic, has a tensile strength of 413 MPa. An elliptical thin 0.25-mm-deep crack is observed before a SIALON plate is tensile tested. The plate fails by brittle fracture at a normal stress of 10 MPa. Calculate the radius of curvature at the crack tip.

SOLUTION

$\sigma_{ut} = \sigma_m = 413$ MPa, Normal stress $= \sigma_0 = 10$ MPa, $a = 0.25$ mm (see Figure 8.3)

By using Equation 10.11,

$$\sigma_m = 2\sigma_0 \sqrt{\frac{a}{\rho_t}}$$

$$413 = 2 \times 10 \times \sqrt{\frac{0.25}{\rho_t}}$$

$$\frac{0.25}{\rho_t} = 426.4$$

$\rho_t = 5.8 \times 10^{-4}$ mm $= 5.8 \times 10^{-7}$ m $= 0.58$ μm
   The radius of curvature at the crack tip $= 0.58$ μm.

EXAMPLE 10.3 COMPUTING THE STRESS CONCENTRATION FACTOR (FOR STATIC LOADING)

By using the data in Example 10.2, calculate the stress concentration factor (for static loading).

SOLUTION

$\sigma_m = 413$ MPa, $\sigma_0 = 10$ MPa

By using Equation 10.10,

$$K_s = \frac{\sigma_m}{\sigma_0} = \frac{413}{10} = 41.3$$

The stress concentration factor (for static loading) $= 41.4$.

EXAMPLE 10.4 DESIGNING A COMPONENT BASED ON THE DESIGN STRESS

The breaking strength of a component's material is 1,100 MPa. The material properties and operating conditions are known in detail. What should be the design stress for the component? Design the cylindrical component for a tensile load of 2 kN.

SOLUTION

For known material properties and operating conditions, FoS = 1.3; σ_{all} = 1,100 MPa, σ_d = ?

By rewriting Equation 10.14, we obtain:

$$\sigma_d = \frac{\sigma_{all}}{FoS_F} = \frac{1,100}{1.3} = 846\,\text{MPa}$$

$$\text{Stress} = \frac{F}{A}$$

$$A = \frac{\pi}{4}d^2 = \frac{F}{\sigma_d}$$

$$d^2 = \frac{4}{\pi}\frac{F}{\sigma_d} = \frac{4}{\pi} \times \frac{2000}{846} = 3\,\text{mm}^2$$

$$d = \sqrt{3} = 1.73\,\text{mm}$$

The design stress = 846 MPa; the cylindrical component design requires diameter = 1.73 mm.

EXAMPLE 10.5 CALCULATING THE STRESS INTENSITY FACTOR FOR A PLATE OF SEMI-FINITE WIDTH CONTAINING AN EMBEDDED CIRCULAR FLAW

A plate of semi-finite width contains an embedded circular flaw of length 1.8 cm. A design stress of 700 MPa is applied normal to the major axis of the crack. Calculate the stress intensity factor.

SOLUTION

By reference to subsection 10.4.2.1, $Y = 2/\pi$; σ = 700 MPa,
Crack length = $2a$ = 1.8 cm = 0.018 m, a = 0.009 m;
By using Equation 10.14,

$$K = Y\sigma\sqrt{\pi a} = \frac{2}{\pi}(700)\sqrt{\pi \times 0.009} = 0.637 \times 700 \times 0.168 = 75\,\text{MPa}\,\sqrt{m}$$

The stress intensity factor = 75 MPa $\sqrt{m}$.

EXAMPLE 10.6 DECIDING IF A DESIGN IS SAFE

A 2.2-mm long central crack was detected in an infinitely wide plate having a fracture toughness of 25 MPa $\sqrt{m}$. Is this plate's design safe to use for a critical application where design stresses are 500 MPa?

SOLUTION

$Y = 1$, $2a$ = 2.2 mm = 2.2×10^{-3} m, a = 0.0011 m; σ = 500 MPa; K_{IC} = 25 MPa $\sqrt{m}$

By using Equation 10.14,

$$K = Y\sigma\sqrt{\pi a} = 1\times 500\sqrt{\pi \, x \, 0.0011} = 29.4$$
$$K = 29.4 \, \text{Mpa}\sqrt{m}$$

By comparing: $K \leftrightarrow K_{IC}$, we find, $\mathbf{K > K_{IC}}$. Hence the **design is unsafe**.

EXAMPLE 10.7 THE APPLICATION OF DESIGN PHILOSOPHY FOR MATERIAL SELECTION

Select a specific polymer for a plate of semi-finite width with an edge crack of length 1.0 mm; which fails at a stress of 45 MPa.

SOLUTION

By reference to subsection 10.4.2.1, Y = 1.1, a_c = 1 mm = 0.001 m, σ_c = 45 MPa
By using Equation 10.16,

$$K_{IC} = Y\,\sigma_c\,\sqrt{\pi a_c} = 1.1\times 45\times\sqrt{\pi \, x \, 0.001} = 2.7\,\text{MPa}\sqrt{m}$$

The polymer to be selected will be the one having $K_{IC} > 2.7$ MPa $\sqrt{m}$
By reference to Table 10.2, we must select polypropylene (PP) for this design application.

EXAMPLE 10.8 DESIGNING A TEST/NDT METHOD TO DETECT A CRACK

A steel plate of semi-finite width has a plane strain fracture toughness of 200 MPa $\sqrt{m}$; and is subjected to a stress of 400 MPa during service. Design a testing method capable of detecting a crack at the edge of the plate before the likely growth of the crack.

SOLUTION

$K_{IC} = 200$ MPa $\sqrt{m}$, $\sigma_c = 400$ MPa, Y = 1.1

In order to compute the minimum size of edge crack that will propagate in the steel under the given condition, we use Equation 10.16, as follows:

$$K_{IC} = Y\,\sigma_c\,\sqrt{\pi a_c}$$
$$200 = 1.1\times 400\sqrt{\pi a_c}$$
$$\sqrt{\pi a_c} = 0.45$$
$$a_c = 0.06\,\text{m} = 6\,\text{cm}$$

Although a 6-cm-long edge-crack can be visually detected, it is safer to conduct non-destructive testing (NDT). The NDT techniques (e.g. dye penetrant inspection or magnetic particle test) are capable of detecting cracks much smaller than this size (6 cm).

EXAMPLE 10.9 DETERMINING THE DESIGN STRESS FOR A STRESSED APPLICATION

A 2.2-mm long central crack was detected in an infinitely wide plate having a fracture toughness of 25 MPa $\sqrt{m}$. What design (working) stresses do you recommend for this application if tensile stresses act normal to the crack length?

SOLUTION

$Y = 1$, $2a = 2.2$ mm $= 2.2 \times 10^{-3}$ m, $a = 0.0011$ m; $K_{IC} = 25$ MPa $\sqrt{m}$

In order to determine the design stress, first we need to know the critical stress, σ_c. By using Equation 10.16,

$$K_{IC} = Y\sigma_c \sqrt{\pi a}$$

$$25 = 1 \times \sigma_c \sqrt{\pi \, x \, 0.0011}$$

$$\sigma_c = \frac{25}{\sqrt{\pi \, x \, 0.0011}} = 425.3 \, \text{MPa}$$

Since design (working) stress must be less than the critical stress, the design stress = 400 MPa.

EXAMPLE 10.10 COMPUTING THE MEAN STRESS AND IDENTIFYING THE STRESS CYCLE

A high steel chimney sways perpendicular to the motion of wind. The bending stress acting on the swaying chimney was determined to be ±82 MPa. Compute the mean stress. Which type of stress cycle exists in the chimney?

SOLUTION

$\sigma_{max} = 82$ MPa, $\sigma_{min} = -82$ MPa

By using Equation 10.17,

$$\sigma_m = \frac{\sigma_{max} + \sigma_{min}}{2} = \frac{82 + (-82)}{2} = 0$$

Since the mean stress is zero, the stress cycle is alternating.

EXAMPLE 10.11 COMPUTING RANGE OF STRESS, STRESS AMPLITUDE, STRESS RATIO, AND R_A

By using the data in Example 10.10, compute the range of stress, stress amplitude, stress ratio, and amplitude ratio for the chimney.

SOLUTION

By using Equation 10.18,

$$\sigma_r = \sigma_{max} - \sigma_{min} = 82 - (-82) = 164 \, \text{Mpa}$$

By using Equation 10.19,

$$\sigma_a = \frac{1}{2}\sigma_r = \frac{1}{2} \times 164 = 82 \, \text{MPa}$$

By using Equation 10.20,

$$R_s = \frac{\sigma_{min}}{\sigma_{max}} = -\frac{82}{82} = -1$$

By using Equation 10.21,

$$R_a = \frac{\sigma_a}{\sigma_m} = \frac{82}{0} = \infty$$

Range of stress = 164 MPa, Stress amplitude = 82 MPa, Stress ratio = –1, Amplitude ratio = ∞.

EXAMPLE 10.12 COMPUTING THE MEAN STRESS WHEN THE STRESS RANGE AND S_{UT} ARE KNOWN

A steel shaft is subjected to a stress range of 340 MPa. The tensile strength of the shaft steel is 480 MPa. Calculate the maximum mean stress the steel shaft is able to withstand.

SOLUTION

$\sigma_r = 340$ MPa, $\sigma_{ut} = 480$ MPa, $\sigma_{FS} = S_e$.

By using Equations 10.19 and 10.22,

$$\sigma_a = \frac{1}{2}\sigma_r = 0.5 \times 340 = 170 \, \text{MPa}$$
$$S_e = \sigma_{FS} = \frac{1}{2}\sigma_{ut} = 0.5 \times 480 = 240 \, \text{MPa}$$

By using Goodman's law,

$$\frac{\sigma_a}{\sigma_{FS}} + \frac{\sigma_m}{\sigma_{ut}} = 1$$

$$\frac{170}{240} + \frac{\sigma_m}{480} = 1$$

$$\sigma_m = 0.71 \times 480 = 140.2\,\text{MPa}$$

The mean stress = 140.2 MPa.

EXAMPLE 10.13 DETERMINING THE FATIGUE LIMIT FOR A NOTCHED SPECIMEN

A material exhibits a smooth (notch-free) bar fatigue limit of 200 MPa. What would be fatigue limit of the same material if a fatigue stress concentration factor of 2 were present due to the notch effect?

SOLUTION

By rewriting Equation 10.25 as:

$$\text{Fatigue limit for notched specimen} = \frac{\textit{Fatigue limit for notch} - \textit{free specimen}}{K_f}$$

$$\text{Fatigue limit for notched specimen} = \frac{200}{2} = 100$$

EXAMPLE 10.14 CALCULATING THE STRESS AND CREEP RATE DURING CREEP TEST

A 450-mm-long cylindrical specimen (having a diameter of 14 mm) is creep tested by hanging a load (weight) of 2.8 kN for a time-duration of 800 h at a temperature of 800°C. Calculate:

(a) the tensile stress acting on the specimen; and (b) creep rate if the strain produced is 0.09.

SOLUTION

A.   Force = 2.8 kN = 2,800 N; Area = $\pi\,r^2 = \pi\,7^2 = 154$ mm²

$$\text{Stress} = \frac{\textit{Force}}{\textit{Area}} = \frac{2800}{154} = 18.18\,\text{N/mm}^2 = 18.18\,\text{MPa}$$

Hence the tensile stress acting on the specimen = 18.18 MPa

B.   Strain = $\Delta\varepsilon$ = 0.09
Time duration = Δt = 800 h
By using Equation 10.26,

$$\text{Creep rate} = \frac{\Delta\varepsilon}{\Delta t} = \frac{0.09}{800} = 0.000112\,h^{-1}$$

The creep rate = $1.12 \times 10^{-4}\,h^{-1}$.

EXAMPLE 10.15 COMPUTING TIME TO CREEP WHEN STRESSES ARE GIVEN

A stainless steel specimen was creep tested at 500°C at a stress of 380 MPa for 400 hours; thereby producing a strain of 0.11. The same specimen was again creep tested at the same temperature at a stress of 300 MPa for 1,000 hours that produced a strain of 0.075. Assuming steady-state creep, calculate the time to produce 0.15% strain in a link bar of the same material when stressed to 70 MPa at the same temperature (500°C).

SOLUTION

At a stress of 380 MPa,

$$\frac{\Delta\varepsilon}{\Delta t} = \frac{0.11}{400} = 2.75 \times 10^{-4}\,h^{-1} \quad \text{(see Equation 10.26)}$$

At a stress of 200 MPa,

$$\frac{\Delta\varepsilon}{\Delta t} = \frac{0.075}{900} = 0.83 \times 10^{-4}\,h^{-1}$$

By using Equation 10.27 ($\frac{\Delta\varepsilon}{\Delta t} = C\sigma^n$), we can write:

2.75 x 10^{-4} = C x 380^n (E-10.15a)
0.83 x 10^{-4} = C x 300^n (E-10.15b)
By combining Equations E-10.15a and E-10.15b, we obtain:

$$\left(\frac{380}{300}\right)^n = \frac{2.75}{0.83}$$
$$\text{or}\quad n = 5.17$$

By substituting the value of n in Equation E-10.15a, we obtain C = 1.26×10^{-17}. For a stress of 70 MPa and strain change of 0.8%, let the strain rate be $x\,h^{-1}$. By using Equation 10.27,

$$x = C \times 70^n = 1.26 \times 10^{-17} \times 70^{5.17} = 1.26 \times 10^{-17} \times 3.4 \times 10^9 = 4.28 \times 10^{-8}$$

Now, creep strain rate $= \frac{\Delta\varepsilon}{\Delta t} = 4.28 \times 10^{-8}$ (see Equation 10.26)

$$\text{or } \Delta t = \frac{\Delta \varepsilon}{creep\ rate} = \frac{0.15\%}{4.28 \times 10^{-8}} = 35,000\,h$$

Time to creep = 35,000 hours.

EXAMPLE 10.16 CALCULATING THE ACTIVATION ENERGY FOR CREEP

Creep testing on a material at a stress of 120 MPa produced the steady-state creep rate of 1.2 x 10^{-4} s^{-1} at 900K and the creep rate of 7.7 x 10^{-9} s^{-1} at 750K. Compute the activation energy for creep for the material.

SOLUTION

By using Equation 10.30 for the conditions T=900K and $\dot\varepsilon = 1.2 \times 10^{-4} s^{-1}$, we can write:

$$ln\,\dot\varepsilon = ln\,A - \frac{Q_c}{RT}$$

$$ln\,1.2 \times 10^{-4} = ln\,A - \frac{Q_c}{8.31 x 900}$$

$$or \qquad ln\,1.2 \times 10^{-4} = ln\,A - \frac{Q_c}{7,479}$$

By using Equation 10.30 for the conditions T=750K and $\dot\varepsilon$ = 7.7 x 10^{-9} s^{-1},

$$ln\,7.7\ x\ 10^{-9} = ln\,A - \frac{Q_c}{8.31 x 750}$$

$$or \qquad ln\,7.7 \times 10^{-9} = ln\,A - \frac{Q_c}{6,232.5}$$

(E-10.16b)

By subtracting Equation E-10.16b from Equation E-10.16a, we obtain:

$$ln\,1.2 \times 10^{-4}\ ln\,7.7 \times 10^{-9} = \frac{Q_c}{6,232.5} - \frac{Q_c}{7,479}$$

$$\frac{Q_c}{6,232.5} - \frac{Q_c}{7,479} = ln\frac{1.2\ x10^{-4}}{7.7\ x10^{-9}}$$

$$Q_c = 3.57 \times 10^5\ J = 357\,kJ/mol$$

The activation energy for creep in the 750–900 K temperature range for the material = 357 kJ.

EXAMPLE 10.17 CALCULATING THE MAXIMUM WORKING TEMPERATURE WHEN Q_C IS KNOWN

By using the data in Example 10.16, calculate the maximum working temperature for the material if the limiting creep rate is 1.5 x 10^{-7} s^{-1}.

SOLUTION

$\dot{\varepsilon} = 1.5$ x 10^{-7} s^{-1}, $Q_c = 3.57$ x 10^5 J/mol

By reference to Equation E-10.16a,

$$ln\,1.2 \times 10^{-4} = ln\,A - \frac{Q_c}{7,479}$$
$$ln\,A - \frac{3.57 \times 10^5}{7,479} = ln\,1.2 \times 10^{-4}$$
$$ln\,A - 47.73 = -9.0$$
$$ln\,A = 38.6$$

By using Equation 10.30,

$$ln\,\dot{\varepsilon} = ln\,A - \frac{Q_c}{RT}$$
$$ln\,1.5 \times 10^{-7} = 38.6 - \frac{3.57 \times 10^5}{8.3145T}$$
$$\frac{3.57 \times 10^5}{8.3145T} = 38.6 - ln\,1.5 \text{ x } 10^{-7}$$
$$T = 790\,K$$

The maximum working temperature = 790 K = 517°C.

EXAMPLE 10.18 CALCULATING THE RUPTURE-LIFE TIME BY USING LM PARAMETER

By reference to the Larson-Miller data in Figure 10.18, calculate the time to rupture the Ti-8-2-9 alloy component that is subjected to a stress of 400 MPa at a temperature of 450°C.

SOLUTION

By reference to Figure 10.18 for the Ti-8-2-9 alloy, at 400 MPa stress, the value of LM parameter is 32; thus we obtain:

$32 = T(20 + log\ t)$ x 10^{-3}

At 450°C or T = 1302R,

$32 = 1302\ (20 + log\ t)$ x 10^{-3}
$log\ t = 4.5$
$t = 10^{4.5} = 31,623$ h =

The creep-rupture time of the component = 31,623 h = 3.6 years.

QUESTIONS AND PROBLEMS

10.1. What are the various mechanisms by which loaded metallic components may fracture/fail? Which type of fracture imposes the greatest threat to society? Give some examples.

10.2. Distinguish between ductile and brittle fractures with the aid of sketches and examples.

10.3. Explain the term ductile-brittle transition (DBT) failure and DBTT with the aid of a sketch. What measures lead to the prevention of DBT failure in steels?

10.4. By drawing a sketch, derive the mathematical form of Griffith's theory; and explain it.

10.5. (a) Define the term: stress concentration factor (K_s); and hence show that the maximum stress at the crack tip can be expressed as: $\sigma_m = 2\sigma_0 \sqrt{\dfrac{a}{\rho_t}}$.

(b) Why are stressed engineering components provided with a radius of curvature?

10.6. Distinguish between the terms: stress intensity factor (K), critical stress intensity factor, (K_c), and the plane strain fracture toughness, K_{IC}.

10.7. Explain fatigue failure with the aid of a sketch. Under which loading conditions do fatigue failures occur? Give three examples of components that generally fail by fatigue.

10.8. What is an S-N curve? How is an S-N curve obtained? Briefly explain the terms: fatigue strength, fatigue limit, and fatigue life.

10.9. Why is a smooth/polished surface important for preventing fatigue? List down the various techniques to control/prevent fatigue failure.

10.10. How is a creep curve obtained? Give some examples of components that fail by creep.

10.11. Show that creep strain rate ($\dot{\varepsilon}$) can be related to the activation energy by: $\ln \dot{\varepsilon} = \ln A - \dfrac{Q_c}{RT}$

P10.12. The specific surface energy and the modulus of elasticity for alumina (Al_2O_3) is 0.90 $J \times m^{-2}$ and 353 GPa, respectively. Calculate the critical stress required for the propagation of a central crack of length 0.6 mm in alumina.

P10.13. An elliptical 0.8-mm-long central crack having a radius of curvature of 6 μm is observed before a plate material is tensile tested. The plate fails by brittle fracture at a normal stress of 150 MPa. Calculate: (a) the theoretical fracture strength of the plate's material, and (b) the stress concentration factor for static loading.

P10.14. The breaking strength of a component's material is 900 MPa. The material, for application under average operating conditions, is untried. What should be the design stress for the component? Design the cylindrical component for a tensile load of 1.7 kN.

P10.15. A plate of semi-finite width contains an edge crack of depth 7 mm. A design stress of 800 MPa is applied normal to the major axis of the crack. Calculate the stress intensity factor.

P10.16. A 2-mm long central crack was detected in an infinitely wide plate having a fracture toughness of 30 MPa $\sqrt{m}$. Is this plate's design safe to use for a critical application where design stresses are 400 MPa?

P10.17. A metallic plate of semi-finite width, containing an embedded circular flaw, has a plane strain fracture toughness of 100 MPa $\sqrt{m}$; and is subjected to a stress of 425 MPa during service. Design a testing method capable of detecting the flaw.

P10.18. A fatigue test was conducted in which the mean stress was 45 MPa and the stress amplitude was 200 MPa. Calculate: (a) the maximum stress level, (b) the minimum stress level, (c) the stress ratio, and (d) the stress range.

P10.19. A steel shaft is subjected to a stress range of 400 MPa. The tensile strength of the shaft steel is 530 MPa. Calculate the mean stress the shaft is able to withstand.

P10.20. The notch sensitivity factor for a component's material is 0.6. The material exhibits a smooth (notch-free) bar fatigue limit of 200 MPa, and its fatigue limit due to the notch effect is 100 MPa. Compute the stress concentration factor for static loading.

P10.21. A 50-cm-long cylindrical specimen, having a diameter of 15 mm, is creep tested by hanging a load of 3 kN for a time-duration of 1,000 h at a temperature of 700°C. Calculate: (a) the tensile stress acting on the specimen; and (b) creep rate if the strain produced is 0.08.

P10.22. An alloy-steel specimen was creep tested at 480°C at a stress of 350 MPa for 370 hours; thereby producing a strain of 0.07. The same specimen was again creep tested at the same temperature at a stress of 270 MPa for 800 hours that produced a strain of 0.05. Assuming steady-state creep, calculate the time to produce 0.14% strain in a link bar of the same material when stressed to 100 MPa at the same temperature (480°C).

P10.23. Creep testing on a material at a stress of 100 MPa produced the steady-state creep rate of 0.9 x 10^{-4} s^{-1} at 890 K and the creep rate of 8.0 x 10^{-9} s^{-1} at 800K. Compute the activation energy for creep for the material for the temperature range (800–890 K) at the same stress.

P10.24. By reference to the Larson-Miller data in Figure 8.18, calculate the time to rupture the Ti-6-2-4-2 alloy component that is subjected to a stress of 200 MPa at a temperature of 400°C.

REFERENCES

Anderson, T.L. (2017) *Fracture Mechanics: Fundamentals and Applications*. 4th ed. CRC Press Inc, Boca Raton.

Ashby, M., Shercliff, H. & Cebon, D. (2010) *Materials: Engineering, Science, and Design*. 2nd ed. Butterworth-Heinemann, Elsevier, Amsterdam.

Blake, A. (1996) *Practical Fracture Mechanics in Design.* Marcel Dekker Inc, New York.

Hertzberg, R.W. (1996) *Deformation and Fracture Mechanics of Engineering Materials.* John Wiley & Sons, Hoboken, NJ.

Huda, Z. & Bulpettt, R. (2012). *Materials Science and Design for Engineers.* Trans Tech Publications, Zurich.

Huda, Z., Bulpett, R. & Lee, K.Y. (2010) *Design Against Fracture and Failure.* Trans Tech Publications, Switzerland.

Huda, Z. & Jie, E.H.C. (2016) A user-friendly approach to the calculation of factor of safety for pressure vessel design. *Journal of King Abdulaziz University (JKAU)*, 27(1), 75–81.

Riemer, A., Leuders, S., Thöne, M., Richard, H.A., Tröster, T. & Niendorf, T. (2014) On the fatigue crack growth behavior in 316l stainless steel manufactured by selective laser melting. *Engineering Fracture Mechanics*, 120, 15–25.

Tawancy, H.M., Hamid, A.U. & Abbas, N.M. (2004) *Practical Engineering Failure Analysis.* CRC Press Inc, Boca Raton.

11 Corrosion and Protection

11.1 CORROSION AND SOCIETY

Corrosion is defined as the environmentally-induced degradation of a material (usually a metal) that involves a chemical reaction. Although, oxidation of iron (rusting) is the most commonly identified form of corrosion, this oxidation process represents only a fraction of material losses. Today, the impacts of corrosion on society are far reaching owing in part to the increased complexity and diversity of material systems. Economic studies on financial losses have shown that premature materials degradation costs industrialized nations around 3% of their gross domestic product (GDP) (Koch *et al.*, 2002).

Corrosion results in the loss of $8–$126 billion annually in the USA alone. Hence, corrosion is a threat to society since it can cause damage to a wide variety of structures/appliances; these include pipelines, bridges, buildings, vehicles, water- and waste-water systems, medical devices, and even home appliances. There are proven techniques to prevent and control corrosion that can reduce or eliminate its impact on public safety, the economy, and the environment. In view of the great economic and technological importance of corrosion, an in-depth scientific explanation and the techniques to combat corrosion are covered in this chapter.

11.2 ELECTROCHEMISTRY OF CORROSION

11.2.1 OXIDATION-REDUCTION (REDOX) REACTIONS

An *oxidation-reduction (REDOX) reaction* is a type of chemical *reaction* that involves transfer of electrons between two species. Consider an apparatus with a zinc plate dipped in a solution of $ZnSO_4$ (zinc sulfate) and a copper plate dipped in a solution of $CuSO_4$ (copper sulfate). This arrangement includes a salt bridge to complete the electric circuit for electrochemical reaction. The electrochemical process can be represented by the following half reactions:

$$\text{Oxidation: } Zn_{(s)} \rightarrow Zn^{2+}_{(aq)} + 2e\text{-} \tag{11.1}$$

$$\text{Reduction: } Cu^{2+}_{(aq)} + 2e\text{-} \rightarrow Cu_{(s)} \tag{11.2}$$

By the addition of Equations 11.1 and 11.2, we obtain the following REDOX reaction:

$$Cu^{2+}_{(aq)} + Zn_{(s)} \rightarrow Cu_{(s)} + Zn^{2+}_{(aq)} \tag{11.3}$$

Equation 11.3 can be used to develop an electrochemical cell; which is explained in the following subsection.

11.2.2 ELECTROCHEMICAL CORROSION AND GALVANIC CELL

It is mentioned in the preceding subsection that placing zinc plate in its sulfate solution results in a *REDOX* reaction that involves transfer of electrons from (highly reactive) zinc to copper (see Equation 11.3); which in turn results from two half reactions. This electrochemical process forms the basis of electrochemical corrosion. *Electrochemical corrosion* of metals occurs when electrons from atoms at the surface of the metal are transferred to a suitable electron acceptor in the presence of water (electrolyte). *Electrochemical corrosion* can be used to build an apparatus where the two half-reactions take place at separate sites. Such as an apparatus, where electrons are transferred indirectly at separate sites, is called an *electrochemical, galvanic,* or *voltaic cell.* Figure 11.1 illustrates the Cu-Zn galvanic cell that involves electrochemical or galvanic corrosion based on *REDOX* reactions.

In the Cu-Zn galvanic cell (Figure 11.1), zinc electrode (anode) is placed in a solution of zinc sulfate whereas copper electrode (cathode) is placed in a solution of copper sulfate. When the electrons flow from the anode to the cathode, zinc corrodes. This electrochemical corrosion of zinc is owing to its negative (or more negative) standard electrode potential value ($E^o_{Zn} = -0.76$) as compared to copper ($E^o_{Cu} = +0.34$) (see subsection 11.2.3).

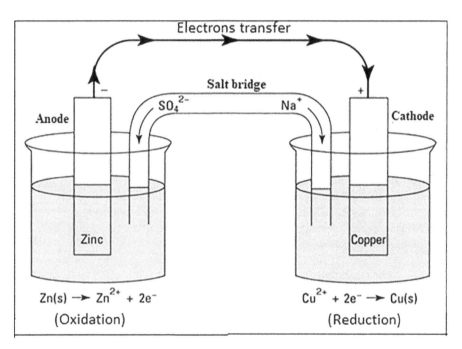

FIGURE 11.1 The Cu-Zn galvanic cell—working principle.

11.2.3 STANDARD ELECTRODE POTENTIAL (E^o)

The standard electrode potential (also called the standard reduction potential), E^o, is the electric potential developed when a pure metal is in contact with its ions at one molar concentration at a temperature of 25°C or 298 K. The standard electrode (half-cell) potentials of some metals are listed in Table 11.1. The voltages in Table 11.1 refer to the half reactions as *reduction* reactions. The metals that are more reactive than hydrogen are assigned negative potentials and are said to be anodic. The anodic metals (e.g. zinc) have a tendency to release electrons; and hence corrode with reference to hydrogen. On the other hand, the metals that are less reactive than hydrogen are assigned positive potentials and are said to be cathodic. The cathodic metals (e.g. copper) have tendency to accept electrons and are reduced to atomic state.

The difference in the standard electrode potentials of the two metals provides the electromotive force; which can be calculated by:

$$E^o = E^o_{red} - E^o_{ox} \qquad (11.4)$$

where E^o is the electromotive force or voltage of the standard galvanic cell, volts (V); E^o_{red} is the standard electrode potential of the cathodic metal (being reduced), V; and E^o_{ox} is the standard electrode potential of the anodic metal (being oxidized), V. For the electrochemical reaction/corrosion to occur spontaneously, the EMF of the cell (E^o) must be positive. By the use of Equation 11.4, the EMF of the Cu-Zn galvanic cell is calculated to be $E^o = +1.10$ V (see Example 11.1). When standard half-cells are coupled together, the metal having more negative potential (E^o) value will oxidize (i.e. corrode), whereas the metal having less negative (or 0 or positive) potential will be reduced (see Example 11.2).

TABLE 11.1
Standard Electrode Potentials (E^o) of Some Metals at 25°C

Cathode half reaction	E^o (volts)
$Li^+_{aq} + e^- \rightarrow Li_{(s)}$	−3.04
$K^+_{aq} + e^- \rightarrow K_{(s)}$	−2.92
$Na^+_{aq} + e^- \rightarrow Na_{(s)}$	−2.71
$Zn^{2+}_{aq} + 2e^- \rightarrow Zn_{(s)}$	−0.76
$Ni^{2+}_{aq} + 2e^- \rightarrow Ni_{(s)}$	−0.25
$2H^+ + 2e^- \rightarrow H_2$	0
$Cu^{2+}_{aq} + 2e^- \rightarrow Cu_{(s)}$	+0.34
$Pt^{2+}_{aq} + 2e^- \rightarrow Pt_{(s)}$	+1.2
$Au^{2+}_{aq} + 3e^- \rightarrow Au_{(s)}$	+1.49

11.3 GALVANIC CORROSION AND GALVANIC SERIES

11.3.1 GALVANIC CORROSION AND CORROSION POTENTIAL

Galvanic corrosion (also called *bi-metallic corrosion*) is an electrochemical corrosion that occurs when dissimilar metals are brought in electrical contact while they are immersed in an electrolyte (see section 11.5.2). In a standard galvanic cell, the metals are immersed in their salt solutions (electrolyte) having $1M$ concentration. However, in real galvanic cells, the electrolytes are usually concentrated much lower than $1M$. This is why the term *standard electrode potential* (for standard cells) is replaced by *corrosion potential* or *cell potential* for real galvanic cells. It means that the corrosion potential of a cell is its electrode potential in real electrolyte. Obviously, owing to the lower electrolyte concentration, the driving force for the oxidation (or corrosion) to occur (in real galvanic cells) is greater as compared to standard galvanic cells (Francis, 2001). It means that *corrosion potential* (E) of a metal is greater than its standard electrode potential (E^o); the relationship between E^o and E is expressed by the *Nernst* equation, as follows (Wahl, 2005):

$$E = E^o - \frac{0.0592}{n} \log Q = - \frac{0.0592}{n} \log \frac{M_1^{n+}}{M_2^{n+}} \qquad (11.5)$$

where E is the corrosion potential of the electrochemical cell, V; E^o is the standard electrode potential of the cell, V; n is the number of electrons transferred; M_1^{n+} is the molar concentration of the ions of the metal being oxidized; and M_2^{n+} is the molar concentration of the ions of the metal being reduced. Nernst equation enables us to calculate the corrosion potential of a galvanic cell for which solution ion concentrations are other than $1M$ (see Example 11.3). It must be noted that the coefficients of the balanced overall reaction equation will appear as exponents in the Q expression in Equation 11.5 (see Example 11.4).

11.3.2 GALVANIC SERIES

The galvanic series is a table reporting the corrosion potentials of metals and commercial alloys in flowing seawater at ambient temperature (see Table 11.2). It is evident in the *galvanic series* that the magnesium and its alloys are the most active in corrosion attack since their corrosion potential are the most negative; it means the metals and alloys at the top are the most anodic. On the other hand, the metals and alloys that are last on the list (e.g. gold, platinum, etc.) are the most cathodic and the least likely to undergo galvanic corrosion.

The galvanic series can be a very beneficial guide for selecting metals to join together so that materials with the least tendency to undergo a galvanic reaction can be chosen. In general, metals and alloys that are located further apart within the series are most likely to undergo galvanic corrosion, which should be prevented through proper materials selection and design (see subsection 11.6.2). In order to avoid possible corrosion reactions brought about by the combination of metals, it is important to select the materials that are near each other in the galvanic series.

TABLE 11.2
The Galvanic Series

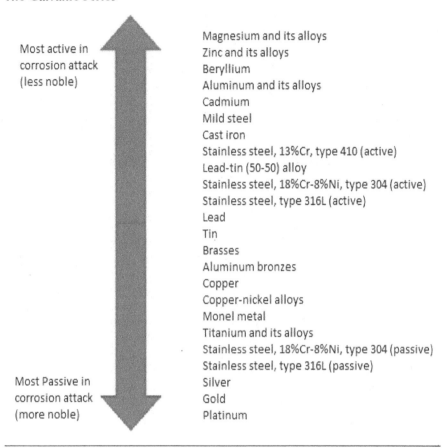

Most active in
corrosion attack
(less noble)

Most Passive in
corrosion attack
(more noble)

Magnesium and its alloys
Zinc and its alloys
Beryllium
Aluminum and its alloys
Cadmium
Mild steel
Cast iron
Stainless steel, 13%Cr, type 410 (active)
Lead-tin (50-50) alloy
Stainless steel, 18%Cr-8%Ni, type 304 (active)
Stainless steel, type 316L (active)
Lead
Tin
Brasses
Aluminum bronzes
Copper
Copper-nickel alloys
Monel metal
Titanium and its alloys
Stainless steel, 18%Cr-8%Ni, type 304 (passive)
Stainless steel, type 316L (passive)
Silver
Gold
Platinum

11.4 RATE OF CORROSION

11.4.1 RATE OF UNIFORM CORROSION

Uniform corrosion involves general thinning of a metallic structure or component owing to the corrosive attack proceeding evenly over the entire or almost entire surface area (see Figure 11.2).

The rate of uniform corrosion or corrosion penetration rate (*CPR*) can be expressed in terms of weight loss in a given time, density, and exposed specimen area, as follows:

$$CPR = \frac{87.6\,W}{\rho\,S\,t}$$

(11.6)

FIGURE 11.2   Uniform corrosion (rusting) of steel screw piles.

where *CPR* is the corrosion penetration rate, mm/yr; *W* is the weight loss, milligrams (mg); ρ is the density of the metal, g/cm³; *S* is the exposed surface area of the specimen, cm²; and *t* is the exposure time of the metal to the corrosive environment, hours (h) (see Example 11.5). A CPR value less than 0.50 mm/yr is acceptable for most applications.

The rate of galvanic corrosion can be calculated by computing the amount of metal uniformly corroded from an anode or electroplated on a cathode in an aqueous solution in a given time. This calculation is governed by Faraday's law of electrolysis; which can be expressed as follow:

$$\text{Corrosion rate (in g/s)} = \frac{w}{t} = \frac{I\,M}{n\,F} \tag{11.7}$$

where *w* is mass of metal corroded, g; *t* is the exposure time, s; *I* is current flowing through the electrolyte, A; *M* is atomic mass of the metal, g/mol; *n* is number of electrons lost by an atom of the corroding metal; and F is Faraday's constant (F= 96,500 Coul./mol) (see Example 11.6).

By rearranging the terms in Equation 11.7, we can obtain an expression for time (*t*), as follows:

$$t = \frac{w\,n\,F}{I\,M} \tag{11.8}$$

where the terms *t*, *w*, *n*, *F*, *I*, and *M* have their usual meanings (see Example 11.7).

An important parameter in electrochemical corrosion is the current density (*J*) defined by:

$$J = \frac{I}{S} \tag{11.9}$$

where J is the current density, A/m²; I is the current, A; and S is the surface area of the electrode, m². If the current density (J) is known, current (I) can be determined and be substituted in Equation 11.7 to compute the rate of corrosion (see Example 11.8).

11.4.2 OXIDATION KINETICS

The oxidation of metals (e.g. rusting of iron) is the common cause of corrosion since the most stable state of most metals is an oxide. The driving force for a metal to oxidize is its *free energy of oxidation* (ΔG_{oxid}); however, the rate of oxidation is determined by the kinetics of the oxidation reactions that largely depend on the nature of the oxide. The oxidation of a metal, M, can be expressed by the following general reaction equation:

$$M+O=MO+\Delta G_{oxid} \tag{11.10}$$

where O is oxygen and MO is the metal oxide; the latter coats the surface of the metal thereby separating it from the corrosive effects of oxygen. The formation of oxide film on the metal surface results in the weight gain of the exposed metallic component. The oxidation kinetics strongly depends on whether the oxide film is porous or compact; in the former case, the kinetic behavior is linear whereas in the latter case, the kinetic behavior is parabolic (see Figure 11.3).

The linear kinetic behavior for oxidation can be expressed as (Ashby *et al.*, 2010):

$$\frac{d(\Delta m)}{dt}=k_l \tag{11.11}$$

where Δm is the weight gain per unit area, kg/m²; and k_l is the *linear kinetic constant for oxidation*, kg/m²×s. By rearranging the terms in Equation 11.11 and then integrating both sides, we obtain:

$$\int d(\Delta m)=\int_0^t k_l\, dt \tag{11.12}$$

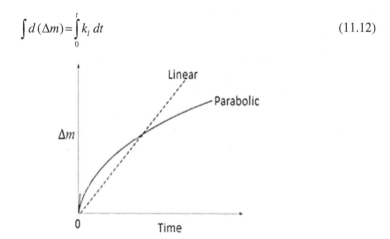

FIGURE 11.3 Two oxidation kinetic behaviors: linear and parabolic (Δm = weight gain per unit area).

$$\Delta m = k_l\, t \qquad\qquad\qquad (11.13)$$

The significance of Equation 11.13 is illustrated in Example 11.9.

The parabolic kinetic behavior for the oxidation of some metals is more favorable as compared to the linear kinetic behavior since in the former case the compact oxide film that develops on their surfaces protects further oxidation/corrosion. The parabolic kinetic behavior is given by:

$$\frac{d(\Delta m)}{dt} = \frac{k_p}{\Delta m} \qquad\qquad\qquad (11.14)$$

Where k_p is the *parabolic kinetic constant for oxidation*, expressed in kg/m^4 × s. By rearranging the terms in Equation 11.14 and then integrating both sides, we obtain:

$$(\Delta m)^2 = k_p\, t \qquad\qquad\qquad (11.15)$$

The significance of Equation 11.15 is illustrated in Example 11.10.

Once the weight gain per unit area (Δm) and the molecular weight of the oxide are known, the weight of the oxide (m_{oxide}) can be determined by:

$$m_{oxide} \frac{Mol.\ wt.\ of\ oxide}{At.\ wt.\ of\ oxygen} \times \Delta m \qquad\qquad\qquad (11.16)$$

Accordingly, the thickness of the oxide film on the metal can be calculated by:

$$x = \frac{m_{oxide}}{\rho} = \frac{Mol.\ wt.\ of\ oxide}{At.\ wt.\ of\ oxygen} \times \frac{\Delta m}{\rho} \qquad\qquad\qquad (11.17)$$

where x is the thickness of the oxide film, m; and ρ is the density of the oxide, kg/m^3 (see Example 11.11).

11.5 FORMS OF CORROSION

11.5.1 The Classification Chart of Corrosion

Metallic corrosion may be classified into six forms: (1) uniform corrosion, (2) galvanic corrosion, (3) localized corrosion, (4) metallurgically-influenced corrosion, (5) mechanically-assisted corrosion, and (6) environmentally-assisted corrosion. These forms of corrosion are illustrated in the classification chart shown in Figure 11.4.

11.5.2 Uniform Corrosion and Galvanic Corrosion

Uniform Corrosion. *Uniform corrosion* is the most common form of corrosion; which has been explained in the preceding section (subsection 11.4.1). Notable examples of uniform corrosion include: rusting of steel and iron and the tarnishing of silverware.

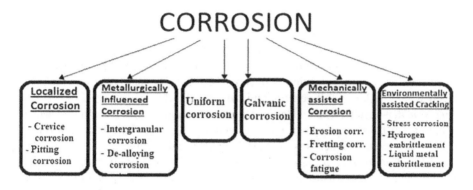

FIGURE 11.4   The classification chart showing various forms of corrosion.

FIGURE 11.5   Galvanic corrosion in a wrought iron support system.

Galvanic Corrosion. Galvanic corrosion involves electrochemical dissolution of a metal when two metals or alloys having different compositions (different electrode potentials) are electrically coupled while exposed to an electrolyte. Figure 11.5 illustrates that a typical structure's wrought iron support system got rusted due to galvanic corrosion because the insulating layer of shellac between the iron (anodic) and copper (cathodic) had failed.

Galvanic corrosion has been discussed in detail in section 11.3. The rate of galvanic corrosion has also been quantitatively analyzed in subsection 11.4.2.

11.5.3   LOCALIZED CORROSION

Localized corrosion involves intense attack at localized sites on the surface of a component while the rest of the surface corrodes at a much lower rate. For example, if corrosion-protective coating breaks down locally, then corrosion may be initiated at these local sites. There are two main types of localized corrosion: (a) *crevice corrosion*, and (b) *pitting corrosion*.

Crevice corrosion. *Crevice corrosion* is a preferential corrosion that occurs in a gap or crevice due to a localized depletion of oxygen. For example, *crevice corrosion* may occur inside a stainless steel piping system owing to a crevice that may be created by lack of a full penetration in an orbital weld (see Figure 11.6). *Crevice corrosion* may result when a material is exposed to stagnant corrosive media; which tend to occur in crevices such as those formed under gaskets, bolt threads, washers, fastener heads, surface deposits, threads, lap joints, and clamps.

Pitting Corrosion. *Pitting corrosion* is a localized form of corrosion by which *pits* or "holes" are produced in the material (see Figure 11.7). *Pitting corrosion* is most likely to occur in the presence of chloride ions, combined with such depolarizers as oxygen or oxidizing salts.

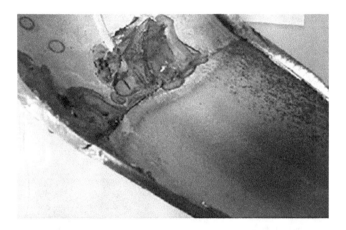

FIGURE 11.6   Crevice corrosion inside a stainless steel piping system due to a crevice.

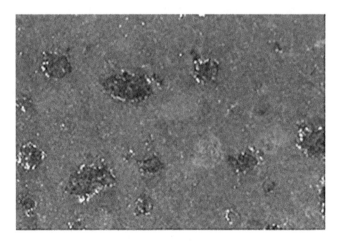

FIGURE 11.7   Pitting corrosion in steel.

Pitting corrosion occurs when localized areas of steel lose their metallic protective layer. These localized areas become anodic, while part of the remaining metal becomes cathodic, resulting in a localized galvanic corrosion (pitting corrosion).

11.5.4 METALLURGICALLY INFLUENCED CORROSION

Metallurgically influenced corrosion is so named due to the significant role of metallurgy played in these forms of attack. There are two forms of metallurgically-influenced corrosion: (a) intergranular (*IG*) corrosion, and (b) de-alloying corrosion.

11.5.4.1 Intergranular Corrosion

Intergranular (IG) corrosion, also called *grain-boundary corrosion*, involves localized corrosion at grain boundaries. In *IG* corrosion, the grain boundaries undergo localized corrosion attack while the rest of the material remains unaffected (see Figure 11.8). For example, small quantities of iron in aluminum or titanium segregate to the grain boundaries where they can induce *IG* corrosion. Certain precipitate phases (e.g. Mg_5Al_8, Mg_2Si, $MgZn_2$, $MnAl_6$, etc.) also cause *IG* corrosion of high-strength aluminum alloys, particularly in chloride-rich media.

11.5.4.2 De-Alloying Corrosion

De-alloying corrosion, also called *selective leaching*, involves the selective removal of one element from an alloy due to corrosion. An example of *de-alloying corrosion* is the graphitic corrosion of gray cast iron, whereby a brittle graphite skeleton remains following preferential iron dissolution. Another example of de-alloying corrosion is the corrosion due to de-zincification of brass.

11.5.5 MECHANICALLY ASSISTED CORROSION

Mechanically assisted corrosion involves corrosion assisted by some mechanical conditions (e.g. the application of cyclic stresses, presence of friction, erosive conditions, etc.). There are three main forms of mechanically-induced corrosion: (a) erosion corrosion, (b) fretting corrosion, and (c) corrosion fatigue.

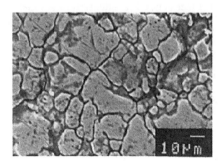

FIGURE 11.8 Intergranular corrosion in steel (IG corrosion has caused dropping of grains resulting in the disintegration of the steel).

11.5.5.1   Erosion Corrosion

Erosion corrosion results from the relative movement between a corrosive fluid and a metal surface. Erosion corrosion leads to the formation of holes, grooves, valleys, wavy surfaces, etc.; which cause damage to the metallic component. Most metals and alloys can be affected by erosion corrosion, particularly those whose corrosion resistance depends on the existence of a surface film (e.g. galvanized steel, stainless steel). Turbulent flow, resulting from relative movement between the corrosive fluid and the coating, can destroy protective films and cause very high corrosion rates in materials otherwise highly resistant under static conditions. The protective designs against erosion corrosion include: selection and design of a more resistant material, improved plant design, and adjustments in the corrosive medium.

11.5.5.2   Fretting Corrosion

Fretting-corrosion is a combined damage mechanism involving corrosion at points where friction is involved i.e. where two moving metal surfaces make rubbing contact. It occurs essentially when the interface is subjected to vibrations and to compressive loads. The protective designs against fretting corrosion include: reduction in friction by lubrication with oils or greases, exclusion of oxygen from the interface; increase in the hardness of one or both materials in contact; selecting a composite showing better friction behavior than others; surface-hardening treatments; use of seals to absorb vibrations and exclude oxygen and/or moisture; reduction of the friction loads in certain cases; and modification of the amplitude of the relative movement between the two contacting surfaces.

11.5.5.3   Corrosion Fatigue

Corrosion fatigue is the corrosion resulting from the combined action of cycling stresses and a corrosive environment. An infamous example of corrosion fatigue is the fatigue failure of the *Aloha 737* aircraft while flying between the Hawaiian Islands in 1988; this disaster cost one life.

Protective design against corrosion fatigue include: minimizing cyclic stresses; reduction in stress concentration; avoiding rapid changes of loading, temperature, or pressure; avoiding internal stress, fluttering, and vibration-producing/transmitting design, increasing natural frequency for reduction of resonance; and limiting corrosion factors either by selecting more resistant material or by using a less corrosive environment.

11.5.6   Environmentally Assisted Corrosion

Environmental assisted corrosion (EAC) refers to the brittle fracture of a normally ductile material in which the *corrosive* effect of the *environment* is a causative factor. *EAC* has presented a significant structural integrity problem in various industries including marine, petrochemical, aerospace, and power generation. *EAC* includes: stress-corrosion cracking, hydrogen embrittlement, and liquid metal embrittlement; which are discussed as follows.

11.5.6.1 Stress-Corrosion Cracking

Stress-corrosion cracking (SCC) occurs as a result of the combined effects of corrosive environment and stress. For *SSC* to occur, three conditions are necessary: (a) the component must be in a corrosive environment, (b) the component must be made of a susceptible material, and (c) the component must be subject to tensile stresses above threshold value. The tensile stresses may be either an externally applied stress or residual stresses in the material. For example, deformation can induce residual stresses in a cold-worked metal thereby causing *SCC* (see Figure 11.9). SCC is restricted to certain alloys subject to specific environments. An example is the failure of age-hardened aerospace aluminum alloys in the presence of chlorides (Talbot and Talbot, 2018).

11.5.6.2 Hydrogen Embrittlement (*HE*)

Hydrogen embrittlement (*HE*) involves ingress of atomic hydrogen into a metallic component. *HE* can seriously reduce the ductility and load-bearing capacity, cause cracking and catastrophic brittle failures at stresses below the yield stress of susceptible materials. Typical examples of *HE* include: cracking of weldments, and cracking of hardened steels when exposed to conditions which inject hydrogen into the component. The alloys that are particularly susceptible to *HE* include: high-strength steels, titanium alloys, and aluminum alloys.

Sources of hydrogen causing *HE* have been encountered in improper practices in steelmaking, in processing parts, in welding, in storage of hydrogen gas, and related to hydrogen as a contaminant in the environment that is often a by-product of general corrosion; it is the latter that concerns the nuclear industry. Hydrogen may be produced by corrosion reactions such as rusting, electroplating, and the like. There are a number of ways for overcoming the *HE* problem. One of the ways is an additional heat treatment of the metal. If in a manufacturing operation (such as welding, electroplating, etc.) significant hydrogen entry is detected, then a final baking heat treatment should be employed to expel hydrogen from the metal.

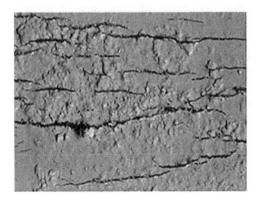

FIGURE 11.9 Stress-corrosion cracks on the external surface of the steel high-pressure gas transmission pipeline.

11.5.6.3 Liquid Metal Embrittlement

Liquid metal embrittlement (*LME*) refers to corrosion and/or embrittlement of a metal resulting from a coating of certain metals (e.g. aluminum, iron, copper, etc.) with a micron-thin layer of certain liquid metals (such as mercury, gallium, cadmium, etc.). *LME* promotes crack growth rate leading to brittle failure. For example, when certain aluminum alloys or brass are coated with mercury, a crack propagation rate of 500 cm/s has been reported (Hertzberg, 1996). *LME* can be avoided by the following techniques: (a) removal of liquid metal from the environment, (b) selection of compatible materials (e.g. *Hg* will embrittle *Al* but not Mg i.e. Hg-Mg is the set of compatible materials), (c) application of resistant coating to act as a barrier between the metal and the environment, and (d) chemical dissolution of the liquid metal (Huda *et al.*, 2010).

11.6 PROTECTION AGAINST CORROSION

11.6.1 Approaches in Protection Against Corrosion

The losses resulting from corrosion have been highlighted in section 11.1. It is, therefore, very important to avoid corrosion by appropriate protective system design. Since almost every form of corrosion is electrochemical in nature, designing against electrochemical corrosion would significantly cover prevention against corrosion in most cases. When designing against corrosion, one of the following six approaches may be adopted: (1) design of metals-assembly system, (2) cathodic protection, (3) surface engineering by coating, (4) inhibitors, (5) material selection, and (6) material treatment.

11.6.2 Design of Metal-Assembly

There are five common techniques in the design of a metal-assembly system: (a) design components with closed fluid systems, (b) avoid the contact of dissimilar metals, (c) avoid unfavorable area effect, (d) avoid crevices between assembled materials, and (e) design assemblies enabling easy replacement of corroded parts. These design techniques are discussed in the following paragraphs.

Design Components with Closed Fluid Systems. If a fluid system is open, it will dissolve gas, providing ions that participate in the cathode reaction leading to galvanic corrosion. It is, therefore, necessary to use *closed fluid systems*; which do not permit accumulation of stagnant pools of liquid thereby preventing dissolution of gas into the fluid system. Hence, for designing components against galvanic corrosion, fluid systems should be closed.

Avoid the Contact of Dissimilar Metals. It is learned in section 11.3.2 that the galvanic series is a very beneficial guide for selecting metals to join together so that materials with the least tendency to undergo a galvanic reaction can be chosen. It means the metallurgist must avoid dissimilar-metals contact by selecting of metals as close together as possible in the galvanic series (see Table 11.2). In case this is impracticable, one must electrically insulate dissimilar metals completely by using intermediate plastic fittings. For instance, the bolted joint between copper and

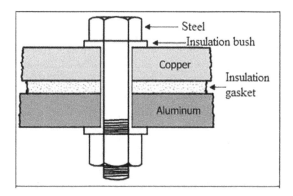

FIGURE 11.10 Design of bolted joint to avoid galvanic corrosion.

aluminum plates can be protected by incorporating an insulating bush and insulating (plastic) gasket (see Figure 11.10).

Avoid Unfavorable Area Effect. In electrochemical corrosion, the cathode accepts electrons released by anode-dissolution reaction. It means that a large cathode area is unfavorable since it accelerates corrosion. Thus, corrosion can be avoided by making the cathode area much smaller than the anode area. This would result in a few electrons-acceptance by the cathode and slow anode-dissolution reaction thereby limiting corrosion. For example, copper rivets (small area) (cathode) can be used to permanently join aluminum sheets (large area) (anode). Example 11.12 illustrates the effects of both favorable and unfavorable areas on galvanic corrosion.

Avoid Crevices between Assembled Materials. Some mechanical fastening operations (e.g. riveting, soldering, or brazing) may result in crevices that form a concentration cell between assembled materials; the latter cause electrochemical corrosion. It is, therefore, important to avoid crevices between assembled materials. In particular, if the materials are apart from each other in the *galvanic series*, they should be assembled by welding; not by riveting or brazing.

Design Assemblies Enabling Easy Replacement of Corroded Parts. In case it is impractical to design metal assembly that fulfills the previously mentioned requirements, the assembly should be so designed as to facilitate easy replacement of the corroded part. For example, in such a case, the assembly should be designed by using bolted joint (see Figure 11.10); not by riveting or brazing.

11.6.3 CATHODIC PROTECTION

Principle. Cathodic protection (CP) refers to the corrosion prevention by converting all of the anodic (active) sites on the metal surface to *cathodic* (passive) sites by supplying free electrons (electrical current) from an alternate source. The function of *CP* is to reduce the potential difference between cathode and anode to a negligibly small value. In general, there are two methods for CP: (a) by using sacrificial anode, and (b) by using impressed voltage. The use of sacrificial anode for CP involves connecting the protected structure to a galvanic anode (or sacrificial anode) (usually

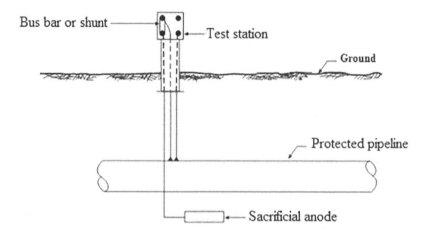

FIGURE 11.11 A typical galvanic/sacrificial anode installation for cathodic protection.

zinc, magnesium, or aluminum), which is placed close to the protected structure (see Figure 11.11). The potential difference between the sacrificial anode and the protected structure (cathode) supplies the free electrons to the cathode for CP.

Applications of CP. *Cathodic protection (CP)* is commonly applied to offshore platform systems with a steel pipe submerged into seawater. It is also applied to buried pipelines (Figure 11.11). Ship hulls are also protected by galvanic cathodic protection by using a sacrificial anode.

11.6.4 SURFACE ENGINEERING BY COATING

11.6.4.1 Coating and Types of Coating Processes

Surface engineering by coating is an important industrial process for corrosion protection. In coating, the part's surface is first cleaned, and then coated by the application of a coating material; finally the coated part is dried either in air or oven. Surface coatings provide protection, durability, and/or decoration to part surfaces. Metals that are commonly used as coating material (for coating on steels) include nickel, copper, chromium, gold, and silver. A good corrosion-resistant surface is produced by chrome plating; however their durability is just fair. The nickel electroplated parts have excellent surface finish and brightness; they are a durable coating, and have fair corrosion resistance.

There are many different surface coating processes; these include electroplating (nickel plating, chrome plating, etc.); hot dip coating, electroless plating, conversion coating, physical vapor deposition (PVD), chemical vapor deposition (CVD), powder coating, painting, and the like. Among these processes, electroplating is the most commonly practiced surface coating process, and is discussed and mathematically modeled in the following section; the other coating processed are described elsewhere (Huda, 2017; Roberge, 2008).

11.6.4.2 Coating by Electroplating

Electroplating process involves the cathodic deposition (plating) of a thin layer of metal on another metal (base metal) or other electrically conductive material. The electroplating process requires the use of direct current (*d.c*) as well as an electrolytic solution consisting of certain chemical compounds that make the solution highly conductive. The positively charged plating metal (anode) ions in the electrolytic solution are drawn out of the solution to coat the negatively charged conductive part surface (cathode). Under an *e.m.f*, the positively charged ions in the solution gain electrons at the part surface and transform into a metal coating (see Figure 11.12).

In modern surface engineering practice, electrolytic solutions usually contain additives to brighten or enhance the uniformity of the plating metal. The amount of plating material deposited strongly depends on the plating time and current levels to deposit a coating of a given thickness. The electroplating process is governed by the Faraday's laws of electrolysis. Since the electroplating time in silver plating is the same as the time of corrosion of silver anode (see Figure 11.12), the electroplating rate and the electroplating time may be computed by using Equations 11.7 and 11.8, respectively (see section 11.4).

An important electroplating parameter is the *cathode current efficiency*; which is defined as "the ratio of the weight of metal actually deposited to that, which would have resulted if all the current had been used for deposition." Numerically, *cathode current efficiency* is given by:

$$Eff_{cathode} = \frac{W}{W_{th}} \tag{11.18}$$

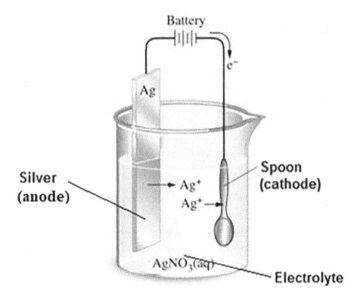

FIGURE 11.12 Electroplating (silver plating) of a steel spoon.

where $Eff_{cathode}$ is the cathode current efficiency; W is the weight of metal actually deposited, g; and W_{th} is the theoretical weight of the deposit, g.

The theoretical weight can be found by:

$$W_{th} = \frac{Q\,M}{F\,n} \tag{11.19}$$

where Q is the electric charge flow, Coul.; M is the atomic mass of the metal, g/mol; F is the Faraday's constant, and n is the valence of the metal deposit (see Example 11.13).

11.6.5 CORROSION INHIBITORS

Corrosion inhibitors are the chemicals that help to slow down or prevent electro-chemical corrosion. Commonly used corrosion inhibitors include organic-based, phosphonate-based, silicate-based, and caustic-based chemicals. For an automotive application, the **Inhibitors** are added to vehicle oils, so that the oil will not corrode the moving parts. Fuels (such as petrol, diesel, and even the renewable fuels) have **corrosion inhibitors** added to them, so that the engine will perform properly, the parts will not rust, and there will be less stress and strain on the engine. However, pipelines and other structures that carry liquids need *corrosion inhibitors* to be able to resist the damage that the liquid can cause. Oil refineries and other storage depots may need to store large quantities of liquids for a long period. These storage tanks are built to withstand corrosion against the stored liquids by using *inhibitors*; this corrosion prevention technique helps to avoid leakage with potentially catastrophic results.

11.6.6 MATERIAL SELECTION

Corrosion can be prevented by selecting a suitable material and/or its heat treatment. Pure metals or their alloys with single-phase microstructure offer better resistance to corrosion as compared to multi-phase alloys. This is because the phases in a multi-phase alloy may form a galvanic cell due to a difference in their electrode potentials. Notable example of corrosion resistant alloys include: 18-8 stainless steels, *Al-7%Mg* alloys, titanium alloys, and the like.

11.6.7 HEAT TREATMENT/COMPOSITION CONTROL

The control of composition and thermal treatment parameters are also important in preventing corrosion. For example, in bending (metal forming), differences in the amount of cold work and residual stresses cause local stress cells; which may cause SCC. This problem can be minimized by a stress-relief anneal (recovery) or a full recrystallization anneal. Additionally, in the casting process, segregation causes the formation of tiny, localized galvanic cells that accelerate corrosion. This galvanic corrosion can be avoided by a homogenization annealing heat treatment.

Heat treatment is particularly important in *austenitic stainless steels* due to its sensitization; the latter may cause *IG* corrosion. When the steel cools slowly from

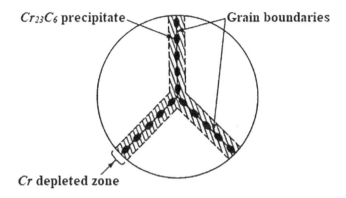

$Cr_{23}C_6$ precipitate Grain boundaries

Cr depleted zone

FIGURE 11.13 Sensitized austenitic stainless steel.

870 to 425°C (e.g. during welding), chromium carbide ($Cr_{23}C_6$) precipitates at the grain boundaries. The austenite at the grain boundaries may contain less than 12% Cr, which is the minimum required to produce a passive oxide layer. The steel is said to have *sensitized* (see Figure 11.13). Since the grain boundary regions are small and highly anodic, rapid corrosion of the austenite at the grain boundaries (*IG* corrosion) occurs. This problem can be overcome by adopting one of the following techniques:

A. Keep the Cr content very high in the steel; this would ensure at least 12% Cr at grain boundaries, even if the $Cr_{23}C_6$ forms.

B. Keep the carbon content less than 0.03%; this composition control would prevent formation of $Cr_{23}C_6$ in the microstructure.

C. *Stabilize* the steel by addition of titanium (Ti) or niobium (Nb) so as to tie up the carbon as TiC or NbC thereby preventing the formation of $Cr_{23}C_6$.

D. Avoid the sensitization temperature range (425–870°C) during tempering, welding, or any other thermal treatment.

E. In case, the preceding techniques (A–D) are impractical, quench heat treat the alloy (sensitized austenitic stainless steel) at above 870°C so as to dissolve the chromium carbide. The resulting microstructure will contain 100% austenite. The rapid cooling (quenching) from above 870°C will prevent formation of $Cr_{23}C_6$ precipitate.

11.7 CALCULATIONS—EXAMPLES ON CORROSION AND PROTECTION

EXAMPLE 11.1 CALCULATING THE EMF OF A GALVANIC CELL

By reference to the standard electrode potential data, calculate the EMF of the standard Cu-Zn cell. Would the electrochemical reaction occur spontaneously?

SOLUTION

In the Cu-Zn galvanic cell (Figure 11.1), copper is reduced and zinc is oxidized. By reference to Table 11.1,

$$E^o_{red} = E^o(Cu^{2+}/Cu) = +0.34 \text{ V}, \text{ and } E^o_{ox} = E^o(Zn^{2+}/Zn) = -0.76 \text{ V}$$

By using Equation 11.4,

$$E^o = E^o_{red} = E^o_{ox} = +0.34 \text{ V} - (-0.76 \text{ V}) = +1.10 \text{ V}$$

Since the EMF value is positive, the electrochemical reaction would occur spontaneously.

EXAMPLE 11.2 IDENTIFYING CATHODE/ANODE AND PREDICTING CORROSION IN A GALVANIC CELL

Consider a Ni-Na galvanic cell. (a) At which electrode does oxidation occur? (b) At which electrode does reduction occur? (c) Which electrode is the anode? (d) Which electrode is the cathode? (e) Which electrode corrodes? (f) Write the spontaneous overall cell reaction. (g) Calculate the EMF of the standard cell.

SOLUTION

$$E^o(Ni^{2+}/Ni) = -0.25 \text{ and } E^o(Na^+/Na) = -2.71$$

A.   Since sodium (Na) has more negative potential, Na will oxidize.
B.   Since nickel (Ni) has less negative potential, Ni will reduce.
C.   Since Na is oxidized, it is the anode.
D.   Nickel (Ni) is the cathode.
E.   Since corrosion occurs at anode, sodium (Na) will corrode.
F.   The half reactions and the overall cell reaction are given in the following:

$$Na \rightarrow Na^+ + e^- \quad E^o = -(-2.7 \text{ V}) = 2.7V$$
$$2Na \rightarrow 2Na^+ + 2e^- \quad E^o = 2.7V$$
$$\underline{Ni^{2+} + 2e^- \rightarrow Ni \qquad E^o = -0.25V}$$
$$Ni^{2+} + Na \rightarrow Na^+ + Ni$$

G.   EMF of the standard cell $= E^o = E^o_{red} - E^o_{ox} = -0.25 - (-2.7) = 2.70 - 0.25$
      $= 2.45 \text{ V}$.

EXAMPLE 11.3 CALCULATING THE CORROSION POTENTIAL OF AN ELECTROCHEMICAL CELL

One-half of an electrochemical cell consists of a pure copper electrode in a solution of Cu^{2+} ions; the other half is a zinc electrode immersed in a Zn^{2+} solution. Calculate the corrosion potential of the cell at 25°C if the Zn^{2+} and Cu^{2+} ions concentrations are $0.3M$ and $0.002M$, respectively.

SOLUTION

By reference to Example 11.1, the standard electrode potential of Cu-Zn cell $= E^o = 1.10$ V

$$[M_1^{n+}]=\left[Zn^{2+}\right]\ =0.3M; \quad [M_2^{n+}]=[Cu^{2+}]=0.002\,M, \qquad\qquad n=2$$

By using Equation 11.5,

$$E=E^o-\frac{0.0592}{n}\log\frac{M_1^{n+}}{M_2^{n+}}=1.10-\frac{0.0592}{2}\log\frac{0.3}{0.002}$$

$$E=1.10-(0.0296\ x\ 2.176)=1.10-0.064\ =\ 1.036\ V$$

The corrosion potential of the specified Cu-Zn cell $= 1.036$ V.

EXAMPLE 11.4 COMPUTING THE CELL EMF WHEN N AND Q NEED TO BE DETERMINED

A zinc electrode is immersed in an acidic 0.75 M Zn^{2+} solution which is connected by a salt bridge to a 1.12 M Ag^+ solution containing a silver electrode. Compute EMF of the cell at 25°C.

Solution

$$[Zn^{2+}] = 0.75 \text{ M}, [Ag^+] = 1.12 \text{ M}$$

By reference to Table 11.1, $E_{ox}^o = E^o(Zn^{2+}\,/\,Zn)=-0.76\,\text{V}$, $E_{red}^o =E^o(Ag^+\,/\,Ag)= 0.8\,\text{V}$. In order to use Nernst equation, we need to know E^o, n, and Q.

$$E^o = E_{red}^0 - E_{ox}^o =0.8-(-0.76)=1.56\,\text{V}$$

Since EMF is positive, the reaction proceeds spontaneously with Zn as anode and Ag as cathode.

The balanced chemical equation for the cell reaction can be obtained by adding half reactions:

$Zn_s \rightarrow Zn^{2+} + 2e\text{-}$
$2Ag^+ + 2e\text{-} \rightarrow 2Ag$

___

$Zn_s + 2Ag^+ \rightarrow Zn^2 + 2Ag$

Since 2 electrons are transferred, $n = 2$
From the balanced chemical equation, $Q=\dfrac{[Zn^{2+}]}{[Ag^+]^2}=\dfrac{0.75}{1.12^2}=\dfrac{0.75}{1.254}=\ 0.598$

$$E = E° - \frac{0.0592}{n} \log Q = 1.56 - \frac{0.0592}{2} \log 0.598 = 1.5622 \text{ V}$$

The EMF of the cell = 1.5622 V.

EXAMPLE 11.5 CALCULATING THE CORROSION PENETRATION RATE AND PREDICTING ITS ACCEPTANCE

A (10 mm x 20 mm) area of a mild steel component was exposed to a corrosive environment for 20 days. The uniform corrosion resulted in a weight loss of 20 mg. Calculate the corrosion penetration rate. Is the CPR value acceptable for a general application?

SOLUTION

S = 10 mm x 20 mm = 1 cm x 2 cm = 2 cm²; W = 20 mg; t = 20 x 24 = 480 h;
ρ = 7.8 g/cm³

By using Equation 11.6,

$$CPR = \frac{87.6\,W}{\rho\,S\,t} = \frac{87.6 \times 20}{7.8 \times 2 \times 480} = 0.2339 \text{ mm/yr}$$

Since the CPR value is less than 0.5 mm/yr, it is acceptable for general applications.

EXAMPLE 11.6 CALCULATING THE CORROSION RATE IN GRAMS PER SECOND

An electroplating process involves dissolution (corrosion) of a copper anode and electroplating a steel cathode by passing a current of 13 A through a copper sulfate solution. The atomic mass of copper is 63.5 g/mol. Calculate the corrosion rate.

SOLUTION

F = 96,500 Coul/mol; M = 63.5 g/mol; I = 13 A; corrosion rate in g/s = ?

For the corrosion of copper at the anode, we can write the following half reaction:

$$Cu_{(s)} \rightarrow Cu^{2+}_{(aq)} + 2e-$$

Thus, the number of electrons given up by an atom of the anodic metal = n = 2
By using Equation 11.7,

$$\text{Corrosion rate (in g/s)} = \frac{I\,M}{n\,F} = \frac{13 \times 63.5}{2 \times 96,500} = 0.00427 \text{ g/s}$$

The corrosion rate = 4.27 mg/s.

EXAMPLE 11.7 CALCULATING THE CORROSION TIME

By reference to the data in Example 11.6, calculate the time required for the galvanic corrosion (or electroplating) if the weight loss of the anodic copper is 9 g.

SOLUTION

$M = 63.5$ g/mol; $I = 13$ A; $w = 9$ g, $n = 2$, $F = 96{,}500$ Coul/mol

By using Equation 11.8,

$$t = \frac{wnF}{IM} = \frac{9 \times 2 \times 96{,}500}{13 \times 63.5} = 2{,}105\,\text{s} = 35\,\text{min}$$

The corrosion or electroplating time = 35 min.

EXAMPLE 11.8 COMPUTING THE RATE OF CORROSION WHEN THE CURRENT DENSITY IS KNOWN

A metal assembly is designed by using a copper-zinc couple. If the current density at the copper cathode is 0.08 A/cm², compute the weight loss of zinc per hour if the cathode area is 140 cm².

SOLUTION

$J = 0.08$ A/cm² $= 800$ A/m², $S = 140$ cm² $= 0.014$ m², $M_{Zn} = 65.4$ g/mol, $w_{Zn}/t = ?$

By using Equation 11.9,

$$I_{Cu} = J_{Cu}\,S_{Cu} = 800 \times 0.014 = 11.2\,\text{A}$$

By using Equation 11.7,

$$\text{Corrosion rate (in g/s)} = \frac{w_{Zn}}{t} = \frac{I_{Cu}\,M_{Zn}}{nF} = \frac{11.2 \times 65.4}{2 \times 96{,}500} = 0.0038 \text{ g/s}$$

The rate of dissolution of zinc or the rate of weight loss of zinc = 0.0038 g/s = 13.66 g/h.

EXAMPLE 11.9 COMPUTING THE *LINEAR KINETIC CONSTANT* FOR THE OXIDATION OF A METAL

The weights of a metallic component (area of exposure = 10 cm x 8 cm) before and after oxidation (corrosion) were measured to be 80 and 82 g, respectively. The time of exposure to oxidizing atmosphere is 300 h; and the oxide film is porous. Compute the *kinetic constant* for the oxidation of the metal.

SOLUTION

Since the oxide film is porous, we have to compute the *linear kinetic constant*.
Area of exposure = 10 cm x 8 cm = 80 cm²; t = 300 h = 300 x 3,600 s = 1,080,000 s
Weight gain in the metallic component = 82–80 = 2 g

$$\Delta m = \frac{2\,g}{80\,cm^2} = 0.025 \; g/cm^2 = 0.25\,kg/m^2$$

By rearranging the terms in Equation 11.13,

$$k_l = \frac{\Delta}{t} = \frac{0.25}{1080000} = 2.31 \times 10^{-7} \; kg/m^2 \times s$$

The *linear kinetic constant* for the oxidation of the metal = 2.31 x 10⁻⁷ kg/m²xs.

EXAMPLE 11.10 COMPUTING THE WEIGHT GAIN FOR OXIDATION WITH PARABOLIC KINETIC BEHAVIOR

Magnesium metal is found to oxidize at 500°C with a *parabolic kinetic constant*
of 3 x 10⁻¹⁰ kg/m⁴ ×s. Calculate the weight gain in a magnesium component
(12 cm x 7 cm) after 800 hours.

SOLUTION

k_p = 3 x 10⁻¹⁰ kg/m⁴xs; t = 800 h = 800 x 3600 = 288 x 10⁴ s;

Exposed area = 0.12 m x 0.07 m = 0.0084 m²
By using a modified form of Equation 11.15,

$$\Delta m = \sqrt{k_p t} = \sqrt{3 \times 10^{-10} \times 288 \times 10^4} = \sqrt{3 \times 10^{-6} \times 288} = 29.4 \; x \; 10^{-3} kg/m^2$$

Weight gain $= \dfrac{\Delta}{exposed\,area} = \dfrac{29.4 \; x10^{-3}}{0.0084} = 3.5 \; kg$

EXAMPLE 11.11 COMPUTING THE THICKNESS OF THE OXIDE FILM

The atomic weight of magnesium is 24.3 g/mol and that of oxygen is 16 g/mol. The
density of magnesium oxide is 3.55 g/cm³. By using the data in Example 11.10, compute the thickness of the magnesium oxide film.

SOLUTION

ρ = 3.55 g/cm³ = 3,550 kg/m³; Δm = 29.4 x 10⁻³ kg/m²

Mol. Wt. of the oxide (MgO) = 24.3 + 16 = 40.3 g/mol = 40.3 kg/kmol
By using Equation 11.17,

$$x = \frac{(Mol.\ wt.\ of\ oxide)}{(At.\ wt.\ of\ oxygen)} \times \frac{\Delta m}{\rho} = \frac{40.3}{16} \times \frac{29.4 \times 10^{-3}}{3550} = 2.1 \times 10^{-5}\ m$$

The thickness of the oxide film = 2.1 x 10^{-5} m = 21 μm.

EXAMPLE 11.12 COMPUTING FAVORABLE AND UNFAVORABLE AREAS FOR CORROSION PROTECTION

A metal assembly is designed by using a copper-zinc couple. If the current density at the copper cathode is 0.08 A/cm², calculate the weight loss of zinc per hour if (a) the copper cathode area is 140 cm² and the zinc anode area is 1.3 cm² and (b) the copper cathode area is 1.3 cm² and the zinc anode area is 140 cm². Which assembly design is good for corrosion protection? (a) or (b)? What conclusion do you draw based on the calculations?

SOLUTION

(a)   J = 0.08 A/cm² = 800 A/m², S = 140 cm² = 0.014 m², M_{Zn} = 65.4 g/mol,
       w_{Zn}/t = ?
By using Equation 11.9,

$$I_{Cu} = J_{Cu} S_{Cu} = 800 \times 0.014 = 11.2\ A$$

By using Equation 11.7,

$$\text{Corrosion rate (in g/s)} = \frac{w_{Zn}}{t} = \frac{I_{Cu} M_{Zn}}{nF} = \frac{0.104 \times 65.4}{2 \times 96,500} = 3.5 \times 10^{-5}\ g/s$$

The rate of dissolution of zinc or the rate of weight loss of zinc = 0.0038 g/s = 13.66 g/h.

(b)   J = 0.08 A/cm² = 800 A/m², S = 1.3 cm² = 0.013 m², M_{Zn} = 65.4 g/mol,
       w_{Zn}/t = ?
By using Equation 11.9,

$$I_{Cu} = J_{Cu} S_{Cu} = 800 \times 0.00013 = 0.104\ A$$

By using Equation 11.7,

$$\text{Corrosion rate (in g/s)} = \frac{w_{Zn}}{t} = \frac{I_{Cu} M_{Zn}}{nF} = \frac{11.2 \times 65.4}{2 \times 96,500} = 0.0038\ g/s$$

The rate of dissolution of zinc or the rate of weight loss of zinc = 3.5 x 10^{-5} g/s = 0.127 g/h.

Since the corrosion rate is significantly lower in case (b), it is good for protection of corrosion. We conclude that the rate of corrosion of zinc is reduced significantly when the copper cathode area is much smaller than the zinc anode. It means that for corrosion protection, the unfavorable area (large cathode area) should be avoided.

EXAMPLE 11.13 CALCULATING THE CATHODE CURRENT EFFICIENCY

A current of 10 A passes for 1.5 h for electroplating nickel on a steel work-piece. Experimentally measured results shows that 15 g nickel is actually deposited on the cathode. Calculate (a) the *theoretical weight of the deposit*, and (b) cathode current efficiency.

SOLUTION

(a) $I = 10$ A, t = 1.5 h = 5400 s, $M = 58.7$ g/mol; $n = 2$, F = 96,500 Coul/equiv.

Electric Charge = $Q = It = 10$ x 5,400 = 54,000 Coul

By using Equation 11.19,

$$W_{th} = \frac{QM}{Fn} = \frac{54000 \times 58.7}{2 \times 96500} = 16.4 \, g$$

The theoretical weight of the nickel deposit = 16.4 g.

(b) $W = 15$ g, $W_{th} = 16.4$ g

By using Equation 11.18,

$$Eff_{cathode} = \frac{W}{W_{th}} = \frac{15}{16.4} = 0.91$$
$$\% \, Eff_{cathode} = \frac{W}{W_{th}} \times 100 = 0.91 \times 100 = 91$$

Cathode current efficiency = 91%

QUESTIONS AND PROBLEMS

11.1. Which of the following statements are true or false?
(a) Almost all corrosion processes are electrochemical in nature (T/F).
(b) *HE* results in ductile failure of the metal (T/F).
(c) Cathode of a galvanic cell releases electrons (T/F).
(d) Anode always corrodes while cathode does not (T/F).
(e) Pure metals are more susceptible to corrosion than alloys (T/F).
(f) IG corrosion involves corrosion of grains (T/F).
(g) SCC requires presence of both stress and corrosive environment and nothing else (T/F).

(h) Austenitic stainless steels should contain low carbon for corrosion resistance (T/F).

(i) One of the reasons of *LME* is Hg environment (T/F).

(j) CP requires supply of free electrons to the metal structure to be protected (T/F).

(k) In metal assembly, the two metals should be close to each other in galvanic series (T/F).

11.2. Corrosion is a threat to society. Justify the statement.

11.3. Explain the working principle of galvanic corrosion with the aid of a sketch.

11.4. Differentiate between the following terms:

(a) standard electrode potential and corrosion potential, (b) HE and LME.

11.5. List (partially) the metals in the galvanic series and highlight its industrial application.

11.6. Diagrammatically illustrate the linear and parabolic oxidation kinetic behaviors. On what factor does the oxidation kinetics depend?

11.7. Draw the classification chart showing various forms of corrosion. Explain SCC.

11.8. Two metals (in plate form), quite apart from each other in galvanic series, are required to be assembled by bolt joint. Draw the sketch showing the design of the metal assembly.

Explain how your design protects corrosion (identifying the type of corrosion involved).

11.9. What is meant by CP? What is the function of CP? Explain CP by use of galvanic anode with the aid of a sketch.

11.10. Illustrate the term sensitization of austenitic stainless steel with the aid of a sketch.

Design a heat treatment to avoid IG corrosion in a sensitized steel.

P11.11. By reference to the standard electrode potential data, calculate the EMF of the standard Pt-Zn cell. Would the electrochemical reaction occur spontaneously?

P11.12. Consider a Pt-K galvanic cell. (a) At which electrode does oxidation occur? (b) At which electrode does reduction occur? (c) Which electrode is the anode? (d) Which electrode is the cathode? (e) Which electrode corrodes? (f) Write the spontaneous overall cell reaction. (g) Calculate the EMF of the standard cell.

P11.13. An aluminum electrode is immersed in an acidic 0.68 M Al^{3+} solution which is connected by a salt bridge to a 1.0 M Ag^+ solution containing a silver electrode.

Compute EMF of the cell at 25°C. Given: $E^o(Al^{3+}/Al) = -1.66$ V, $E^o(Ag^+/Ag) = 0.8$ V

P11.14. A (15 mm x 23 mm) area of a mild steel component was exposed to a corrosive environment for 30 days. The uniform corrosion resulted in a weight loss of 25 mg. Calculate the corrosion penetration rate.

P11.15. An electroplating process involves dissolution (corrosion) of a silver anode and electroplating a steel spoon by passing a current of 15 A through a suitable electrolyte. The atomic mass of silver is 107.87 g/mol. (a) Identify the used electrolyte, (b) calculate the corrosion rate.

P11.16. A metal assembly is designed by using a platinum-iron couple. If the current density at the platinum cathode is 0.1 A/cm^2, compute the weight loss of iron per hour if the cathode area is 140 cm^2. The atomic mass of iron is 55.8 g/mol.

P11.17. By using the data in P11.15, calculate the time required for the galvanic corrosion (or electroplating) if the weight loss of the anodic silver is 7 g.

P11.18. Magnesium metal is found to oxidize at 500°C with a *parabolic kinetic constant* of 3×10^{-10} kg/m × s Calculate the weight gain in a magnesium component (9 cm x 6 cm) after 1,000 hours.

P11.19. The weights of a metallic component (area of exposure = 14 cm x 10 cm) before and after oxidation were measured to be 82 and 85 g, respectively. The time of exposure to oxidizing atmosphere is 400 h; and the oxide film is porous. Compute the *kinetic constant* for the oxidation of the metal.

P11.20. A metal assembly is designed by using a silver-sodium couple. If the current density at the silver cathode is 0.07 A/cm^2, calculate the weight loss of sodium per hour if (a) the silver cathode area is 120 cm^2 and the sodium anode area is 1.1 cm^2 and (b) the silver cathode area is 1.1 cm^2 and the sodium anode area is 120 cm^2. Which assembly design is good for corrosion protection? (a) or (b)? What conclusion do you draw.

P11.21. The atomic weight of potassium is 39.1 g/mol and that of oxygen is 16 g/mol. The density of potassium oxide is 2.35 g/cm^3. Compute the thickness of the potassium oxide film if an oxidation phenomenon results in weight gain of 40 x 10^{-3} kg/m^2.

P11.22. A current of 12 A passes for 2 h for electroplating chromium on a steel work-piece. Experimental results shows that 18 g chromium is actually deposited on the cathode. Calculate (a) the *theoretical weight of the deposit*, and (b) cathode current efficiency.

REFERENCES

Ashby, M., Shercliff, H. & Cebon, D. (2010) *Materials: Engineering, Science, Processing, and Design*. Butterworth-Heinemann, Elsevier, Oxford and Burlington.

Francis, R. (2001) *Galvanic Corrosion: A Practical Guide for Engineers*. NACE International, Houston.

Hertzberg, R.W. (1996) *Deformation and Fracture Mechanics of Engineering Materials*. John Wiley & Sons, Inc, Hoboken, NJ.

Huda, Z. (2017) *Materials Processing for Engineering Manufacture*. Trans Tech Publications, Switzerland.

Huda, Z., Bulpett, R. & Lee, K.Y. (2010) *Design Against Fracture and Failure*. Trans Tech Publications, Switzerland.

Koch, G.H., Brongers, M.P.H., Thompson, N.G., Virmani, Y.P. & Payer, J.H. (2002) *A Report on: Corrosion Cost and Preventive Strategies in the United States*. National Academy of Sciences, Washington.

Roberge, P.R. (2008) *Corrosion Engineering: Principles and Practice*. McGraw Hill Publishers, New York.

Talbot, D.E.J. & Talbot, J.D.R. (2018) *Corrosion Science and Technology*. 3nd ed. CRC Press, Boca Raton, FL.

Wahl, D. (2005) A short history of electrochemistry. *Galvanotechnik*, 96(8), 1820–1828.

12 Ferrous Alloys

12.1 FERROUS ALLOYS—CLASSIFICATION AND DESIGNATION

Ferrous alloys are the metallic materials that contain a large percentage of iron. Examples of ferrous alloys include: carbon steels, alloy steels, cast irons, and wrought iron. In general, ferrous metals/alloys possess good strength and durability. In particular, alloy steels are the engineering alloys that are noted for their excellent strength and toughness; these steels find wide applications in automotive and aerospace components, ship structures, rails, pressure vessels, boilers, cutting tools, and the like.

The classification of·ferrous alloys is presented in Figure 12.1; which shows that *ferrous alloys* may be classified into three main groups: (a) plain carbon steels, (b) alloy steels, and (c) cast irons. Additionally, wrought iron is also a ferrous metal. The three categories of ferrous alloys are explained in the following sections.

The distinctive point between "steels" and "cast irons" is 2.1 wt% C; a ferrous alloy containing greater than 2.1 wt% C is called cast iron whereas ferrous alloys containing less than 2 wt% C are called steels. Steels are generally designated based on their AISI (American Iron and Steel Institute) and SAE (Society of Automotive Engineers) numbering systems. These designation systems generally use four-digit numbers; the first two digits indicate the major alloying elements, and the last two digits refer to the wt% C in the steel. The digit series 1xxx indicate carbon steels. The 10xx series refers to carbon steels containing 1.00% Mn maximum; the 11xx series represent resulfurized carbon steels; 12xx series refer to resulfurized and rephosphorized carbon steels; and 15xx series indicate non-resulfurized high-manganese (up-to 1.65 wt%) carbon steels which are produced for applications requiring good machinability. For example, an AISI 1050 steel is a plain-carbon steel with 0.50 wt% C. The AISI-SAE system classifies alloy steels by using first digit for a specific alloying element (see section 12.3).

12.2 PLAIN CARBON STEELS

12.2.1 Classification and Applications

Plain carbon steels, also called *carbon steels*, are the steels whose properties mainly depend on their carbon content, and which do not contain more than 0.5% wt% Si and 1.5% wt% Mn. Steel is made by refining pig iron by oxidation of impurities in a steelmaking furnace (Turkdogan, 1996). The carbon content in plain carbon steels varies from 0.05% to 1.5%. Depending upon their carbon content, carbon steels are divided into the following four types: (a) mild steels (MS), (b) low-carbon steels, (c) medium-carbon steels, and (d) high-carbon steels.

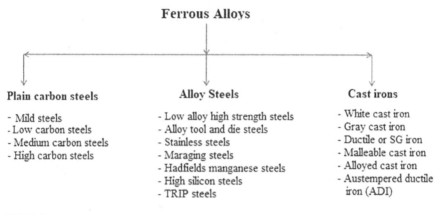

FIGURE 12.1   Classification of ferrous alloys.

Mild Steels. A *mild steel* is a very low-carbon steel that contains up to 0.05 wt% C. Mild steels are ductile and have properties similar to iron itself. Owing to their low carbon content, they cannot be modified by heat treatment. They are cheap, and find applications in noncritical components and general paneling and fabrication work.

Low-Carbon Steels. Low-carbon steels contain carbon in the range of 0.05–0.2 wt%. Examples of low-carbon steels in the AISI (American Iron and Steel Institute) standard include: AISI 1006 (0.06% C steel), AISI 1009, AISI 1020 steels, and the like. Low-carbon steels cannot be effectively heat treated; consequently there are no heat affected zones (HAZ) in welding. They are used in applications requiring high ductility, such as screws, nails, paper-clip wires, etc.

Medium-Carbon Steels. Medium-carbon steels contain carbon in the range of 0.2–0.8% wt%. Examples include: AISI 1030, AISI 1055, AISI 1077, and the like. Hardenability of medium-carbon steels increase with increasing carbon content. They are designed both for general purpose applications and stressed applications (such as gears, pylons, pipelines, etc.).

High-Carbon Steels. High-carbon steels contain from 0.8 to 1.5% wt% C. They are highly sensitive to heat treatments. They find applications requiring high hardness (e.g. files, drills, etc.).

12.2.2   MICROSTRUCTURES OF CARBON STEELS

The pearlitic microstructure of eutectoid steel (a medium-carbon steel) has well been discussed and analyzed in Chapter 7 (see Figure 7.8). The microstructures of typical samples of mild steel, low-carbon steel, and high-carbon steel are shown in Figures 12.2(a, b, and c), respectively.

The microstructural phases and microconstituents of carbon steels can be deduced from the steel portion of the iron-carbon phase diagram (Figure 12.3). The phases and microconstituents in the four types of steels are listed in Table 12.1. According

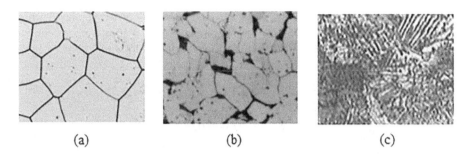

FIGURE 12.2 The microstructures of typical samples of mild steels (a), low-carbon steel (b), and high-carbon steel (c); (see Table 12.1).

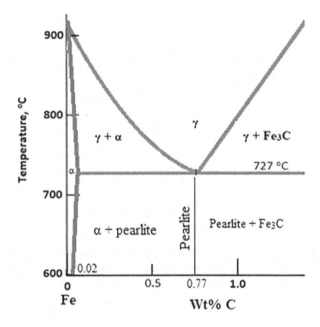

FIGURE 12.3 Steel portion of the iron-carbon phase diagram.

TABLE 12.1

The Phases and Microconstituents in the Four Types of Steels

Ferrous alloy	Mild steel	Low-carbon steel	Eutectoid steel	High-carbon steel
Microstructure	Ferrite (α)	α + Pearlite	Pearlite (P)	P + cementite

to Figure 12.3, when a low-carbon steel is heated above 727°C and slowly cooled to ambient temperature, the resulting microstructure will comprise of ferrite and pearlite (Figure 12.2b). Similar thermal treatment of a high-carbon steel will result in the formation of pearlite and cementite (Figure 12.2c).

By reference to Figure 12.3 and Table 12.1, the amounts of ferrite and pearlite in a low-carbon steel (Fe-C alloy), at a temperature just below the eutectoid, can be determined by using lever rule (Equations 6.15 and 6.16), as follows:

$$W_\alpha = \frac{C_{Pearlite} - C_0}{C_{Pearlite} - C_\alpha} = \frac{0.77 - C_0}{0.77 - 0.02} \tag{12.1}$$

$$W_{Pearlite} = \frac{C_0 - C_\alpha}{C_{Pearlite} - C_\alpha} = \frac{C_0 - 0.02}{0.77 - 0.02} \tag{12.2}$$

where $C_{pearlite}$ is the wt% carbon in pearlite (=0.77), C_o is the wt% carbon in the low-carbon steel, and C_α is the wt% carbon in ferrite (=0.02).

Similarly, the amounts of pearlite and cementite (Fe$_3$C) in a high-carbon steel (Fe-C ally), at a temperature just below the eutectoid, can be computed by:

$$W_{Pearlite} = \frac{C_{Fe3C} - C'_o}{C_{Fe3C} - C_{Pearlite}} = \frac{6.67 - C'_o}{6.67 - 0.77} \tag{12.3}$$

$$W_{Fe3C} = \frac{C'_o - C_{Pearlite}}{C_{Fe3C} - C_{Peralite}} = \frac{C'_o - 0.77}{6.67 - 0.77} \tag{12.4}$$

where C'_o is the wt% C in the Fe-C alloy (high-carbon steel). The significance of Equations 12.1–12.4 are illustrated in Examples 12.1–12.3.

12.2.3 The Effects of Carbon and Other Elements on the Mechanical Properties of Steel

In the preceding subsection, we learned that carbon is the main impurity in plain carbon steels. In addition to carbon, there are other impurities in steels; which include silicon (Si), manganese (Mn), sulfur (S), and phosphorous (P). The effects of these elements on the mechanical properties of steels are discussed in the following paragraphs.

Effects of Carbon. In general, an increase in the carbon content from 0.01 to 1.5% in steel results in an increase in its hardness and strength. However, an increase beyond 1.5% carbon causes an appreciable reduction in the ductility and toughness of the steel. The effects of carbon content on the tensile strength, % elongation, and toughness of annealed carbon steel is shown in Figure 12.4. It is evident in Figure 12.4 that the tensile strength (ρ_{ts}) of steel varies linearly as its carbon content increases up to 0.8 wt% C; beyond which the rise in ρ_{ts} follows a curve. However, impact toughness significantly decreases as carbon content increases from 0.2 to 0.7%. The optimum composition for a compromise in tensile strength without loss of impact toughness is in the range of 0.3–0.5 wt% C in the steel. The significance of Figure 12.4 is illustrated in Examples 12.4–12.5.

Effects of Silicon. The silicon (Si) content in the finished steel is usually in the range of 0.05–0.4 wt%. The addition of Si prevents porosity and blow holes in low-carbon steels. Si also removes the gases and oxides thereby making the steel tougher and stronger.

Effects of Sulfur. The sulfur (S) content in carbon steels is usually up to 0.05 wt%. In absence of manganese, sulfur occurs as iron sulfide (FeS) in steel; the *FeS*,

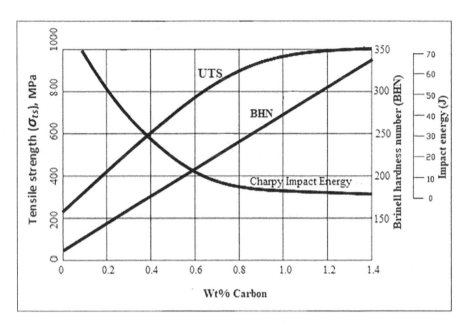

FIGURE 12.4   The effects of carbon on the mechanical properties of annealed carbon steel.

during hot working, may segregate along grain boundaries causing *hot shortness* (brittleness at a high temperature). However, in the presence of manganese (Mn), a sulfur content of up to 0.35% in steel is desirable since it improves machinability.

Effects of Manganese. Carbon steels may contain up to 1.2% Mn; the latter is added to steel for de-oxidizing. Manganese preferentially combines with sulfur to form MnS so as to decrease the harmful effect of sulfur in the steel. When used in low-carbon steels, Mn makes the metal stronger and tougher.

Effects of Phosphorous. Phosphorous (P) renders carbon steels brittle at ambient temperature thereby causing *cold shortness* in steel. Hence, P content is usually less than 0.04 wt% in steels. However, in low-carbon steels, P raises the yield strength and improves the resistance to uniform corrosion (rusting). For good ductility, the sum of carbon and phosphorus contents should be up to 0.25 wt% in the steel.

12.3   ALLOY STEELS

12.3.1   DEFINITION, AISI DESIGNATIONS, AND APPLICATIONS

An *alloy steel* is a type of steel to which one or more elements besides carbon have been intentionally added, for obtaining the desired mechanical properties. An *alloy steel* may contain up to 50 wt% of alloying elements. The main alloying elements used in alloy steels include: silicon (Si), manganese (Mn), nickel (Ni), chromium (Cr), molybdenum (Mo), tungsten (W), vanadium (V), cobalt (Co), boron (B), copper (Cu), aluminum (Al), titanium (Ti), and niobium (Nb).

The *AISI* system classifies alloy steels by using the first digit as follows: 2 for Ni steels, 3 for Ni-Cr steels, 4 for Mo steels, 5 for Cr steels, 6 for Cr-V steels, 7 for W-Cr steels, and 9 for Si-Mn steels. For example, a typical alloy steel with AISI-SAE

TABLE 12.2

AISI Desig. No. of Steels (Oil Quenched and Tempered) and Applications

Steel AISI desig.	4063	4140	4340	5150	6150	410 SS*	440A SS
Application	Springs, hand tools	Shafts, gears, forgings	Bushings, gears, aircraft tubes	Heavy duty shafts, gears	Pistons, shafts, gears	Gun barrel, jet- engine parts	Bearings, surgical tools, cutlery

*SS = stainless steel

designation number 4140 has the following composition ranges (wt%): 0.38–0.43C, 0.15–0.25Mo, 0.75–1.00Mn, 0.15–0.35 Si, 0.8–1.1Cr, balance: Fe (Degarmo *et al.*, 2003). The application of an alloy steel strongly depends on the nature and the amount of alloying elements and the treatment given to the steel (see Table 12.2).

12.3.2 Effects of Alloying Element on Steel

The microstructure and mechanical properties of alloy steels strongly depend on the type and amount of alloying elements in the steel. The effects of the addition of alloying elements on alloy steels are listed and described in the following paragraphs:

A. Alloying elements (e.g. Ni, Nb, V, etc.) generally result in an increase in hardness and strength of solid solution (ferrite) without decreasing the ductility. They also alter the eutectoid composition and eutectoid temperature.

B. The addition of an alloying element reduces the critical cooling rate (CCR) (except Co) by making the transformation to the martensite slower (see Chapter 7, section 7.6) (see also Table 12.3). Thus, alloy steels may be hardened at a lower CCR by an oil or even air quench thereby reducing the risk of cracking or distortion that may result from a rapid water quench.

C. An alloying element either increases or decreases α to γ transition temperature. Elements are either α stabilizer (e.g. *Cr, W, V, Mo, Al*, Si, etc.) or γ stabilizer (e.g. Ni, Mn, Co, Cu, and C).

D. Some elements (e.g. Cr, W, V, Mo, Nb, Mn) form hard, stable carbides (e.g. Cr_7C_3, W_2C).

E. Some elements (e.g. Si, Ni, Co, Al, etc.) cause graphitization of Fe_3C. This is why these elements are not added to high-carbon steels unless counteracted by a carbide former.

F. Alloying elements confer the characteristic property of the element on the steel. For example, Cr confers corrosion resistance when wt% Cr > 12, rendering it a stainless steel.

G. An alloying element may either promote or inhibit the rate of grain growth. For example, Cr promotes grain growth rate thereby reducing the strength of steel, if overheated. On the other hand, grain refining elements (e.g. V, Ti, Nb, Al, Ni, etc.) slow down grain growth rates and so they are used in case hardening steels.

TABLE 12.3
Effects of Alloying Element on CCR of Cteel

Wt% carbon	Wt% alloying element	CCR (ºC/s)
0.42	0.55 Mn	550
0.40	1.60 Mn	500
0.42	1.12 Ni	450
0.40	4.80 Ni	85
0.38	2.64 Cr	10

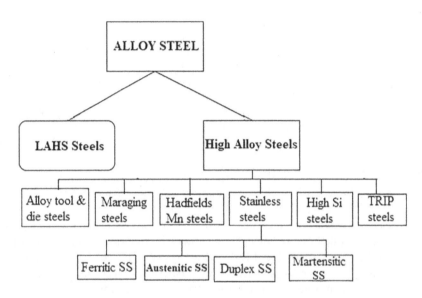

FIGURE 12.5 Classification of alloy steels (LAHS = low-alloy high strength, SS = stainless steels).

12.3.3 CLASSIFICATION OF ALLOY STEELS

Depending on the of alloying additions, alloy steels may be broadly classified into: (1) low-alloy high strength steels, and (2) high-alloy steels; the former contain up to 4% alloying additions whereas the latter contain from 4 to 45% alloying additions. Figure 12.5 presents the classification of alloy steels; the various types of alloy steels are discussed in the following subsections.

12.3.4 LOW ALLOY HIGH STRENGTH (LAHS) STEELS

High-strength low-alloy (HSLA) steels are the low-carbon steels strengthened by the addition of alloying elements in the range of 1–4 wt% (with up to 1.65 wt% Mn), and sometimes strengthened by special rolling and heat-treatment techniques. A typical LAHS steel contains about 0.15% C, 1.65% Mn and P and S (each < 0.035%), and small additions of Cu, Ni, Cr, Mo, Nb, and V.

Nickel (Ni) contributes the most to improve steel's tensile strength since it strengthens ferrite solid solutions, but it also causes graphitization of carbides. This is why Ni is usually accompanied by strong carbide formers (e.g. Cr, V, etc.); the latter also increase hardenability. Air hardening LAHS steel contain around 4.25% Ni and 1.25% Cr. An additions of 0.3% Mo renders "nickel-chrome-moly" steels suitable for applications in axles, shafts, gears, connecting rods, and the like. LAHS steels are notable for their use in fabricating large structures such as truck bodies, construction equipment, off-highway vehicles, mining equipment, and other heavy-duty vehicles. LAHS steel sheets and plates also find applications in chassis components, buckets, grader blades, and structural members outside the vehicle body.

12.3.5 TOOL AND DIE STEELS

Tool and die steels are alloy steels having high hardness, toughness, hardenability, and the resistance to wear and corrosion. They are designed by incorporating carbides that are harder than cementite, while retaining strength and toughness. Alloying elements include Cr, W, Mo, and V; which are carbide formers and stabilize ferrite and martensite. A typical composition of tool steel is Fe-0.8%C-18%W-4%Cr-1%V. Such steels are heat treatable (see Chapter 15).

12.3.6 HADFIELD MANGANESE STEELS

Hadfield manganese steels are high-Mn alloy steels that contain Mn contents in the range of 12%–14% Mn and 1% C. These steels are austenitic at all temperatures and therefore non-magnetic. Hadfield steels have a unique property in that when the surface is abraded or deformed, their surface hardness is significantly increased, while the core still remains tough. For this reason these steels are used in pneumatic drill bits, excavator bucket teeth, rock crusher jaws, ball mill linings, and railway points and switches. When water quenched from 1050°C to retain carbon in solution, the soft core has a tensile strength of 849 MPa, a ductility of 40%, and a Brinell hardness of 200, but after abrasion the surface hardness increases to 550 BHN.

12.3.7 MARAGING STEELS

Maraging steels are the high-alloy steels containing high Ni, Co, and Mo. A typical composition of maraging steel is: Fe-18%Ni-8%Co-5%Mo-0.4%Ti-0.1%Al and up to 0.05% C. Heat treatment involves solution treatment followed by quenching of the austenite to give a BCC martensitic structure (see Chapter 15). Aging heat treatments can produces finely dispersed precipitates of complex intermetallics (e.g. $TiNi_3$) resulting in high tensile strengths of around 2,000 MPa. These steels are relatively tough; with good corrosion resistance and good weldability. Maraging steels find applications in aircraft undercarriage components, wing fittings, extrusion-press rams, and the like.

12.3.8 STAINLESS STEELS

Stainless steels (SS) are the corrosion resistant steels containing more than 12% Cr. However, Cr promotes grain growth during thermo-mechanical processing; which is why grain-refining elements (chiefly Ni) are added to SS. Stainless steels can be classified into three types: (a) ferritic SS, (b) martensitic SS, (c) austenitic SS, and (d) duplex SS.

Ferritic stainless steels (FSS). FSS are the SS containing 12% to 25% Cr and up to 0.1% C. A typical composition is Fe-18% Cr-8% Ni-0.1%C, called *18-8 stainless steel.* FSS are ferritic up to the melting point; they cannot be quench hardened to produce martensite. They can be work hardened but are only ductile above the ductile-brittle transition temperature found in BCC metals. Prolonged overheating can cause the precipitation of an embrittling sigma (σ) phase.

Martensitic stainless steels (MSS). MSS are the SS containing 12% to 25% Cr and 0.1% to 1.5%C. The higher carbon content restores the α-to-γ transition temperature by making the γ loop larger. Thus, the steel can be heated into the austenite region and quenched to give a martensitic structure (Figure 12.6). Hardenability is generally high enough so that hardening can be achieved by air-cooling. Due to their high strength and wear resistance in combination with some corrosion resistance, *MSS* are suitable for applications in hydroelectric turbines, knives, cutlery, etc.

Austenitic stainless steels (ASS). ASS are the non-magnetic SS that contain 16 to 26% Cr and up to 35% Ni. Since Ni has a greater effect on the α-to-γ transition temperature, and this can be reduced to below room temperature, the austenitic FCC phase is retained. *ASS* is more ductile than other SS, and can be cold-worked to produce deep-drawn shapes used in chemical plants, kitchenware, and architectural work. The welding or heat treatment of ASS in the temperature range of 870 to 425°C can lead to the problem of "sensitization" of SS; the solution to this problem has been discussed in the preceding chapter (see section 11.6.8).

Duplex stainless steel (DSS). DSS are so named because they have a two-phase (duplex) microstructure consisting of ferrite and austenite in the stainless steel (see Figure 12.7). DSS are about twice as strong as ASS or FSS. They have significantly better impact toughness and ductility than FSS; however, they do not reach the excellent toughness values of ASS. For chloride pitting and crevice corrosion resistance, the Cr, Mo, and N contents of *DSS* are the most important. DSS show excellent resistance to stress-corrosion cracking (*SCC*).

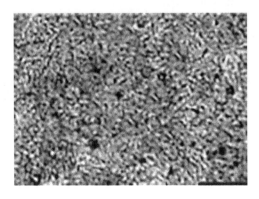

FIGURE 12.6   The microstructure of MSS.

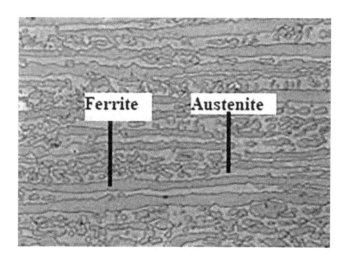

FIGURE 12.7   The two-phase microstructure of DSS.

12.3.9   HIGH-SILICON ELECTRICAL STEELS

When low-carbon steel is alloyed with silicon (usually up to 4 wt% Si), it helps to reduce eddy current losses; which is beneficial for electrical applications. Si electrical steels are widely used in transformer cores and electrical drive components; in such applications, silicon steels are generally specified and selected on the basis of allowable core loss in watts/kg. The performance of silicon electrical steels strongly depends on the processing technique applied to manufacture the steel. Depending on the processing technique, silicon electrical steels may be classified into: (a) non-grain oriented electrical steels (NGOES), (b) grain-oriented electrical steels (GOES), and (c) low-carbon electrical steels (LCES). *GOES* are produced by a complex processing technique starting from a hot-rolled coil of thick steel sheet that goes through trimming, annealing, de-scaling, and cleaning processes followed by cold rolling-annealing-cold rolling processes to obtain a sheet thickness of 0.20–0.35 mm (Mazurek, 2012). GOES possess magnetic properties that are strongly oriented with respect to the rolling direction (*RD*).

12.3.10   TRIP STEELS

Transformation induced plasticity (TRIP) steels are the high-alloy steels having high strength and enhanced formability. Owing to their excellent formability, TRIP steels can be used to produce more complicated parts than other high-strength steels, thus allowing the automotive engineer more freedom in parts design to optimize weight and structural performance. The excellent mechanical properties arise from a martensitic transformation of metastable retained austenite, induced by external stress and/or plastic deformation. The TRIP steels possess a multi-phase microstructure, consisting typically of α-Fe (ferrite), bainite, and retained (metastable) γ-Fe (austenite).

12.3.11   MATHEMATICAL MODELS FOR ALLOY STEELMAKING (SECONDARY STEELMAKING)

12.3.11.1   Effects of Different Processes on the Steel-Holding Ladle Temperature

In traditional steelmaking and casting, liquid steel is tapped from a steelmaking furnace (e.g. basic oxygen furnace [BOF], electric arc furnace [EAF], or the like) into an open-topped cylindrical *ladle*, and then transferred to the casting bay where the steel is cast into either ingots or a final castings (Fruehan, 1998). The *ladle* is used not only for holding and transporting liquid steel, but also for de-slagging, re-slagging, alloy addition, heating, and the like. In order to ensure that the ladle arrives at the casting bay at the correct temperature, it is important to calculate the effects of different processes on the ladle temperature. During tapping, the steel temperature will decrease by around 60°C. For most alloy additions, each 1,000 kg added to the ladle results in an additional temperature drop of about 6°C. It is necessary to prevent the steel bath temperature falling below the liquidus temperature (i.e. the temperature at which the steel starts to solidify).

12.3.11.2   Mathematical Models for Liquidus Temperature for Alloy Steels

An alloy steel's liquidus temperature, T_{liq}, is strongly dependent on its composition; and it can be calculated from the following equations (World Steel Association, 2019):

For steels containing C < 0.5%,

$$
\begin{aligned}
T_{liq}\left(\text{in}^{\circ}\text{C}\right) &= 1537 - \left(73{\cdot}1\%\text{C}\right) - \left(4{\cdot}\%\text{Mn}\right) - \left(14{\cdot}\%\text{Si}\right) \\
&- \left(45{\cdot}\%\text{S}\right) - \left(30{\cdot}\%\text{P}\right) - \left(1.5{\cdot}\%\text{Cr}\right) - 2.5{\cdot}\%\text{Al} - \left(3.5{\cdot}\%\text{Ni}\right) \\
&- \left(4{\cdot}\%\text{V}\right) - \left(5{\cdot}\%\text{Mo}\right)
\end{aligned}
\tag{12.5}
$$

For steels with C > 0.5%,

$$
\begin{aligned}
T_{liq}\left(\text{in}^{\circ}\text{C}\right) &= 1531 - \left(61{\cdot}5\%\text{C}\right) - \left(4{\cdot}\%\text{Mn}\right) - \left(14{\cdot}\%\text{Si}\right) \\
&- \left(45{\cdot}\%\text{S}\right) - \left(30{\cdot}\%\text{P}\right) - \left(1.5{\cdot}\%\text{Cr}\right) - 2.5{\cdot}\%\text{Al} \\
&- \left(3.5{\cdot}\%\text{Ni}\right) - \left(4{\cdot}\%\text{V}\right) - \left(5{\cdot}\%\text{Mo}\right)
\end{aligned}
\tag{12.6}
$$

The significance of Equations 12.5–12.6 is illustrated in Examples 12.8–12.9.

12.3.11.3   Mathematical Models for Additions to Achieve the Aim Composition

Addition of Elemental Metal. If a pure element is added to the steel-holding ladle, the mass of the additive required ($m_{additive}$) can be calculated by (World Steel Association, 2019):

$$
m_{additive} = \frac{\Delta\% X \times m_{ladle}}{100\%}
\tag{12.7}
$$

where m_{ladle} is the mass of the molten steel in the ladle, kg, and $\Delta\%X$ is the required increase in $wt\% \, X$ i.e. $\Delta\%X = \% \, X_{aim} — \% \, X_{actual}$ (see Example 12.10).

Addition of Ferroalloys. It is economical to make additions through ferroalloys rather than by pure elements. In such cases, the wt% of the desired element in the ferroalloy must be considered. It is also important to include the "%recovery" (the wt% of the element that actually increases the liquid steel composition rather than being lost to the slag, etc.) in the calculation.

The mass of the ferroalloy required to added ($m_{Fe-alloy}$) can be calculated by:

$$m_{Fe-alloy} = \frac{100\% \times \Delta\%X \times m_{ladle}}{\%X \, in \, ferroalloy \times \% \, recovery \, X} \tag{12.8}$$

The significance of Equation 12.8 is illustrated in Example 12.11.

Pick-up of Carbon due to Ferroalloy Addition. Some high-carbon ferroalloys can cause an increase in the C contents of the alloy steel. In such cases, the amount of carbon pick up by steel ($\Delta\%$ C) can be computed by:

$$\Delta\%C = \frac{m_{ferroalloy} \times \% \, C \, in \, ferroalloy \times \% \, recovery \, of \, C}{100\% \times m_{ladle}} \tag{12.9}$$

The significance of Equation 12.9 is illustrated in Example 12.12.

12.4   CAST IRONS

12.4.1   Metallurgical Characteristics and Applications

Cast iron (CI) is a ferrous alloy that contains more than 2% carbon and 0.5% or more silicon. Cast irons are the least expensive of all engineering metallic materials. The design and manufacturing advantages of cast irons include: low tooling and production cost, ready availability, good machinability without burring, easy to cast into complex shapes, excellent wear resistance and high hardness (particularly white irons), and high damping ability. CI finds wide applications in automotive components (e.g. clutch plates, brake discs, cylinder blocks and heads, piston, liner, and the like) and other engineering components (e.g. gears, pipes, water pipes, flywheels, etc.).

The distinctive point between the two principal types of cast iron (gray CI and white CI) is the carbon equivalent (CE); which is numerically defined as:

$$CE \, (in \, wt\%) = C + \frac{Si + P}{3} \tag{12.10}$$

where C is wt% carbon, Si is wt% silicon, and P is wt% phosphorous. For optimum metallurgical characteristics, the CE value should be about 4.3 in the gray CI.

A low CE and a high cooling rate during solidification favors the formation of white CI whereas a high CE and a low cooling rate promotes gray CI (see Example 12.13). During solidification, the major proportion of the carbon precipitates in the form of either graphite or cementite. In gray CI, if the cooling rate through the eutectoid temperature is sufficiently slow, *graphitization* occurs involving the decomposition of the cementite (in the pearlite) into α-ferrite and graphite, according to the reaction:

$$Fe_3C \rightarrow 3Fe_{(\alpha)} + C_{(graphite)} \tag{12.11}$$

Besides slow cooling rate, graphite formation is also promoted by the presence of silicon (Si) in concentration exceeding 1 wt% (see Equation 12.10).

12.4.2 Types of Cast Irons

There are six types of cast iron: (1) white iron, (2) gray iron, (3) ductile or spheroidal graphitic (SG) iron, (4) malleable iron, (5) alloyed cast iron, and (6) austempered ductile iron (ADI). Each of these types of cast iron is explained with references to their microstructures and applications.

12.4.3 White Cast Iron

White cast iron is so named because when it is sheared, the fractured surface appears white due to the absence of graphite. A high cooling rate and a low CE favor the formation of white CI (see Equation 12.11). They are hard and brittle with poor machinability; however their compressive strength and wear resistance are good. White CI finds applications in pumps, grinding tools, conveying equipment, and the like.

12.4.4 Gray Cast Iron

Gray cast iron is so named because when it is sheared, the fractured surface appears gray due to the presence of graphite. Gray CIs are characterized with carbon as graphite flakes in their microstructures (see Figure 12.8). They possess good

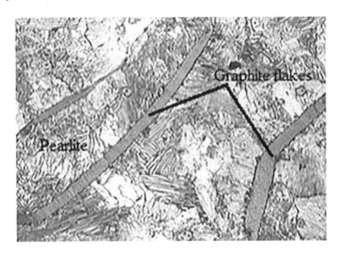

FIGURE 12.8 Microstructure of gray cast iron showing graphite flakes.

machinability and wear resistance as well as excellent damping capacity to absorb vibrational energy. These metallurgical characteristics render gray CI suitable for applications in pump housings, IC engines' cylindrical blocks, valve bodies, and the base structures for machines and heavy equipment that are exposed to vibrations.

12.4.5 DUCTILE OR NODULAR CAST IRON

Ductile cast iron, also called spheroidal graphitic (SG) iron or nodular cast iron, is so named because its microstructure shows graphite in the form of nodules/spheroids (see Figure 12.9). Ductile CI is basically gray CI into which small amounts of inoculants (magnesium, Mg, or cesium, Cs) have been added to nodulates the graphite. A typical composition of SG iron is as follows: Fe-3.3%C-2.2%Si-0.3% Mn-0.04%Mg-0.01%P-0.01% S. Ductile CIs possess moderately high tensile strength and high ductility; which renders them suitable for applications in pipes for water and sewerage lines.

The mechanical and thermal properties of ductile iron strongly depend on the number, size, and shape of graphite nodules. The nodules to be considered should be round or close to it. As the number of the nodule counts increases, the structure and properties of the SG iron become more uniform. The nodule count per unit area (N_A) can be computed by (Fras and Lopez, 2010):

$$N_A = \frac{(N_i + 0.5N_u + 1)M^2}{A} \tag{12.12}$$

where N_A is the nodule count, nod./mm²; N_i is the number of nodules inside a measured rectangle U; N_u is the number of nodules that intersect a U side excluding the corners; and A is the surface area of the rectangle U, mm²; and M is the magnification of the micrograph (see Example 12.14). In general, N_A values lie in the range of 100–500.

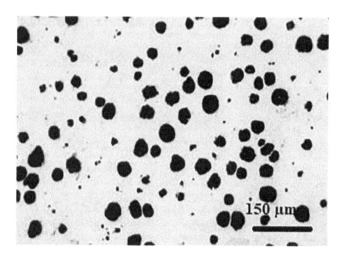

FIGURE 12.9 Microstructure of ductile CI showing graphite nodules.

The graphite nodule count/volume (N_V) can be computed by (Pedersen and Tiedje, 2008):

$$N_V = \sqrt{\frac{\pi}{6 f_g}} \left(\alpha N_A \right)^{3/2} \tag{12.13}$$

where N_V is the volume nodule count, nod/mm^3; f_g is the fraction of graphite (usually 0.1); α is a parameter whose value depends on the width of the size distribution of the nodules ($\alpha = 1$–1.25) (see Example 12.15).

12.4.6 MALLEABLE CAST IRON

Malleable cast iron (CI) is basically a white CI that has been heat treated to improve ductility without sacrificing its tensile strength. The heat treatment involves heating white CI at 870°C for a prolonged period and then slow cooling at a controlled rate. As a result, cementite decomposes to free nodules of graphite (see Equation 12.11). The microstructure of black-heart malleable CI shows ferrite and graphite free nodules (Figure 12.10).

There are three types of malleable CI: (a) white-heart CI, (b) black-heart CI, and (c) pearlitic CI (see Table 12.4). Malleable CIs find applications in thin-section castings, and in components that require good formability, machinability, and impact toughness.

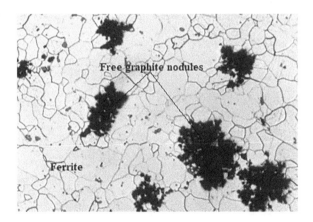

FIGURE 12.10 Microstructure of black-heart malleable iron.

TABLE 12.4

Microstructural Phases and Mechanical Properties of Malleable Cast Irons

Type of malleable iron/ phases/properties	White-heart malleable iron	Black-heart malleable iron	Pearlitic malleable iron
Microstructural phases	Ferrite (case) + pearlite (core)	Ferrite + graphite free nodules	pearlite
Tensile strength, MPa	250–400	290–340	400–450
% elongation	4–10%	6–12%	6–12%

12.4.7 ALLOYED CAST IRON

Alloyed cast irons are the CI alloyed with Ni, Cr, and Mo; the microstructure shows carbides with graphite flakes in a matrix of either pearlite or bainite. They find applications in mill rolls; which require a rich and deep layer of chill with a minimal reduction in hardness. Alloyed CIs having high hardness (where the matrix is bainitic instead of pearlitic) and are widely used in finishing stands of rolling mills.

12.4.8 AUSTEMPERED DUCTILE IRON

Austempered ductile irons (ADI) are the cast irons that have been heat treated by austenizing at 950°C followed by austempering at 350°C for 64 min to obtain bainitic structure. A typical composition of ADI is: Fe-3.52C-2.51Si-0.49Mn-0.15Mo-0.31Cu (wt%). ADIs possess excellent damping capacity (to reduce engine noise) and fatigue resistance for automotive applications.

12.5 CALCULATIONS—WORKED EXAMPLES ON FERROUS ALLOYS

EXAMPLE 12.1 CALCULATING THE AMOUNTS OF PHASE/MICROCONSTITUENT IN A LOW-CARBON STEEL

Calculate the amounts of phase/microconstituent in a low-carbon steel having composition of Fe-0.18 wt% C at a temperature just below the eutectoid.

SOLUTION

By using Equations 12.1 and 12.2,

$$W_\alpha = \frac{C_{Pearlite} - C_0}{C_{Pearlite} - C_\alpha} = \frac{0.77 - 0.18}{0.77 - 0.02} = \frac{0.60}{0.75} = 0.80 = 80\%$$

$$W_{Pearlite} = \frac{C_0 - C_\alpha}{C_{Pearlite} - C_\alpha} = \frac{0.18 - 0.02}{0.77 - 0.02} = \frac{0.16}{0.75} = 0.20 = 20\%$$

Hence, the steel is comprised of 80% ferrite and 20% pearlite.

EXAMPLE 12.2 CALCULATING THE AMOUNTS OF PHASE/MICROCONSTITUENT IN A HIGH-CARBON STEEL

Calculate the amounts of phase/microconstituent in a high-carbon steel (Fe-1.2 wt% C alloy) at a temperature just below the eutectoid.

SOLUTION

By using Equations 12.3 and 12.4,

$$W_{Pearlite} = \frac{C_{Fe3C} - C'_o}{C_{Fe3C} - C_{Pearlite}} = \frac{6.67 - 1.2}{6.67 - 0.77} = 0.927 = 92.7\%$$

$$W_{Fe3C} = \frac{C_o' - C_{Pearlite}}{C_{Fe3C} - C_{Peralite}} = \frac{1.2 - 0.77}{6.67 - 0.77} = 0.073 = 7.3\%$$

The Fe-C alloy (high-carbon steel) comprises of 92.7% pearlite and 7.3% cementite.

EXAMPLE 12.3 DETERMINING AISI NUMBER OF A CARBON STEEL WHEN WT% C IS UNKNOWN

The composition/grade of an unalloyed steel sheet used for making car bodies is unknown. A metallurgist heated a sample of the steel above 727°C to obtain austenite, held it for a sufficiently long time, and then slowly cooled it in a furnace to room temperature. The resulting microstructure comprises of 85% ferrite and 15% pearlite. What is the AISI-SAE designation number of the steel?

SOLUTION

Since unalloyed steel sheets used in car bodies are made of low-carbon steel, we can determine the composition of the steel by using either Equation 12.1 or Equation 12.2, as follows:

$$W_\alpha = \frac{0.77 - C_0}{0.77 - 0.02}$$

$$85\% = \frac{0.77 - C_0}{0.77 - 0.02}$$

$$C_0 = 0.13$$

The plain carbon steel contains 0.13 wt% C. Hence, the AISI-SAE number is AISI 1013.

EXAMPLE 12.4 DETERMINING THE MECHANICAL PROPERTIES OF STEELS WHEN WT% C IS KNOWN

By reference to Figure 12.4, determine the tensile strength (ρ_{ts}), Brinell hardness number (BHN), and Charpy impact energy of a plain carbon steel containing 1.0 wt% C. Is this steel suitable for manufacturing car bodies? Justify your answer.

SOLUTION

By reference to Figure 12.4, a carbon steel containing 1.0 wt% C possesses the following properties:

$\rho_{ts} = 970$ MPa, BHN = 680, and Impact energy = 6 J.

The steel is unsuitable for manufacturing car bodies which require high ductility and impact toughness. Unfortunately the impact energy of the steel (mentioned here) is too low.

EXAMPLE 12.5 ESTIMATING THE P_{TS} RANGES FOR THE FOUR TYPES OF CARBON STEELS

By using Figure 12.4, estimate the tensile strength (ρ_{ts}) ranges for the following types of annealed steels: (a) MS, (b) low-carbon steel, (c) medium-carbon steel, (d) high-carbon steel.

SOLUTION

(a) By reference to the subsection 12.2.1, MS (mild steel) contains 0.01–0.05 wt% C. By reference to Figure 12.4, the tensile strength (ρ_{ts}) range for mild steel is 210–230 MPa.

(b) Low-carbon steel contains carbon in the range of 0.05–0.2 wt%. By reference to Figure 12.4, the ρ_{ts} range for low-carbon steel is 250–400 MPa.

(c) Medium-carbon steel contains carbon in the range of 0.2–0.8% wt%. By reference to Figure 12.4, the ρ_{ts} range for medium-carbon steel is 250–400 MPa.

(d) High-carbon steel contains carbon in the range of 0.8–1.5% wt%. By reference to Figure 12.4, the ρ_{ts} range for high-carbon steel is 900–1,000 MPa.

EXAMPLE 12.6 ESTIMATING THE BHN RANGES FOR THE FOUR TYPES OF CARBON STEELS

By using Figure 12.4, estimate the Brinell hardness number (BHN) ranges for the following types of annealed steels: (a) MS, (b) low-carbon steel, (c) medium-carbon steel, (d) high-carbon steel.

SOLUTION

(a) By reference to Figure 12.4, the BHN range for mild steel is 40–60.

(b) By reference to Figure 12.4, the BHN range for low-carbon steel is 60–140.

(c) By reference to Figure 12.4, the BHN range for medium-carbon steel is 150–240.

(d) By reference to Figure 12.4, the BHN range for high-carbon steel is 240–340.

EXAMPLE 12.7 ESTIMATING IMPACT ENERGY RANGES FOR THE FOUR TYPES OF CARBON STEELS

By using Figure 12.4, estimate the impact energy ranges for the following types of annealed carbon steels: (a) MS, (b) low-carbon steel, (c) medium-carbon steel, (d) high-carbon steel.

SOLUTION

(a) By reference to Figure 12.4, the impact energy range for mild steel is 65–80 J.

(b) By reference to Figure 12.4, the impact energy range for low-carbon steel is 50–60 J.

(c) By reference to Figure 12.4, the impact energy range for medium-carbon steel is 8–50 J.

(d) By reference to Figure 12.4, the impact energy range for high-carbon steel is 2–7 J.

EXAMPLE 12.8 CALCULATING THE LIQUIDUS TEMPERATURE OF ALLOY STEEL CONTAINING C < 0.5%

A maraging steel contains the following weight percentages of alloying elements: 18%Ni-8%Co-5%Mo-0.4%TI-0.1%Al-0.05% C. Calculate the temperature at which the steel starts to solidify.

Solution

%C = 0.05, %Ni = 18, %Co = 8, %Mo = 5, %Ti = 0.4, %Al = 0.1
Since C < 0.5%, we must use the modified form of Equation 12.5 as follows:

$$T_{liq} \left(in°C \right) = 1537 - \left(73.1 \times \%C \right) - \left(2.5 \times \%Al \right) - \left(3.5 \times \%Ni \right) - \left(5 \times \%Mo \right)$$

$$= 1537 - \left(73.1 \times 0.05 \right) - \left(2.5 \times 0.1 \right) - \left(3.5 \times 18 \right) - \left(5 \times 5 \right)$$

$$= 1445°C$$

The temperature at which the steel starts to solidify or the liquidus temperature = 1,445°C.

EXAMPLE 12.9 CALCULATING THE LIQUIDUS TEMPERATURE OF ALLOY STEEL CONTAINING C > 0.5%

A tool steel (high-speed steel) contains the following weight percentages of alloying elements: 0.8%C-18%W-4%Cr-1%V. Calculate the temperature at which the steel starts to solidify.

Solution

%C = 0.8, %W = 18, %Cr = 4, %V = 1

Since C > 0.5%, we must use the modified form of Equation 12.6 as follows:

$$T_{liq} \left(in°C \right) = 1531 - \left(61.5 \times \%C \right) - \left(1.5 \times \%Cr \right) - \left(4 \times \%V \right)$$

$$= 1531 - \left(61.5 \times 0.8 \right) - \left(1.5 \times 4 \right) - \left(4 \times 1 \right)$$

$$= 1531 - 59.2 = 1472°C$$

The temperature at which the steel starts to solidify or the liquidus temperature = 1,472°C.

EXAMPLE 12.10 CALCULATING THE MASS OF AN ELEMENTAL METAL REQUIRED TO BE ADDED TO LADLE

A 20-tons ladle of steel actually contains 0.03% Ni. Calculate the mass of the elemental nickel (Ni) that must be added to achieve an aim composition of 1.2% Ni.

SOLUTION

m_{ladle} = 20 tons = 20,000 kg, $\%Ni_{actual}$ = 0.03%, $\%Ni_{aim}$ = 1.2%,
$\Delta\%Ni = \%Ni_{aim} - \%Ni_{actual}$ = 1.2% − 0.03% = 1.17%

By using Equation 12.7,

$$m_{additive} = \frac{\Delta\%Ni \times m_{ladle}}{100\%} = \frac{1.17\% \times 20,000}{100\%} = 234 \text{ kg}$$

The mass of the elemental nickel required to be added = 234 kg.

EXAMPLE 12.11 CALCULATING THE MASS OF THE FERROALLOY REQUIRED TO BE ADDED TO THE LADLE

An 8-tons ladle of steel contains 2% Cr at tap. Calculate the amount of ferrochrome required to be added to achieve a composition of 13% Cr. Ferrochrome contains 66.5% Cr. The typical %recovery for Cr is 99%.

SOLUTION

m_{ladle} = 8 ton = 8,000 kg, $\Delta\%Cr = \%Cr_{aim} - \%Cr_{actual}$ = 13%−2% = 11%

By using Equation 12.8,

$$m_{ferrochrome} = \frac{100\% \times \Delta\%Cr \times m_{ladle}}{\%Cr \text{ in ferrochrome} \times \% \text{ recovery}Cr}$$

$$= \frac{100\% \times 11\% \times 8,000}{66.5\% \times 99\%}$$

$$m_{ferrochrome} = \frac{100 \times 11 \times 8,000}{66.5 \times 99} = 1,337 \text{ kg}$$

The amount of ferrochrome required to be added to the ladle = 1,337 kg.

EXAMPLE 12.12 CALCULATING THE WT% OF CARBON PICKUP DUE TO FERROALLOY ADDITION

By using the data in Example 12.11, calculate the wt% of carbon pickup. Ferrochrome (FeCr) contains 6.4% carbon with 99% recovery

SOLUTION

m_{ladle} = 8 ton = 8,000 kg, m_{FeCr} = 1,337 kg, % C in FeCr = 6.4%

By using Equation 12.9,

$$\Delta\%C = \frac{m_{FeCr} \times wt\%\,C\,in\,FeCr \times \%\,recovery\,of\,C}{100\% \times m_{ladle}} = \frac{1,337 \times 6.4\% \times 99\%}{100\% \times 8,000}$$
$$= 1.06\ wt\%$$

The amount of carbon pick-up = 1.06 wt%.

EXAMPLE 12.13 CALCULATING CARBON EQUIVALENT AND IDENTIFYING CAST IRONS

Two samples of cast irons (A and B) are provided to you. The sample A has the composition:

C: 3.2, Si: 1.8, P: 1.4, S: 0.1, Mn: 0.6, Fe: balance (wt%). The composition of sample B is: C: 2.5, Si: 1.3, P: 0.15, Mn: 0.4, Fe: balance (wt%). Calculate CE for each sample; and hence identify which one is gray CI and the other white CI.

SOLUTION

By using Equation 12.10 for sample A,

$$CE\,(in\ wt\%) = C + \frac{Si+P}{3} = 3.2 + \frac{1.8+1.4}{3} = 4.26$$

By using Equation 12.10 for sample B,

$$CE\,(in\ wt\%) = C + \frac{Si+P}{3} = 2.5 + \frac{1.3+0.15}{3} = 2.98$$

The high CE value of sample A confirms it to be gray CI. The sample B, having a low CE value, is white CI.

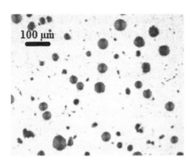

FIGURE E-12.14(A)   Micrograph of SG iron.

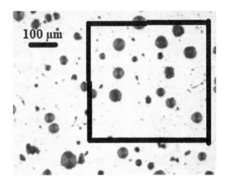

FIGURE E-12.14(B)

EXAMPLE 12.14 COMPUTING THE NODULE COUNTS PER UNIT AREA FOR SG IRON

By using the micrograph in Figure E-12.14, calculate the graphite nodule count per unit area, in the upper right region, for the ductile iron.

SOLUTION

In order to determine the nodule count, we select a measured area of 3 cm x 3 cm. rectangle, as shown in Figure E-12.14(B).
By reference to Figure E-12.14(B),

$N_i = 16$, $N_u = 6$, A = 30 mm x 30 mm = 900 mm^2

$$The\ magnification = M = \frac{Measured\ length\ of\ the\ magnification\ bar}{Length\ shown\ on\ the\ bar}$$

$$= \frac{7\ mm}{10\ \mu m} = 70$$

By using Equation 12.12,

$$N_A = \frac{(N_i + 0.5N_u + 1)M^2}{A} = \frac{(16 + 0.5 \times 6 + 1)70^2}{900} = 109 \text{nod/mm}^2$$

Nodule counts per unit area = N_A = 109 nod/mm^2.

EXAMPLE 12.15 CALCULATING THE GRAPHITE NODULE COUNT PER UNIT VOLUME FOR DUCTILE CI

By using the data in Example 12.14, calculate the graphite nodule count per unit volume for the ductile cast iron. Given: the parameter α = 1.2, and the fraction of graphite is 0.1.

SOLUTION

N_A = 109 nod/mm^2, f_g = 0.1, α = 1.2

By using Equation 12.13,

$$N_V = \sqrt{\frac{\pi}{6f_g}}(\alpha N_A)^{3/2} = \sqrt{\frac{\pi}{6 \times 0.1}}(1.2 \times 109)^{1.5} = 2.288 \times 1496$$

$$= 3,423 \text{ nod/mm}^3$$

The graphite volume nodule count = 3,423 nod/mm^3.

QUESTIONS AND PROBLEMS

12.1. What is the meaning of each of the following acronyms used in ferrous alloys metallurgy?
(a) DSS, (b) ADI, (c) AISI, (d) SAE, (e) CCR, (f) TRIP steel, (g) CE.

12.2. Encircle the most appropriate answers for the following statements:
(a) Which steel is the most suitable for application in cutting tools? (i) MS, (ii) low-carbon steel, (iii) medium-carbon steel, (iv) high-carbon steel.
(b) Which steel offers the best compromise in the tensile strength and impact toughness? (i) MS, (ii) low-carbon steel, (iii) medium-carbon steel, (iv) high-carbon steel.
(c) Which steel has the highest ductility? (i) MS, (ii) low-carbon steel, (iii) medium-carbon steel, (iv) high-carbon steel.
(d) Which steel has either ferrite or ferrite+pearlite in its microstructure? (i) MS, (ii) low-carbon steel, (iii) medium-carbon steel, (iv) high-carbon steel.
(e) Which steel is the most suitable for application in wires for cloth-hangers? (i) low-carbon steel, (ii) medium-carbon steel, (iii) high-carbon steel.
(f) Which type of steel is the Fe-3.5 wt% C alloy? (i) low-carbon steel, (ii) medium-carbon steel, (iii) high-carbon steel, (iv) cast iron.

(g) Which steel has pearlite+cementite in its microstructure? (i) MS, (ii) low-carbon steel, (iii) medium-carbon steel, (iv) high-carbon steel.

(h) Which steel is the Fe-1.0 wt% C alloy? (i) MS, (ii) low-carbon steel, (iii) medium-carbon steel, (iv) high-carbon steel.

(*i*) Which of the following elements causes *hot shortness* in carbon steels? (i) silicon, (ii) manganese, (iii) sulfur, (iv) phosphorous.

(j) Which of the following elements causes *cold shortness* in carbon steels? (i) silicon, (ii) manganese, (iii) sulfur, (iv) phosphorous.

(k) Which of the following elements prevents porosity and blow holes in carbon steels? (i) manganese, (ii) silicon, (iii) sulfur, (iv) phosphorous.

(*l*) Which of the following element eliminates the harmful effects of sulfur in carbon steels? (i) silicon, (ii) manganese, (iii) sulfur, (iv) phosphorous.

(m) Which element improves yield strength and resistance to rusting in low-carbon steels? (i) silicon, (ii) manganese, (iii) sulfur, (iv) phosphorous.

(n) What is the main role of the alloy steel containing around 4.25% Ni and 1.25% Cr? (i) corrosion resistance, (ii) air hardening, (iii) high thermal stability, (iv) high hardness.

(o) Which type of alloy steel is suitable for applications in axles, shafts, and gears? (i) Cr-W-V, (ii) Ni-Cr-Mo, (iii) Mn-Cr-C, (iv) high nickel.

(p) Which alloy steel has ferrite, bainite, and retained austenite in its microstructure? (i) TRIP steel, (ii) DSS, (iii) MSS, (iv) Hadfield Mn steel.

(q) Which element is added to transform gray CI to ductile CI? (i) Ni, (ii) Mo, (iii) Mg, (iv) Cr.

(r) Which type of cast iron is suitable for application in mill rolls? (i) malleable CI, (ii) gray CI, (iii) nodular CI, (iv) alloyed CI.

(s) Which type of CI has good formability, machinability, and impact toughness? (i) malleable CI, (ii) gray CI, (iii) nodular CI, (iv) alloyed CI.

12.3. Arrange the four types of carbon steels in the order of increasing ductility.

12.4. Draw the classification chart for various types of: (a) ferrous alloys, and (b) alloy steels.

P12.5. Calculate the amounts of phase/microconstituent in a high-carbon steel (98.8% Fe-1.4% C alloy) at a temperature just below the eutectoid.

P12.6. By using the micrograph of the cast iron in Figure 12.9, calculate the (a) graphite nodule count per unit area, in the central region, and (b) the nodule count per unit volume, if $\alpha = 1.1$ and graphite fraction is 0.1.

P12.7. A ferritic stainless steel has the following composition: Fe-18% Cr-8% Ni-0.1%C. Calculate the alloy steel's liquidus temperature.

P12.8. A 10-tons ladle of steel contains 0.3% Mn at tap. Calculate the amount of high-purity ferromanganese required to be added to achieve a composition of 1.6% Mn. Ferromanganese contains 49% Mn. The typical %recovery for Mn is 95%.

P12.9. A 17-tons ladle of steel actually contains 0.5% Cr. Calculate the mass of the elemental chromium (Cr) that must be added to achieve an aim composition of 15% Cr.

P12.10. Calculate the carbon equivalent of the cast iron having composition: Fe-2.8%C-1.4%Si- 0.18%P-0.4%Mn.

REFERENCES

Degarmo, E.P., Black, J.T. & Kohser, R.A. (2003) *Materials and Processes in Manufacturing.* 9th ed. John Wiley & Sons Inc, New York.

Fras, E. & Lopez, H. (2010) Eutectic cells and nodule count—An index of molten iron quality. *International Journal of Metal Casting*, 4(July 3), 35–61.

Fruehan, R.J. (1998) *The Making, Shaping, and Treating of Steel, Volume. 2, Steelmaking and Refining.* 11th ed. AISE Steel Foundation, New York.

Mazurek, R. (2012) *Effects of Burrs on a Three Phase Transformer Core Including Local Loss, Total Loss and Flux Distribution.* PhD Thesis, Cardiff School of Engineering, Cardiff University.

Pedersen, K.M. & Tiedje, N.S. (2008) Graphite nodule count and size distribution in thin walled ductile cast iron. *Materials Characterization*, 59, 1111–1121.

Turkdogan, E.T. (1996) *Fundamentals of Steelmaking.* CRC Press, Boca Raton, FL.

World Steel Association (2019) *Secondary Steelmaking Simulation—User Guide.* Online: file:///C:/Users/Welcome/Desktop/SSM_User_Guide_EN_v02.pdf. Accessed on 4 February 2019.

13 Nonferrous Alloys

13.1 NONFERROUS ALLOYS

Nonferrous alloys are the metallic materials that do not contain iron as a base element. Notable examples of nonferrous metallic materials include alloys of aluminum (*Al*), copper (*Cu*), nickel (*Ni*), and titanium (*Ti*). In particular, aluminum alloys, superalloys, brasses, and bronzes are commonly used engineering materials. Nonferrous alloys find a broad range of decorative and engineering applications, especially in aerospace and automotive industries. Aluminum and titanium alloys are specified for aerospace structural applications requiring reduced weight, higher specific strength, nonmagnetic properties, and resistance to corrosion. Copper and aluminum are also specified for electrical and electronic applications.

13.2 ALUMINUM ALLOYS

13.2.1 PROPERTIES AND APPLICATIONS

Aluminum and it alloys are the world's most widely used nonferrous materials. Pure aluminum has good ductility, corrosion resistance, and electrical conductivity; these properties render metallic aluminum suitable for applications as foils and conductor cables. However, alloying of aluminum with other elements is necessary to provide the higher strengths needed for engineering applications. For example, the tensile strength of pure aluminum is around 30 MPa; but that of *Al*-2%M*g* is about 75 MPa. Owing to their light weight, strength, and corrosion resistance, aluminum alloys applications range from aircraft structures through automotive engine components and construction materials to packaging foils (Kaufman, 2000; Huda, 2016, Huda *et al.*, 2009).

13.2.2 DESIGNATIONS AND APPLICATIONS

The strength of an aluminum alloy strongly depends on its alloying contents, degree of cold working, and heat treatment. Aluminum alloys usually contain Cu, Mg, Si, Mn, Zn, and lithium (Li) as the chief alloying elements. Owing to a broad range of *Al* alloys, classification systems have been developed for wrought and cast alloys. Wrought *Al* alloys cannot be heat-treated; they are strengthened by cold working. The wrought *Al* alloys have a four-digit system, and the cast alloys have a three-digit and one-decimal place system.

Table 13.1 presents the designations systems for wrought, cast, and heat-treated *Al* alloys. In each designation system, the first digit indicates the principal alloying element. In the *wrought alloy designation* system, the second digit, if different from

TABLE 13.1

The Designations Systems for Wrought, Cast, and Heat-treated *Al* Alloys

Wrought alloys series	Principal alloying elements	Cast alloys series	Principal alloying elements	Heat-treated alloys suffix codes	Heat treatment, temper, and post-process
1xxx	99.xx%*Al*	1xx.x	99%*Al*	xxxx-T3	Solution heat-treated (SHT), then cold-worked (CW)
2xxx	Copper	2xx.x	Copper	xxxx-T36	SHT, and CW (controlled)
3xxx	Manganese	3xx.x	Si+Cu and/or Mg	xxxx-T351	SHT, stress-relieved stretched (SRS), then CW
4xxx	Silicon	4xx.x	Silicon	xxxx-T4	SHT, then naturally aged
5xxx	Magnesium	5xx.x	Magnesium	xxxx-T451	SHT, then SRS
6xxx	Mg + Si	6xx.x	Unused series	xxxx-T5	Artificially aged only
7xxx	Zinc	7xx.x	Zinc	xxxx-T6	SHT, artificial aging
8xxx	Other elements (e.g. Li)	8xx.x	Other elements	xxxx-T6	SHT, then artificially aged

0, indicates a modification of the specific alloy (see Table 13.1's columns 1 and 2). In the *cast alloy designation*, the second and third digits are arbitrary numbers given to identify a specific alloy in the series. The digit following the decimal point indicates whether the alloy is a casting (.0) or an ingot (.1 or .2) (see columns 3 and 4). The heat treated *Al* alloys have suffix T3, T-36, T-351, and the like (see columns 5 and 6 in Table 13.1). Example 13.1 illustrates the significance of Table 13.1.

The series 1xxx *Al* alloys are used in heat exchangers and chemical-handling equipment. The series 2xxx *Al* alloys (containing 1.9–6.8% Cu and some Mn, Mg, and Zn) (e.g. 2024-T4) are used in forgings, extruded parts, and liquefied-gas storage tanks in commercial and military aircrafts. The series 7xxx Al-Zn-Mg alloys offer the potential for precipitation strengthening (see Chapter 9) though Cu is often added to improve stress-corrosion cracking. In particular, 7075-T6 *Al* alloy is noted for aircraft structural parts and other highly stressed components (Tajally *et al.*, 2010). The series 6xxx alloys (e.g. 6061-T4) find applications in pipelines and automotive engines. The *Al* alloy 356.0 is used in automotive transmission cases and cylinder heads.

13.2.3 ALUMINUM-SILICON CASTING ALLOYS

The metallurgical characteristic of *Al*-Si alloys can be studied by the Al-Si phase diagram (Figure 13.1). The maximum solid solubility of *Si* in *Al* is 1.6 wt%, which

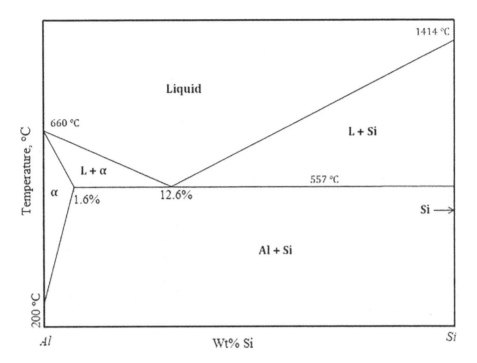

FIGURE 13.1   Aluminum-silicon phase diagram.

decreases to zero at 200°C. It is evident in Figure 13.1 that *Al-Si* system forms a eutectic at 12.6 wt% Si at 557°C; thus *Al-12%Si* is a good casting alloy. However, the presence of Si in *Al* imparts brittleness; this is why, in industrial practice, a lower Si content (1.1–7.0 wt% Si) is used in *Al* casting alloys. The eutectic *Al*-Si alloy, at a temperature just below the eutectic, contains around 88.8 wt% α-*Al* solid solution and 11.2 wt% Si crystals (see Example 13.2). The large Si crystals in the microstructure are detrimental for mechanical properties; this is why a small amount of sodium is added for the refinement of the microstructure of the casting eutectic *Al* alloys (see Figure 13.2).

The study of the kinetics of *Si* precipitation in *Al-Si* casting alloys is interesting to physical metallurgists. It has been reported that the Ω parameter for the recovery process induced by dislocation glide and annihilation can be expressed as (Zamani, 2015):

$$\Omega = A\exp\left(-\frac{Q_{dislocation-glide}}{RT}\right) \tag{13.1}$$

where Ω is the recovery function that depends on temperature and strain rate; A is the empirical constant, R is the gas constant (= 8.31 J/K·mol); and $Q_{dislocation-glide}$ is the activation energy for dislocation glide, J/mol (see Example 13.3). The addition of *Si* to *Al* causes a decrease in the vacancy formation energy, and in an increase in the concentration of vacancies near the Si solute atoms (Roswell and Nowick,

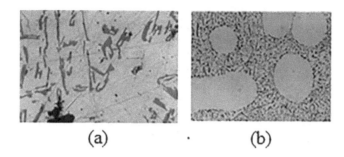

FIGURE 13.2 Microstructures of Al-12%Si alloy; (a) before Na addition, and (b) after Na addition.

1953; Lomer, 1958). In order to provide a better description of the recovery process expressed in Equation 13.1, it is necessary to include a correction term by considering the effects of vacancy formation. The interaction between Si solute atoms and vacancies can be expressed as (Zamani, 2015):

(13.2)

$$\frac{X_{vac}^{Al-Si}}{X_{vac}^{Al}} = 1 - Z \cdot X_{si} + Z \cdot X_{si} \exp\left(\frac{E}{kT}\right) \tag{13.2}$$

where Z is the coordination number; X_{si} is the concentration of the solute Si in the Al alloy; E is the vacancy-Si binary interaction energy in the alloy, J; and k is the Boltzmann constant ($k = 1.38$ x 10^{-23} J/K) (see Example 13.4).

13.2.4 ALUMINUM-COPPER ALLOYS AND AEROSPACE APPLICATIONS

Aluminum-copper alloys are strong and lightweight; they are noted for aerospace applications. In particular, the age-hardened 2024-T3 Al-Cu alloy is used in fuselage, wing tension members, shear webs, and structural areas where stiffness, fatigue performance, and good strength are required. The selection of heat treatment approaches for the aerospace Al-Cu alloys play an important role in strengthening the alloys (see Chapter 15). For example, the aerospace 2024-T81 Al alloy has a high yield strength and ultimate tensile strength of 450 MPa and 485 MPa, respectively. In aerospace alloys, for tensile loading below the yield limit, the applied stress on a structural component (e.g. fuselage) should be considered regarding its specific weight (weight per unit length). In comparing specific weights of two materials, the design relationship is expressed as:

$$\frac{W_a}{W_b} = \frac{\rho_a \cdot \sigma_{ys(b)}}{\rho_b \cdot \sigma_{ys(a)}} \tag{13.3}$$

where W_a is the specific weight of the part using material (a), W_b is the specific weight of the part using material (b); ρ_a and ρ_b are the densities of materials (a) and (b), respectively; and $\sigma_{ys(a)}$ and $\sigma_{ys(b)}$ are the yield strengths of materials (a) and (b), respectively (see Example 13.9).

13.3 COPPER ALLOYS

13.3.1 THE APPLICATIONS OF COPPER ALLOYS

Copper is an excellent electrical conductor, and finds wide applications in electrical industry. Examples of copper alloys include: brasses (copper-zinc), bronzes (copper-tin), gun metal (Cu-Sn-Zn-Pb), Cu-Ni alloys, nickel silver (Cu-Ni-Zn), and the like. In particular, brasses are very important engineering materials; these include red brass, gilding metal, cartridge brass, clock brass, and *Muntz metal*. Copper alloys for high temperature applications (e.g. in boilers) include: silicon bronzes, aluminum brasses, and copper nickels. Copper and its alloys find applications in architecture, automotive, electrical, telecommunication, and mechanical engineering (tubes, marine, pipe, and fittings, fuel gas piping system, etc.) fields. For example, copper tubes are used as refrigerant lines in HVAC systems (see Figure 1.3). Copper alloys are specified/graded depending on their compositions, processing, and heat-treatment conditions (see Chapter 15, Table 15.2).

13.3.2 COPPER-ZINC ALLOYS

The copper-zinc alloys system is shown in the Cu-Zn phase diagram (Figure 13.3). It is evident in Figure 13.3 that the α phase (FCC crystal structure) is stable for

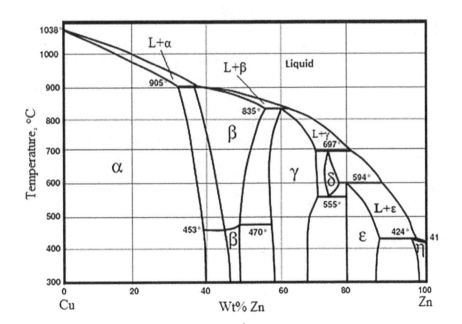

FIGURE 13.3 Copper-zinc phase diagram.

TABLE 13.2

Equivalent Zinc Data for Copper Alloys

Element	Si	Sn	Pb	As	Fe	Ni	Mn
Equivalent Zn	10	2	1	1	0.9	0.8	0.5

concentrations up to 38 wt% Zn. Alpha (α) brasses are relatively ductile with good formability; and find applications in deep drawn and cold-worked components (e.g. automotive radiator caps, battery terminals, etc.). Cartridge brass is 70%Cu-30%Zn. In addition to α phase, there are other phases (β, γ, ε, η) that are stable at various temperatures corresponding to various compositions of Cu-Zn alloys (see Examples 13.5–13.8). The β-brasses are suitable for applications where high tensile strength is required but must be hot-worked because of their restricted ductility below 450°C (if cold worked). The γ-brass is hard and brittle both at low and high temperatures and cannot be worked.

Cold working significantly enhances the strength of brass. The tensile strength of a cold -worked cartridge brass wire may be as high as 900 MPa. Various alloying additions (e.g. lead, arsenic, nickel, iron, gold, silver, etc.) are made to the basic copper-zinc alloys for a variety of reasons; which include: (a) improving machinability, (b) improving tensile strength and wear resistance, and (c) improving corrosion resistance. In copper industry, the calculation of *zinc equivalent* is of great technological importance. The *zinc equivalent* may be calculated by:

$$\text{Zinc Equivalent } \% = \frac{X}{Y} \times 100 \qquad (13.4)$$

where

$$X = \Big[\text{sum of } \big(\text{Equivalent Zn} \times \text{ wt \% of each alloying element} \big) \Big] + \text{wt\% Zn} \quad (13.5)$$

$$\text{and } Y = X + \text{wt\% Cu} \qquad (13.6)$$

The values of equivalent zinc for various alloying elements (for Cu-Zn base alloys) are given in Table 13.2. The significance of Equations (13.4)–(13.6) is illustrated in Example 13.10.

Once the zinc equivalent % has been computed by using Equation 13.4, the stability of phase(s) (e.g. α, $\alpha+\beta$, β, etc.) in the alloy can be determined by using Figure 13.3 (see Example 13.11).

13.3.3 Copper-Tin Alloys

Copper-tin (Cu-Sn) alloys, or bronzes, are known for their remarkable corrosion resistance. They have higher strength and ductility than red brass and semi-red brass. Bronzes, with up 15.8% Sn, retain the FCC structure of alpha copper. Bronzes find applications in bearings, gears, piston rings, valves, and fittings. Lead (Pb) is added to bronzes to improve machinability. Since Pb decreases the tensile strength of the bronzes, the composition is usually adjusted to balance machinability and strength requirements. High leaded bronzes are primarily used in sleeve bearings.

13.4 NICKEL AND ITS ALLOYS

13.4.1 General Properties and Applications

Nickel and its alloys have excellent high-temperature strength and corrosion resistance. Nickel metal has a low coefficient of thermal expansion; and finds application in bimetallic (iron-nickel) thermocouple sensors. *Monel metal* is a nickel-copper alloy that is resistant to corrosion in many environments. Monel 400 contains at least 63% Ni, 30% Cu, 2.3% Fe, and 1.8% Mn. It is one of the few alloys that maintains its strength at sub-zero temperatures. The most important engineering Ni alloy is the superalloy; which refers to Ni-base alloys containing Cr, Fe, Mo, W, Ti, Al, C, Ta, Zr, Nb, B, etc. Superalloys are known for their exceptional high-temperature strength and resistances to creep and hot corrosion; these remarkable properties render Ni-base superalloys suitable for applications in hot-section components (vanes, disks, and blades) of gas-turbine (GT) engines as well as in nuclear reactors. In particular, directionally solidified (DS) superalloy blades with exceptional creep resistance are used in modern aircraft engines (see Figure 2.9, Chapter 2). Several Ni-base superalloys have been developed for application in the GT blades; these include: the *GTD-111*, the Allvac 718Plus, and the single crystal superalloys Rene-N6 and MC-534 (Huda, 2017).

13.4.2 Ni-Base Superalloys and Creep Behavior

Designing an alloy against creep involves analyzing the variables on which creep strength depends; these variables include: crystal structure, solutes, precipitates, dispersoids, grain size, and grain-boundary structure (Huda, 1995). Besides the FCC crystal lattice of Ni, many alloying elements (solutes) contribute favorably to impart solid-solution strengthening, precipitation strengthening, and dispersoids effects to achieve a good tensile and creep strength in Ni-base superalloys (see also Figure 6.6). For example, the IN-718 superalloy (53Ni-18.4Cr-3Mo-0.56Al-1.0Ti-0.16Co-0.003B-0.054C-0.09Si-balanceFe) is extensively used in hot sections of gas-turbine engines (Huda *et al.*, 2014). In particular, the fine dispersed γ' (having formula: $Ni_3[Al,Ti]$) that precipitates in the FCC γ matrix of the microstructure significantly enhances the creep strength (see Figure 13.4).

The study of the creep behavior of Ni-base superalloys is of great technological importance. It has been explained in Chapter 10 that an increase in temperature has a pronounced effect on creep strain rate (see Equation 10.28). Harrison and co-workers have studied the creep behavior of Ni-base superalloy 720Li for application in gas turbine disk; they have reported that the stress σ (in Equation 10.28) can be normalized by σ_{ut} according to (Harrison *et al.*, 2013):

$$\dot{\varepsilon}_{ss} = A^* \left(\frac{\sigma}{\sigma_{ut}} \right)^n \exp\left(-\frac{Q_C^*}{RT} \right) \tag{13.7}$$

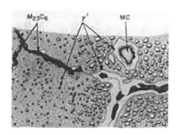

FIGURE 13.4   Schematic of the phases in the microstructure of a Ni-base superalloy.

where $\dot{\varepsilon}_{ss}$ is the steady-state strain rate, /s; A^* is the material constant; σ is the stress, Pa; σ_{ut} is the ultimate tensile strength, Pa; n is the stress exponent (n varies from $n \cong 14$ for the tests with conditions giving short lives to $n \cong 5$ for the long duration tests).; T is the temperature, K; R = 8.3145 J/mol·K; and Q_c^* is the normalized activation energy for creep, J/mol (see Example 13.12). Besides the γ' phase in the microstructure, an increase in the average grain size strongly enhances the creep strength of Ni-base superalloys at elevated temperatures. It has been experimentally shown that at a high temperature of 850°C, the creep strain rates in a coarse- grained and a fine-grained superalloy at 250 MPa may be co-related as (Thebaud *et al.*, 2018):

$$\dot{\varepsilon}(FG) = 100 * \dot{\varepsilon}(CG) \tag{13.8}$$

where $\dot{\varepsilon}(FG)$ and $\dot{\varepsilon}(CG)$ are the creep strain rates in a fine-grained (FG) and the coarse-grained (CG) superalloy, respectively (see Example 13.13).

13.5   TITANIUM AND ITS ALLOYS

13.5.1   PROPERTIES AND APPLICATIONS

Commercially pure (99–99.5% purity) titanium has a low ultimate tensile strength (UTS = 330–650 MPa); however, when alloyed (with *Al*, Cu, Mn, Mo, Sn, V, or Zr), its strength significantly increases to 800–1450 MPa. The elevated-temperature strength properties of titanium alloys are also good. The modulus of elasticity for titanium alloys is excellent (E = 125 GPa for the Ti-6Al-4V alloy). The fatigue endurance strength at 10^7 cycles for Ti alloys is about 0.6 times the UTS. When these exceptional mechanical properties are considered together with a low density of 4500 kg/m^3 (slightly over half that of steel), titanium alloys are the material of choice for applications in aircraft and spacecraft structures and gas turbine components (e.g. compressor blades), naval ships, guided missiles, and lightweight armor plate for tanks (see Example 13.14). Owing to its excellent corrosion resistance and bio-compatibility, titanium alloys are used in artificial hips and other orthopedic bioengineering applications.

13.5.2   MICROSTRUCTURE AND PROPERTIES OF TITANIUM ALLOYS

Pure titanium (Ti) exists in two allotropic forms: (a) the α-Ti, having hexagonal close-packed (HCP) crystal structure, is stable from room temperature to 882°C; (b) the β-Ti, with a body-centered cubic (BCC) structure, is stable above 882°C. The microstructural phases in a titanium alloy depend on the presence and amount of alloying elements (*Al, V, Sn, Mo, Mn*, etc.) and processing. The stability of α, β, α+β, and other phases in a *Ti-Al* alloy at various temperatures can be observed in the alloy phase diagram (Figure 13.5).

For many stressed applications requiring high strength, an alpha-beta titanium alloy is the material of choice; this microstructure can be achieved by alloying with *Al* and *V* (see Figure 13.6). In particular, the Ti-6%Al-4%V alpha-beta alloy, in the annealed condition, finds applications in prosthetic implants, airframe structural components, and chemical-processing equipment.

Titanium alloys may be classified into three groups: (a) α Ti alloys, (b) β Ti alloys, and (c) α+β Ti alloys. Alpha Ti alloys have low-to-medium strength but good

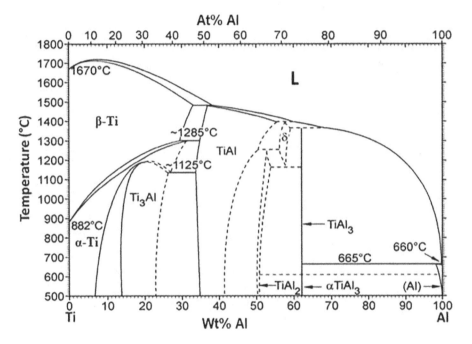

FIGURE 13.5   The *Ti-Al* phase diagram.

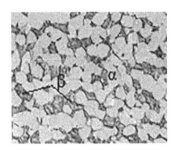

FIGURE 13.6   The microstructure of α+β titanium alloy.

TABLE 13.3

Types, Mechanical Properties, and Applications of Ti Alloys

Alloy type	Composition	S_{ut}, MPa	0.2% Yield strength, MPa	% elongation	Fracture toughness, MPa$\sqrt{m}$	Applications
Un-Alloyed: Not heat-treatable	commercially pure (>99%) Ti	330–650	220–550	20–30	70 or more	Parts requiring high corrosion resistance
Alpha and near alpha: High strength and creep resistance	Ti-2.25Al-11Sn-1Mo-5Zr-0.2Si, Ti-6Al-5Zr-0.5Mo-0.25Si	1,000	900	N/A	50–60	Aircraft frames and jet engine components (compressor blades)
Alpha-Beta: Heat treatable, good hot formability	Ti-6Al-V4, Ti-4Al-4Mo-2Sn-0.5Si	900–1,100	800–1,000	14–17	40–60	Prosthetic implants, artificial hips and other orthopedic applications; airframe structural components, and processing equipment
Beta: Can be cold worked in solution and quenched condition	Ti-11.5Mo-6Zr-4.5Sn, Ti-3Al-8V-6Cr-4Mo-4Zr	1,400	1,300	11	N/A	Parts requiring good weldability and strength

toughness and ductility. They are non-heat treatable and also have a good weld-ability. They have excellent creep strength, resistance to oxidation, and excellent properties at cryogenic temperatures. Alpha-Beta alloys are heat treatable to varying extents; their strength levels are medium to high. Hot working formability is good but cold forming often presents difficulties. Creep strengths of α+β Ti alloys are just fair. Beta Ti alloys are readily heat treatable, weldable, and offer high strength up to intermediate temperature levels. Cold formability of solution-treated β Ti alloys is generally excellent. Table 13.3 presents the types, properties, and applications of commonly used titanium alloys.

13.6   PRECIOUS METALS

The precious metals, also called noble metals, are the set of expensive metals that are oxidation resistant, soft, and ductile. The precious metals are gold (Au), silver (Ag), platinum (Pt), palladium (Pd), rhodium (Rh), ruthenium (Ru), iridium (Ir), and osmium (Os). Gold, silver, and platinum are extensively used in jewelry. Gold jew-elry is usually made of commercially pure gold since pure gold is too soft to be formed/bent. Hence, for making a gold jewelry, gold is alloyed with copper and zinc;

the resulting alloy is harder and more workable for making jewelry. In order to calculate the actual amount of pure gold in a jewelry, it is important to know its weight and karat. The karat marks on gold jewelry are usually stamped as "K" or "Kt." The standard karat numbers for gold are: **10K**, **14K**, **18K**, **20K**, and **22K**.

Gold karats can be converted to percentage by the following formula:

$$\%\text{gold} = \frac{K}{24} \times 100 \tag{13.9}$$

where K is the karat number of the gold (see Example 13.14).

The weight of gold in a jewelry can be determined by:

$$W_g = \% \text{ gold} \times W_j \tag{13.10}$$

where W_g is the weight of gold, grams; and W_j is the weight of the jewelry, grams (see Example 13.15).

Besides their application in jewelry, alloys of gold and alloys of silver are used as dental restoration materials. Platinum is used in thermocouples for measuring elevated temperatures; it is also used as a catalyst in petrochemical processing. Each of Ru, Rh, Pd, and Os alloys is also used as a dental alloy.

13.7   OTHER NONFERROUS ALLOYS

Magnesium and Its Alloys. Magnesium is the lightest structural known metal; it is a silvery-white metal weighing only two-thirds as much as aluminum. When alloyed with zinc, aluminum, and manganese, the magnesium alloy produces a material with the highest specific strength. Magnesium parts can be severely damaged in corrosive environments unless galvanic couples are avoided by proper design or surface protection. The poor corrosion resistance of magnesium alloys is of primary concern. Another threat (hazard) associated with the use of magnesium is its high flammability. It is, therefore, important to follow the safety precautions (e.g. keeping container tightly closed and dry, using CO_2 or dry powder in case of fire, etc.). Magnesium alloys possess good casting characteristics and strength. For example, the Mg-5.5%Zn-0.45%Zr alloy, in the artificially aged condition, having a tensile strength of 350 MPa and a good ductility of 11% elongation (in 50 mm), is used in high-strength forgings for aircraft components.

Zinc and Zinc Die-Casting Alloys. Zinc is widely used in the galvanizing industry since a protective layer is easily produced on the surface of zinc; which prevents further atmospheric corrosion. Galvanizing is primarily used on steel sheets for automobiles, construction, shipbuilding, light, and some other industries. Zinc alloys are the easiest to die cast. They have high ductility and impact strength. Zinc alloys can be cast with thin walls and excellent surface finish. Low limits on lead, tin, and cadmium ensure the long-term integrity of the alloy's strength and dimensional stability. The applications of the die cast products are varied, including automobile, construction, and electronic industries, as well as clothing, furniture, and toys.

Rare Earth Metals. Rare earth metals are a group of 17 chemical elements that occur together in the periodic table; these elements are: yttrium and the 15 lanthanide series elements (lanthanum, cerium, praseodymium, neodymium, promethium, samarium, europium, gadolinium, terbium, dysprosium, holmium, erbium, thulium, ytterbium, and lutetium); additionally scandium is found in most rare earth element deposits and is sometimes classified as a rare earth metal. Rare earth metals and alloys that contain them are used in many devices, such as computer memory, DVDs, rechargeable batteries, cell phones, catalytic converters, magnets, and fluorescent lighting.

13.8 CALCULATIONS—WORKED EXAMPLES ON NONFERROUS ALLOYS

EXAMPLE 13.1 IDENTIFYING THE ALLOYING ELEMENT AND HEAT TREATMENT IN AN *AL* ALLOY

The heat-treated 2024-T4 aluminum alloy finds applications in aircraft fuselage structure and other aerospace components. (a) What is the principal alloying element in the alloy? (b) What heat treatment has been given to the alloy?

SOLUTION

(a) By reference to Table 3.1's columns 1 and 2, the principal alloying element in the 2024-T4 aluminum alloy is copper (Cu).
(b) By reference to Table 3.1's columns 5 and 6, the heat treatment given to the 2024-T4 alloy involves solution heat treatment following by natural aging.

EXAMPLE 13.2 CALCULATING THE AMOUNTS OF A-*AL* AND *SI* PHASES IN A EUTECTIC *AL-SI* ALLOY

A eutectic *Al* alloy is cooled from liquid state to just below the eutectic temperature. Calculate the amounts of α-*Al* and Si phases at the specified temperature.

SOLUTION

By reference to Figure 13.1, we may consider the specified temperature to be around 550°C; at this temperature the alloy contains 1.6 wt% α-*Al*.

$$C_o = 12.6 \text{ wt\% Si}, C_{\alpha Al} = 1.6 \text{ wt\%} \alpha - Al, C_{si} = 100 \text{ wt\% Si}$$

By using Equation 6.15 and Equation 6.16, we obtain:

$$W_{\alpha Al} = \frac{C_{Si} - C_o}{C_{Si} - C_{\alpha Al}} = \frac{100 - 12.6}{100 - 1.6} = 0.888 = 88.8\%$$

$$W_{si} = \frac{C_0 - C_{\alpha Al}}{C_{Si} - C_{\alpha Al}} = \frac{12.6 - 1.6}{100 - 1.6} = 0.108 = 11.2\%$$

The amounts of α-*Al* and Si phases at the specified temperature are 88.8% and 11.2%, respectively.

EXAMPLE 13.3 COMPUTING THE EMPIRICAL CONSTANT USING Ω PARAMETER FOR THE RECOVERY PROCESS IN AN *AL-SI* CASTING ALLOY

The Ω parameter for the recovery process in an *Al-Si* casting alloy is determined to be 75. The activation energy for dislocation glide is 6,500 J/mol at a temperature of 400K. Calculate the empirical constant for the Arrhenius equation.

SOLUTION

$$\Omega = 75, \ T = 400K, Q_{dislocation-glide} = 6500 \text{ J/mol.},$$
$$R = 8.31 J/K \times gmol, A = ?$$

By using Equation 13.1,

$$75 = A \exp\left(-\frac{6500}{8.31 \times 400}\right)$$

$$75 = A \exp(-1.955)$$

$$A = \frac{75}{0.1415} = 530$$

The empirical constant for the Arrhenius equation = 530

EXAMPLE 13.4 CALCULATING THE CORRECTION TERM: $\dfrac{X_{vac}^{Al-Si}}{X_{vac}^{Al}}$ IN AN *AL-SI* ALLOY

The concentration of solute Si in an Al-Si alloy matrix was measured (by WDXS) to be 0.015. The vacancy-Si binary interaction energy in the alloy is 1.6 x 10^{-19} J. Calculate the correction term for the interaction between Si solute atoms and vacancies at 400K.

SOLUTION

$X_{si} = 0.015$, Z = coordination number for the FCC Al = 12 (see Chapter 2),
    E = 1.6 x 10^{-19} J

By using Equation 13.2,

$$\frac{X_{vac}^{Al-Si}}{X_{vac}^{Al}} = 1 - Z \cdot X_{si} + Z \cdot X_{si} \cdot \exp\left(\frac{E}{kT}\right)$$

$$= 1 - (12 \times 0.015) + 0.18 \cdot \exp\left(\frac{1.6 \times 10^{-19}}{1.38 \times 10^{-23} \times 400}\right)$$

$$\frac{X_{vac}^{Al-Si}}{X_{vac}^{Al}} = 1 - 0.18 + (0.18 \times 3.85 \times 10^{12}) = 6.94 \times 10^{11}$$

The correction term for the interaction between Si solute atoms and vacancies $= 6.94 \times 10^{11}$.

EXAMPLE 13.5 IDENTIFYING THE STABILITIES OF PHASES IN CU-ZN ALLOYS

What phases are stable at room temperature in the following alloys: (a) Cu-42 wt% Zn, (b) Cu-55 wt% Zn, (c) Cu-75 wt% Zn, and (d) Cu-90%Zn?

SOLUTION

By reference to the Cu-Zn phase diagram (Figure 13.3),

A. the stable phases in the Cu-42 wt% Zn alloy are: $\alpha + \beta$,
B. the stable phases in the Cu-55 wt% Zn alloy are: $\beta + \gamma$,
C. the stable phases in the Cu-75 wt% Zn alloy are: $\gamma + \varepsilon$,
D. the stable phases in the Cu-90%Zn alloy are: $\varepsilon + \eta$.

EXAMPLE 13.6 IDENTIFYING PERITECTIC PHASE TRANSFORMATIONS IN CU-ZN ALLOYS

Refer to Figure 13.3. What peritectic phase-transformation reactions occur at the following peritectic temperatures: (a) 835°C, and (b) 697°C?

SOLUTION

(a) $L + \beta \rightarrow \gamma$
(b) $L + \gamma \rightarrow \delta$

EXAMPLE 13.7 IDENTIFYING PERITECTIC COMPOSITIONS IN CU-ZN ALLOYS

What peritectic compositions correspond to the following peritectic temperatures: (a) 835°C, and (b) 697°C?

SOLUTION

By reference to Figure 13.3,

(a) the temperature of 835°C corresponds to the peritectic composition: 40%Cu-60%Zn.
(b) the temperature of 697°C corresponds to the peritectic composition: 25%Cu-75%Zn.

EXAMPLE 13.8 CALCULATING THE AMOUNTS OF α AND β PHASES IN CU-43 WT% ZN ALLOY

Calculate the amounts of α and β phases in Cu-43wt%Zn alloy at room temperature.

SOLUTION

By reference to Figure 13.2, $C_o = 0.43$, $C_\beta = 0.48$, $C_\alpha = 0.40$
By using Equation 6.15,

$$W_\alpha = \frac{C_\beta - C_o}{C_\beta - C_\alpha} = \frac{0.48 - 0.43}{0.48 - 0.40} = 0.625 = 62.5\%$$

$$W_\beta = 100 - 62.5 = 37.5\%$$

EXAMPLE 13.9 COMPARING SPECIFIC WEIGHTS OF STEEL AND ALUMINUM ALLOY

The densities of the 2024-T81 *Al* alloy and steel are 2.87 and 7.87 g/cm³, respectively. The yield strengths of the same alloys are 450 MPa and 483 MPa, respectively. Compare the specific weights of the two aerospace alloys.

SOLUTION

$$\rho_a = \rho_{Al\ alloy} = 2.87 g/cm^3 \qquad \rho_b = \rho_{steel} = 7.87 g/cm^3$$

$$\sigma_{ys(a)} = \sigma_{ys(Al\ alloy)} = 450 MPa \quad \sigma_{ys(b)} = \sigma_{ys(steel)} = 483 MPa$$

By using Equation 13.3,

$$\text{Ratio of the specific weights} = \frac{W_a}{W_b} = \frac{\rho_a \cdot \sigma_{ys(b)}}{\rho_b \cdot \sigma_{ys(a)}} = \frac{2.87 \times 483}{7.87 \times 450} = 0.39$$

$$= \frac{39}{100} \cong \frac{1}{3}$$

It means that the 2024-T81 *Al* alloy weighs about one-third as compared to steel without a significant sacrifice of yield strength.

EXAMPLE 13.10 CALCULATING THE ZINC EQUIVALENT % FOR A CU-ZN BASE ALLOY

In a Cu-Zn base alloy, the weight percentages of each element are listed as follows:

Cu = 69.55, Zn = 22.52, Sn = 2.02, As = 1.03, Fe = 2.76, Pb = 2.11.

Calculate the zinc equivalent % for the alloy.

SOLUTION

By using Equation 13.5,

$$X = \left[(2 \times wt\% \text{ Sn}) + (1 \times wt\% \text{ As}) + (0.9 \times wt\% \text{ Fe}) + (1 \times wt\% \text{ Pb}) \right]$$
$$+ wt\% \text{ Zn}$$
$$= \left[(2 \times 2.02) + (1 \times 1.03) + (0.9 \times 2.76) + (1 \times 2.11) \right] + 22.52$$
$$X = 32.18$$

By using Equation 13.6,

$$Y = X + wt\% \ Cu = 32.18 + 69.55 = 101.73$$

By using Equation 13.4,

$$\text{Zinc Equivalent } \% = \frac{X}{Y} \times 100 = \frac{32.18}{101.73} \times 100 = 31.63$$

The zinc equivalent for the Cu-Zn base alloy = 31.63 wt%.

EXAMPLE 13.11 IDENTIFYING THE MICROSTRUCTURAL PHASE WHEN THE ZN EQUIVALENT IS KNOWN

By using the data in Example 13.10, identify the microstructural phase in the Cu-base alloy.

SOLUTION

In Example 13.10, we calculated the zinc equivalent for the Cu-Zn base alloy = 31.63 wt%. By reference to Figure 13.3, the stable phase in the microstructure of the alloy at room temperature (for 31.63 wt% Zn) is the **α-phase**.

EXAMPLE 13.12 CALCULATING THE CREEP CONSTANT IN S⁻¹ FOR A NI- BASE SUPERALLOY

A steady-state strain rate of 5×10^{-5} s^{-1} is observed in the Ni-base superalloy *720Li* at temperatures in the range of 550–750°C. For the alloy 720Li, a value of $Q_c^* \cong 330 \text{KJ mol}^{-1}$ is established. The yield stress and the ultimate tensile strength at 550°C are determined to be 1,088 and 1,500 MPa, respectively. Take $n = 8$. Calculate the constant A^* (in Equation 13.7) for the alloy 720Li.

SOLUTION

$$\dot{\varepsilon}_{ss} = 5 \times 10^{-5} s^{-1,} T = 550 + 273 = 823K, Q_c^* = 330,000 \text{ J mol}^{-1},$$
$$\sigma = 1088 \text{ MPa}, \sigma_{ut} = 1500 \text{ MPa}$$

By using Equation 13.7,

$$\dot{\varepsilon}_{ss} = A^* \left(\frac{\sigma}{\sigma_{ut}} \right)^n \exp\left(-\frac{Q_c^*}{RT} \right)$$

$$5 \times 10^{-5} = A^* \left(\frac{1088}{1500} \right)^8 \exp\left(-\frac{330,000}{8.3145 \times 823} \right)$$

$$5 \times 10^{-5} = A^* (0.0766) \exp(-48.225)$$

$$A^* = 5.73 \times 10^{17} \, s^{-1}$$

The creep constant $A^* = 5.73 \times 10^{17} \, s^{-1}$.

EXAMPLE 13.13 ESTIMATING THE CREEP STRAIN RATE IN FG SUPERALLOY AT HIGH TEMPERATURE

A creep strain rate of 8.33×10^{-9}/s was observed at 850°C/250 MPa for the AD730 superalloy with a grain size of 350 μm. Estimate the creep strain rate in the superalloy with a grain size of 10 μm at the same testing conditions.

SOLUTION

$$\dot{\varepsilon}(CG) = 8.33 \times 10^{-9} / s, \dot{\varepsilon}(FG) = ?$$

By using Equation 13.8,

$$\dot{\varepsilon}(FG) = 100 * \dot{\varepsilon}(CG) = 100 \times 8.33 \times 10^{-9} = 8.3 \times 10^{-7} / s$$

$$\dot{\varepsilon}(FG) = 8.3 \times 10^{-7} / s$$

EXAMPLE 13.14 COMPARING SPECIFIC WEIGHTS OF STEEL AND TITANIUM ALLOY

The densities of the Ti-6Al-4V alloy and carbon steel are 4.42 and 7.87 g/cm³, respectively. The yield strengths of the same alloys are 800 MPa and 483 MPa, respectively. Compare the specific weights of the two aerospace alloys.

SOLUTION

$$\rho_a = \rho_{Ti \, alloy} = 4.42 g/cm^3 \qquad \rho_b = \rho_{steel} = 7.87 g/cm^3$$

$$\sigma_{ys(a)} = \sigma_{ys(Ti \, alloy)} = 800 MPa \quad \sigma_{ys(b)} = \sigma_{ys(steel)} = 483 MPa$$

By using Equation 13.3,

$$\text{Ratio of the specific weights} = \frac{W_a}{W_b} = \frac{\rho_a \cdot \sigma_{ys(b)}}{\rho_b \cdot \sigma_{ys(a)}} = \frac{4.42 \times 483}{7.87 \times 800} = 0.34 = \frac{34}{100} \cong \frac{1}{3}$$

It means that the Ti-6Al-4V alloy weighs about one-third as compared to carbon steel with a significantly higher yield strength.

EXAMPLE 13.15 CALCULATING THE PERCENTAGE OF GOLD WHEN ITS KARAT IS KNOWN

A gold jewelry is made of 18-karat gold. Calculate the percentage of gold.

SOLUTION

By using Equation 13.9,

$$\% \text{ gold} = \frac{K}{24} \times 100 = \frac{18}{24} \times 100 = 75$$

It means that 18 karat is 75% gold content in the jewelry.

EXAMPLE 13.16 CALCULATING THE WEIGHT OF PURE GOLD WHEN ITS KARAT AND THE JEWELRY WEIGHT ARE KNOWN

An 18K necklace weighs 20 grams. Calculate the weight of pure gold in the necklace.

SOLUTION

By reference to Example 13.15, the percentage of gold in 18K jewelry = 75%
By using Equation 13.10,

$$W_g = \% \text{ gold} \times W_j = 75\% \times 20 = 0.75 \times 20 = 15 \text{ g}$$

It means that the pure gold in the necklace is 15 grams.

QUESTIONS AND PROBLEMS

13.1. Encircle the most appropriate answers for the following statements.
  (a) Which metal has the poorest corrosion resistance?
     (i) wrought iron, (ii) magnesium, (iii) copper, (iv) titanium.
  (b) Which metal's alloy is the most extensively used bio-material?
     (i) iron, (ii) magnesium, (iii) copper, (iv) titanium.
  (c) Which metal's alloy is the most commonly used in hot sections of gas turbines?
     (i) nickel, (ii) magnesium, (iii) copper, (iv) titanium.
  (d) Which alloy has the best ductility?
     (i) superalloy, (ii) bronze, (iii) brass, (iv) Monel.
  (e) Which metal's alloy is the most commonly die cast?
     (i) aluminum, (ii) zinc, (iii) copper, (iv) titanium.

(f) Which metal's alloys are the most widely used in aerospace structures?
(i) aluminum, (ii) zinc, (iii) copper, (iv) titanium.

(g) Which metal's alloy is used in landing gears and compressor blades of aircraft?
(i) aluminum, (ii) zinc, (iii) copper, (iv) titanium.

(h) Which metal's alloys are the most widely used nonferrous material in the world?
(i) aluminum, (ii) zinc, (iii) copper, (iv) titanium.

(i) Which metal has the best specific strength (strength to weight ratio)?
(i) iron, (ii) magnesium, (iii) copper, (iv) titanium.

(j) Which type of superalloy is the best for application in GT blades?
(i) equiaxed, (ii) single crystal, (iii) columnar grains.

(k) Which alloy is the most suitable for application in nuclear reactors?
(i) superalloy, (ii) bronze, (iii) brass, (iv) Monel.

(l) Which metal is extensively used in electrical industry?
(i) aluminum, (ii) zinc, (iii) copper, (iv) titanium.

(m) Which designation series of *Al* alloys refers to solution heat treated, then naturally aged?
(i) xxxx-T3, (ii) xxxx-T4, (iii) xxxx-T5, (iv) xxxx-T36.

(n) Which designation series of *Al* alloys refers to zinc as the main alloying element?
(i) 2xxx, (ii) 4xxx, (iii) 5xxx, (iv) 7xxx.

(o) Which microstructural phase in Ni-base superalloy mainly imparts creep strength?
(i) gamma, γ, (ii) gamma prime, γ', (iii) sigma, σ, (iv) carbide.

(p) Which nonferrous metal is not used as a dental implant?
(i) zinc, (ii) gold, (iii) titanium, (iv) silver.

(q) Which titanium alloy is the most extensively used in engineering applications?
(i) Ti-4Al-6V, (ii) Ti-6Al-6V, (iii) Ti-6Al-4V, (iv) Ti-4Al-4V.

(r) Which nonferrous metal is used as a catalyst in petrochemical processing?
(i) silver, (ii) polonium, (iii) gold, (iv) platinum.

13.2. Draw *Al-Si* phase diagram; and identify the eutectic temperature and composition.

13.3. Why is a small amount of sodium added to the eutectic *Al-Si* alloy in industrial practice?

13.4. Do you recommend a high Si content in Al-Si alloy for engineering applications? Explain!

13.5. List engineering applications of copper and its alloys.

13.6. (a) By reference to Figure 13.3, identify the composition ranges and temperatures for α, $\alpha+\beta$, and β brasses.
(b) Discuss the characteristics and applications of α, $\alpha+\beta$, and β brasses.

13.7. What are bronzes? List the general characteristics and applications of bronzes.

13.8. (a) What is meant by a Ni-base superalloy?

(b) Draw a sketch showing the phases in the microstructure of a Ni-base superalloy.

(c) List the applications of a Ni-base superalloy.

13.9. Highlight the general properties and applications of titanium alloys.

13.10. What are the two allotropic forms of titanium? Specify their stability temperatures.

13.11. Discuss the characteristics of α, $\alpha+\beta$, and β titanium alloys.

P13.12. The 7075-T6 is an engineering Al alloy. (a) What is the principal alloying element in the alloy? (b) What heat treatment has been given to the alloy? (c) Specify the applications.

P13.13. A eutectic Al alloy is cooled from liquid state to 300°C. Calculate the amounts of α-Al and Si phases at the specified temperature.

P13.14. The concentration of solute Si in an Al-Si alloy matrix was measured (by WDXS) to be 0.014. The vacancy-Si binary interaction energy in the alloy is 1.8×10^{-19} J. Calculate the correction term for the interaction between Si solute atoms and vacancies at 500K.

P13.15. The densities of the 7075-T6 Al alloy and carbon steel are 2.81 and 7.87 g/cm^3, respectively. The yield strengths of the same alloys are 500 MPa and 483 MPa, respectively. Compare the specific weights of the two aerospace alloys.

P13.16. Calculate the amounts of α and β phases in Cu-41wt%Zn alloy at 500°C.

P13.17. (a) In a Cu-Zn base alloy, the weight percentages of each element are listed as follows:

Cu = 70, Zn = 22.97, Sn = 2.00, As = 1.05, Fe = 2.80, Pb = 2.07.

Calculate the zinc equivalent % for the alloy.

(b) Identify the microstructural phase in the Cu-Zn base alloy.

P13.18. A steady-state strain rate of 5×10^{-5} s^{-1} is observed in the Ni-base superalloy *720Li* at temperatures in the range of 550–750°C. For the alloy 720Li, a value of $Q_c^* \cong 330$ kJ mol^{-1} is established. The yield stress and the ultimate tensile strength at 700°C are determined to be 927 and 1254 MPa, respectively. Taking n=8, calculate the constant A^* (in Equation 13.7) at 700°C for the alloy 720Li.

P13.19. A gold ring made of 22-karat gold weighs 12 grams. Calculate the weight of pure gold in the ring.

REFERENCES

Harrison, W., Whittaker, M. & Williams, S. (2013) Recent advances in creep modelling of the nickel base superalloy, alloy 720Li. *Materials (Basel)*, 6(3), 1118–1137.

Huda, Z. (1995) Development of design principles for a creep-limited alloy for turbine blades. *Journal of Materials Engineering and Performance*, 4(1), 48–53. DOI: https://doi.org/10.1007/BF02682704

Huda, Z. (2016) Materials selection in design of structures of subsonic and supersonic aircrafts. In: Liu, Y (ed) *Frontiers in Aerospace Science*, Vol. 1. Bentham Science Publishers, Sharjah, pp. 457–481.

Huda, Z. (2017) Energy-efficient gas-turbine blade-material technology—*A review. Materiali in Tehnologije (Materials and Technology)*, 3(51), 355–361. Online: http://mit.imt.si/Revija/izvodi/mit173/huda.pdf.

Huda, Z., Taib, N.I. & Zaharinie, T. (2009) Characterization of 2024-T3: An aerospace aluminum alloy. *Materials Chemistry and Physics*, 113, 515–517.

Huda, Z., Zaharinie, T., Metselaar, I.H.S.C., Ibrahim, S. & Min, G.J. (2014) Kinetics of grain growth in 718 Ni-base superalloy. *Archives of Metallurgy and Materials*, 59(3), 847–852. DOI: https://doi.org/10.2478/amm-2014-0143

Kaufman, J.G. (2000) *Introduction to Aluminum Alloys and Tempers*. ASM International, Materials Park, OH.

Lomer, W.M. (1958) Point defects and diffusion in metals and alloys. In: *Vacancies and Other Point Defects in Metals and Alloys, Institute of Metals*, Monograph and report series, 23. The Institute of Metals, London, pp. 79–86.

Roswell, A.E. & Nowick, A.S. (1953) Decay of lattice defects frozen into an alloy by quenching. *Transactions, Metals Society, American Institute of Mining, Metallurgical, and Petroleum Engineers (AIME)*, 197, 1259–1266.

Tajally, M., Huda, Z. & Masjuki, H.H. (2010) A comparative analysis of tensile and impact-toughness behavior of cold-worked and annealed 7075 aluminum alloy. *International Journal of Impact Engineering*, 37, 425–432.

Thebaud, L., Villechaise, P., Coraline, C. & Cormier, J. (2018) Is there an optimal grain size for creep resistance in Ni-based disk superalloys? *Materials Science & Engineering A*, 716, 274–283.

Zamani, M. (2015) *Al-Si Cast Alloys—Microstructure and Mechanical Properties at Ambient and Elevated Temperatures*. Licentiate Thesis, School of Engineering, Jönköping University, Sweden.

Part IV

Thermal Processing and Surface Engineering

14 Recrystallization and Grain Growth

14.1 THE THREE STAGES OF ANNEALING

It is learned in the preceding chapter that cold working results in an increase in hardness and reduction in ductility. A cold-worked specimen, being in a state of higher energy, is thermodynamically unstable. In order to restore ductility and to transform the cold-worked metal to a lower-energy state, it is heated through a sequence of *annealing* processes: (1) *recovery*, (2) *recrystallization*, and (3) *grain growth* (Rollett *et al.*, 2017). These three stages involve microstructural changes that occur during annealing after cold working or during hot working.

Recovery, the first stage of annealing, involves the relief of internal stresses due to cold working. Recovery is followed by *recrystallization*; which occurs by the nucleation and growth of new grains, which are essentially strain-free, at the expense of the polygonized matrix (Figure 14.1). Once recrystallization is complete, further annealing increases the average size of the grains; this process is called *grain growth*. The grain size and the mechanical properties of an annealed metal strongly depend on the annealing temperature (see Figure 14.1).

14.2 RECRYSTALLIZATION

It is the recrystallization stage of annealing that mainly restores the ductility of CW metal. Recrystallization refers to the rearrangement and reorientation of the elongated CW grains to refined equiaxed grains when a metal is heated at its recrystallization temperature, T_r. Recrystallization temperature, T_r, is the minimum temperature at which complete recrystallization occurs in an annealed cold-worked metal within a specified time.

The recrystallization temperature, T_r, is related to melting temperature, T_m, of a metal by:

$$T_r \sim 0.4 T_m \qquad (14.1)$$

where the temperatures are measured in Kelvin (K).

Recrystallization is a time-dependent process; and the time dependence of the fraction of statically recrystallized material is commonly described through Avrami general equation (Tajjaly and Huda, 2011; Callister, 2007):

$$X = 1 - exp\left(-kt^n\right) \qquad (14.2)$$

where X is the fraction of recrystallization; t is the annealing time; and k and n are time-independent constants for the particular process. The kinetics of static recrystallization are often characterized in terms of the time required for recrystallization of 50% of the material, i.e. $t_{0.5}$.

The rate of recrystallization ($Rate_{ReX}$) can be computed as the reciprocal of the time required for recrystallization of 50% of the material, i.e. $t_{0.5}$., as follows:

$$Rate_{reX} = \frac{1}{t_{0.5}} \qquad (14.3)$$

In metallographic practice, it is often necessary to determine the fraction recrystallized. The determination of fraction recrystallized usually involves quenching the material and mapping the locations of grains, and their sizes and shapes by use of a metallurgical microscope; this quantitative metallographic technique enables us to find the ratio of recrystallized grains. However, this is a complicated technique. Horsinka and co-researchers (2012) have reported a convenient stress-relaxation testing method to compute the recrystallization fraction. This testing method involves heating to a required temperature, cooling down to deformation temperature, deformation at a prescribed rate, and then holding the strain constant while recording the force (or stress) and time (from the start of the hold) (Horsinka et al., 2012). By using the stress-relaxation test, it is possible to calculate the recrystallization fraction, X, by the following formula:

$$X = \frac{[\sigma_{01} - (\alpha_1 \log t)] - \sigma}{(\sigma_{01} - \sigma_{03}) - (\alpha_1 - \alpha_3)\log t} \qquad (14.4)$$

where α_1 is the slope of the stress-relaxation curve (recovery stage); α_3 is the slope of the curve (final stage); σ_{01} and σ_{03} are the stress values during the recovery and final stage, respectively; σ is the stress value in the second part of the curve, MPa; and t is the relaxation time, s.

14.3   GRAIN GROWTH

Grain growth is the third stage of annealing; it refers to the increase in the average grain size of the metal when the metal is heated beyond the completion of recrystallization (see Figure 14.1). The control of grain growth is great technological importance because many material properties strongly depend on the grain size and its distribution (Huda et al., 2014; Huda and Ralph, 1990).

Many polycrystalline materials follow the grain-growth kinetic behavior as follows (Huda, 2004):

$$D^m - D_0{}^m = Kt \qquad (14.5)$$

where D is the average grain diameter at the end of isothermal annealing at a specified temperature for a time duration of t; K is a temperature-dependent constant, D_0 is the pre-growth grain diameter; and m is the reciprocal of the grain-growth exponent, n. For most metals, the value of $n=0.5$ i.e. $m=2$ (Huda et al., 2014); thus Equation 14.5 may be rewritten as:

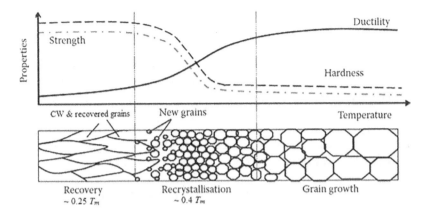

FIGURE 14.1 The effects of annealing temperature on microstructure and mechanical properties (CW = cold worked; T_m = melting temperature).

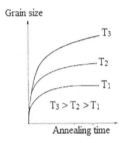

FIGURE 14.2 Effects of time and temperature on grain growth (T_1, T_2, and T_3 are the annealing temperatures).

$$D^2 - D_0^{\,2} = Kt \tag{14.6}$$

For a polycrystalline material with extremely fine grain size, the original grain diameter may be taken as zero i.e. $D_0 = 0$. Thus Equation 14.5 takes the form:

$$D^m = Kt \tag{14.7}$$

Since m is the reciprocal of the grain-growth exponent, n, i.e. $m = 1/n$; Equation 14.7 becomes:

$$D = kt^n \tag{14.8}$$

where k is a new constant. This equation (Equation 14.8) is valid only for fine-grained materials.

The kinetics of grain growth depends on three main factors: (a) time, (b) temperature, and (c) the presence of second-phase particles. During isothermal annealing, the grains first grow fast but the grain-growth becomes slower with increase in annealing time (see any curve in Figure 14.2). The rate of grain growth increases with increasing temperature (Figure 14.2). This is why overheating of a metal during its hot working must be avoided else extensive grain growth would "burn" the metal. The presence of second-phase particles also strongly influences the rate of grain growth by particle pinning i.e. fine second-phase particles inhibit the rate of grain growth (Huda, 2004). This is why many alloy steels are strengthened by adding nickel; the latter acts as a grain refiner.

14.4   CALCULATIONS—WORKED EXAMPLES ON RECRYSTALLIZATION AND GRAIN GROWTH

EXAMPLE 14.1 COMPUTING THE RECRYSTALLIZATION TEMPERATURE

The melting temperature of aluminum is 660°C and that of lead is 327°C. Compute the recrystallization temperatures of the two metals.

SOLUTION

Melting temperature of aluminum = T_m = 660°C = 660 + 273 = 933K
By using Equation 14.1,

$$T_r = 0.4T_m = 0.4 \times 933 = 373.2K = 373.2 - 273 = 100.2°C$$

Recrystallization temperature of aluminum = 101°C.
   Melting temperature of lead = T_m = 327°C = 327 + 273 = 600K
   Recrystallization temperature of lead = T_r = 0.4 T_m = 0.4 x 600 = 240K = 240–273 = –33°C.

EXAMPLE 14.2 DECIDING WHETHER OR NOT A METAL NEEDS RECRYSTALLIZATION ANNEALING

Two different samples of aluminum and lead metals were formed at room temperature (25°C). Which of the two metals does not require a furnace for recrystallization annealing? Explain!

SOLUTION

We refer to the data obtained in the Solution of Example 14.1. For aluminum, the working temperature (25°C) is below the recrystallization temperature of Al (101°C). It means aluminum was cold worked at 25°C; hence Al needs to be recrystallized in a furnace.
   For lead, the working temperature (25°C) is above the T_r of lead (–33°C). It means that lead was hot worked at 25°C; hence lead does NOT need to be recrystallized in a furnace.

EXAMPLE 14.3 APPLICATION OF THE AVRAMI EQUATION TO DETERMINE THE CONSTANT K

The kinetics of recrystallization for some alloy follow the Avrami relationship (Equation 7.2), and that the value of n in the exponential is 2.3. If, after 120 s, 50% recrystallization is complete, compute the constant k.

SOLUTION

For the Avrami relationship, $n = 2.3$, $t = 120$ s, $X = 50\% = 0.50$, $k = ?$
By using Equation 14.2,

$$X = 1 - \exp\left(-kt^n\right)$$

$$\text{or} \quad \exp\left(-kt^n\right) = 1 - X$$

$$\text{or} \quad -kt^n = \ln\left(1 - X\right)$$

$$\text{or} \quad k = -\left\{\left[\ln\left(1 - X\right)\right]/t^n\right\}$$

$$k = -\left\{\left[\ln\left(1 - 0.5\right)\right]/120^{2.3}\right\} = -\left\{\left[\ln\left(0.5\right)\right]/60550.3\right\} = 1.14 \times 10^{-5}$$

$$k = 1.14 \times 10^{-5}$$

EXAMPLE 14.4 USE OF THE AVRAMI EQUATION TO DETERMINE THE RECRYSTALLIZATION TIME

By using the data in Example 14.3, determine the total time to complete 95% recrystallization.

SOLUTION

For the Avrami relationship, $n = 2.3$, $X = 95\% = 0.95$, $k = 1.14 \times 10^{-5}$, $t = ?$

$$X = 1 - \exp\left(-kt^n\right)$$

$$\text{or} \quad t^n = -\left\{\left[\ln\left(1 - X\right)\right]/k\right\}$$

$$t = \left\{-\left[\ln\left(1 - X\right)\right]/k\right\}^{1/n}$$

$$t = \left\{-\left[\ln\left(1 - 0.95\right)/\left(1.14 \times 10^{-5}\right)\right]\right\}^{1/2.3} = \left\{-\left[\ln\left(0.05\right)/\left(1.14 \times 10^{-5}\right)\right]\right\}^{0.434}$$

$$t = \left\{-\left[\left(-2.62 \times 10^5\right)\right]\right\}^{0.434} = \left(2.62\right)^{0.434} \times 10^{2.17} = 1.52 \times 148 = 225 s$$

The total time to complete 95% recrystallization = 225 s = 3.74 min.

EXAMPLE 14.5 CALCULATING THE RATE OF RECRYSTALLIZATION

Determine the rate of recrystallization of the alloy with the parameters stated in Example 7.3.

SOLUTION

In Example 14.3, time to complete 50% recrystallization = $t_{0.5}$ = 120 s. By using Equation 14.3,

$$Rate_{Re\,X} \frac{1}{t_{0.5}} = \frac{1}{120} = 0.008\,/\,s$$

The rate of recrystallization of the alloy = 0.008 s^{-1}.

EXAMPLE 14.6 COMPUTING THE FRACTION OF RECRYSTALLIZATION WHEN N AND K ARE UNKNOWN

A cold-worked aluminum sample was recrystallized at 350°C for 95 min; this annealing treatment resulted in 30% recrystallized microstructure. Another sample of the same CW Al metal was recrystallized for 127 min at the same temperature; which resulted in 80% recrystallized structure. Determine the fraction recrystallized after an annealing time of 110 min, assuming that the kinetics of this recrystallization process obey the Avrami equation.

SOLUTION

By using Equation 14.2,

$$X = 1 - \exp\left(-kt^n\right)$$
$$or$$
$$k = -\left\{\left[ln\left(1 - X\right)\right]/t^n\right\}$$

X = 0.3 when t = 95 min

$$k = -\left\{\left[ln\left(1 - 0.3\right)\right]/95^n\right\} \tag{E-7.6a}$$
$$k = 0.35/95^n$$

X = 0.8 when t = 110 min

$$k = -\left\{\left[ln\left(1 - 0.8\right)\right]/110^n\right\} \tag{E-7.6b}$$
$$k = 1.6/127^n$$

By solving Equations (E-14.6a) and (E-14.6b) simultaneously, we obtain: n = 5.2, k = 1.8 x 10^{-11}.

In order to determine the fraction recrystallized after an annealing time, t = 110 min,

$$X = 1 - \exp\left(-kt^n\right) = 1 - \left\{\exp[-\left(1.8 \times 10^{-11}\right) 110^{5.2}\right]\}$$

$$= 1 - \left[\exp\left(-0.74\right)\right] = 1 - 0.47 = 0.52$$

$$X = 0.52$$

The fraction recrystallized after an annealing time of 110 min = 0.52.

EXAMPLE 14.7 COMPUTING THE STRESS FOR RECRYSTALLIZED FRACTION BASED ON STRESS-RELAXATION TEST

A metal sample was recrystallized by using the stress-relaxation testing method. Calculate the stress value in the second part of the stress-relaxation curve based on the following data:

The slope of the stress-relaxation curve (recovery stage) = 25 MPa;
The slope of the stress-relaxation curve (final stage) = 8 MPa;
Stress during the recovery stage = 65 MPa;
Stress during the final stage = 20 MPa;
Time to complete 50% recrystallization = 2.3 s; n = 1.3.

SOLUTION

$\alpha_1 = 25$ MPa, $\alpha_3 = 8$ MPa, $\sigma_{01} = 65$ MPa, $\sigma_{03} = 20$ MPa, $n = 1.3$, $X = 0.5$, $t_{0.5} = 2.3$ s, $\sigma = ?$

By using Equation 14.4,

$$X = \left\{\left[\sigma_{01} - \left(\alpha_1 log t\right)\right] - \sigma\right\} / \left\{\left(\sigma_{01} - \sigma_{03}\right) - \left(\alpha_1 - \alpha_3\right) log t\right\}$$

$$0.5 = \left\{\left[65 - \left(25 log\left(2.3\right)\right)\right] - \alpha\right\} / \left\{\left(65 - 20\right) - \left(25 - 8\right) log\left(2.3\right)\right\}$$

$$0.5 = \left\{\left[65 - \left(25 \times 0.36\right)\right] - \alpha\right\} / \left\{\left(45\right) - \left(17 \times 0.36\right)\right\}$$

$$\sigma = 36.5 \text{ MPa}$$

The stress value in the second part of the stress-relaxation curve = 36.5 MPa.

EXAMPLE 14.8 COMPUTING ANNEALING TIME FOR GRAIN-GROWTH PROCESS FOR A METAL

It takes 3,000 minutes to increase the average grain diameter of brass from 0.03 to 0.3 mm at 600°C. Assuming that the grain-growth exponent of brass is equal to 0.5, calculate the time required to increase the average grain diameter of brass from 0.3 to 0.6 mm at 600°C.

SOLUTION

For $D_0 = 0.03$ mm and $D = 0.3$ mm, $t = 3,000$ min

By using Equation 14.6,

$$D^2 - D_0^2 = Kt$$
$$(0.3)^2 - (0.03)^2 = K\,(3000)$$
$$K = 2.9 \times 10^{-5}$$

Now, we can compute t for $D_0 = 0.3$ mm and $D = 0.6$ mm; and $K = 2.9 \times 10^{-5}$ (since the temperature is the same) by using Equation 14.6, as follows:

$$(0.6)^2 - (0.3)^2 = \left(2.9 \times 10^{-5}\right) t$$
$$t = 9300 \text{ min}$$

Hence, the time required to increase the average grain diameter of brass from 0.3 to 0.6 mm at 600°C = 9,300 minutes.

EXAMPLE 14.9 COMPUTING ANNEALING TIME WHEN THE GRAIN- GROWTH EXPONENT IS UNKNOWN

An extremely fine-grained material was annealed for 2,000 minutes to obtain an average grain diameter of 0.02 mm at temperature T. In another experiment, the same fine-grained material was annealed at T for 6,000 minutes and the resulting average grain diameter was 0.03 mm. Compute: (a) the grain-growth exponent for the material, and (b) the annealing time to obtain an average grain diameter of 0.04 mm at the same temperature T.

SOLUTION

Since this problem deals with a fine-grained material ($D_0=0$), Equation 14.7 is applicable.

(a) By using Equation 14.7 for the first experiment ($D = 0.02$ mm, $t = 2,000$ min),

$$D = k\,t^n,$$
$$0.02 = k\,(2000)^n \qquad\qquad\qquad \text{(E-14.9a)}$$

By using Equation 14.7 for the second experiment ($D = 0.04$ mm, $t = 6,000$ min),

$$0.03 = k\,(6000)^n \qquad\qquad\qquad \text{(E-14.9b)}$$

By solving the two equations simultaneously, we get: $n = 0.37$, and $k = 0.0012$
    Hence, the grain-growth exponent of the material = 0.37.

(b) In order to compute the annealing time for D=0.04 mm, we again use Equation 14.7:

$$D = kt^n$$
$$0.04 = 0.0012 \, t^{0.37}$$
$$\log 0.04 = \log (0.0012) + (0.37) \log t$$
$$\log t = 4.13, \quad t = 10^{4.13} = 13{,}490$$

The required annealing time = t = 13,490 min.

QUESTIONS AND PROBLEMS

14.1. Explain the effects of annealing temperature on the microstructure and mechanical properties of a metal with the aid of a diagram.

14.2. Compare the metallographic technique of determining the fraction recrystallized with stress-relaxation technique.

14.3. Show that grain growth behavior in a fine-grained material may be expressed as: $D = kt^n$.

14.4. Why do hot-worked metals not generally require recrystallization annealing treatment?

14.5. Do you recommend heating a metal well above its recrystallization temperature during the metal's hot working? Justify your answer.

14.6. Why is it important to add nickel to stainless steels?

P14.7. The melting temperature of tin is 232°C and that of copper is 1080°C. Compute the recrystallization temperatures of the two metals.

P14.8. A cold-worked aluminum alloy was recrystallized at 370°C for 100 min; this annealing treatment resulted in 35% recrystallized microstructure. Another sample of the same CW *Al* metal was recrystallized for 120 min at the same temperature; which resulted in 75% recrystallized structure. Determine the fraction recrystallized after an annealing (total) time of 140 min (assume that the kinetics of the recrystallization obeys the Avrami equation).

P14.9. A fine-grained metal was annealed for 1,500 minutes to obtain an average grain diameter of 0.010 mm at temperature *T*. Another sample of the same material was annealed at *T* for 5,000 minutes and the resulting average grain diameter was 0.017 mm. Compute the grain- growth exponent.

P14.10. By using the data in P14.9, calculate the annealing time to obtain an average grain diameter of 0.02 mm for the metal sample at the same temperature *T*.

REFERENCES

Huda, Z., Zaharinie, T., Metselaar, I.H.S.C., Ibrahim, S. & Min, G.J. (2014) Kinetics of grain growth in 718 Ni-base superalloy. *Archives of Metallurgy and Materials*, 59(3), 847–852.

320 Metallurgy for Physicists and Engineers

Callister, W.D. (2007) *Materials Science and Engineering: An Introduction*. John Wiley & Sons Inc, Hoboken, NJ.

Horsinka, J., Kliber, J., Knapinski, M. & Mamuzic, L. (2012) Study of the kinetics of static recrystallization using Avrami Equation and stress relaxation method. *Metal*, 23(May 25).

Huda, Z. (2004) Influence of particle mechanisms on the kinetics of grain growth in a P/M superalloy. *Materials Science Forum*, 467–470, 985–990.

Huda, Z., Zaharinie, T., Metselaar, I.H.S.C., Ibrahim, S. & Min, G.J. (2014), Kinetics of grain growth in 718 Ni-base Superalloy. *Archives of Metallurgy and Materials*, 59(3), 847–852. DOI: https://doi.org/10.2478/amm-2014-0143

Huda, Z. & Ralph, B. (1990) Kinetics of grain growth in powder formed IN-792 superalloy. *Materials Characterization*, 25, 211–220.

Rollett, A., Rohrer, G.S. & Humphreys, J. (2017) *Recrystallization and Related Annealing Phenomena*. Elsevier Publishers, New York.

Tajjaly, M. & Huda, Z. (2011) Recrystallization kinetics for 7075 aluminum alloy. *Metal Science and Heat Treatment*, 53(5–6), 213–217.

15 Heat Treatment of Metals

15.1 HEAT TREATMENT OF METALS—AN OVERVIEW

Heat treatment is the process that involves controlled heating and cooling of metals to improve their physical and mechanical properties by controlling the microstructure. The purpose of heat treatment may be to: (a) increase the strength, (b) increase/decrease hardness, (c) improve machinability, (d) improve formability, or (e) restore ductility after a cold-working operation. Heat-treatment processes include: full annealing, normalizing, hardening, tempering, recrystallization annealing, case hardening, precipitation strengthening, solution treatment, aging treatment, and the like (Zakharov, 1962; Huda, 2007). Many engineering components are given heat treatment for specific applications; these include: automotive components, aerospace components, and the like.

15.2 HEAT TREATMENT OF STEELS

15.2.1 THE PROCESSES IN HEAT TREATMENT OF CARBON STEELS

Heat treatment of steel is an important manufacturing process since it improves the mechanical properties of the metal by controlling its microstructure. In particular, hardness, ductility, and strength properties can be significantly improved by heat treatment. Heat-treatment processes in steels include: full annealing, normalizing, hardening, tempering, process (recrystallization) annealing, case hardening, and the like (Totten and Howes, 1997). These heat-treatment processes are described in sections 15.2.3–15.2.8, the recrystallization thermal process has been explained in Chapter 14. The understanding of heat treatment of carbon steels requires an ability to interpret the Fe-Fe$_3$C phase diagram, which is explained in the following subsection.

15.2.2 EQUILIBRIUM COOLING OF CARBON STEELS

The equilibrium cooling of plain-carbon steel containing 0.77 wt% C (eutectoid steel) has been explained with reference to eutectoid reaction in Chapter 7 (see section 7.5). It was established that the microstructure of slowly cooled eutectoid steel (Fe-0.77 wt% C) is pearlite; which is comprised of 88.7 wt% ferrite and 11.3 wt% cementite. The equilibrium cooling of hypo-eutectoid and hyper-eutectoid steels can be illustrated with reference to the steel portion of the iron-carbon phase diagram, as shown in Figure 15.1.

Let us consider the slow cooling of a hypo-eutectoid steel. When a hypo-eutectoid steel (say Fe-0.4 wt% C alloy) is slowly cooled from austenitic range (say 850°C) to

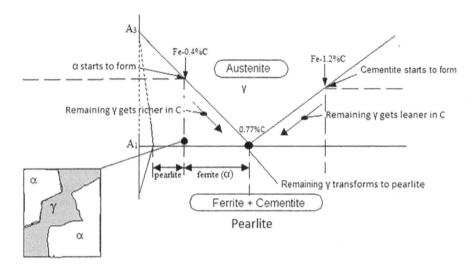

FIGURE 15.1   Equilibrium cooling of hypo- and hyper-eutectoid steels.

a temperature about 770°C, **_proeutectoid ferrite_** starts to form at the austenite (γ) grain boundaries (see the microstructure in Figure 15.1). On further cooling to a temperature just above A_1, the remaining γ becomes enriched in carbon so that the carbon content of the remaining γ will be increased from 0.4 to 0.8 wt%C. At the eutectoid (A_1) temperature (723°C), the remaining γ will transform to pearlite; here the ferrite in the pearlite is called **_eutectoid ferrite_**. Thus, the weight percentage of total ferrite in the microstructure of carbon steel can be determined as follows:

$$Wt\% \text{ total ferrite} = wt\% \text{ eutectoid ferrite}$$
$$+ wt\% \text{ proeutectoid ferrite} \qquad (15.1)$$

The significance of Equation 15.1 is illustrated in Example 15.1.

Now, we consider the equilibrium cooling of a hyper-eutectoid steel. When a hyper-eutectoid steel (say Fe-1.2 wt% C alloy) is slowly cooled from austenitic range to a temperature about 770°C, **_proeutectoid cementite_** starts to form at the γ grain boundaries. At the temperature A_1, the remaining γ (which has become leaner in carbon from 1.2 to 0.8 wt%C) transforms into pearlite; here the cementite in the pearlite is called **_eutectoid cementite_**. Hence, we can write:

$$Wt\% \text{ total cementite} = wt\% \text{ eutectoid cementite}$$
$$+ wt\% \text{ proeutectoid cementite} \qquad (15.2)$$

15.2.3   HEAT-TREATMENT TEMPERATURES RANGES AND THERMAL CYCLES FOR CARBON STEELS

In general, heat-treatment processes in carbon steels include: full annealing, normalizing, hardening, tempering, process (recrystallization) annealing, spheroidizing annealing, and stress relief annealing. The heating temperatures for these

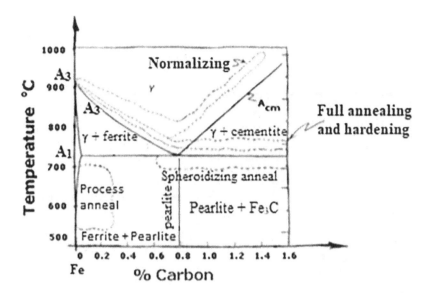

FIGURE 15.2 Partial Fe-Fe$_3$C phase diagram showing temperature ranges for various heat-treatment processes in carbon steels.

heat-treatment processes in steel are illustrated in the steel portion of Fe-Fe$_3$C phase diagram (Figure 15.2). The phase diagram in Figure 15.2 enables us to specify the heating temperature ranges for various heat-treatment processes; the latter are explained in the following subsection.

Since heat treatment of a metal involves controlled heating and cooling, it is useful to specify thermal cycles for the heat-treatment processes in steels. The thermal cycles for various heat-treatment processes in carbon steels are illustrated in Figure 15.3. At this stage, it is important to refer to the phase transformations in steel as discussed in Chapter 7 (section 7.5).

15.2.4 FULL ANNEALING OF CARBON STEEL

Full annealing of steel involves the following steps:

1. Slowly raising the temperature of steel about 30°C above the line A$_3$ (in case of hypo-eutectoid steels) or 30°C above line A$_1$ (in case of hypereu-tectoid steels) (see Figure 15.2);
2. Holding steel at this temperature for sufficient time to allow complete phase transformation into austenite or austenite-cementite as the case may be;
3. The steel is then slowly cooled at the rate of about 20°C/h in a furnace to about 50°C into the ferrite-cementite range (see Figure 15.3). At this point, it may be cooled in room temperature in air with natural convection.

Full annealing of eutectoid steel (Fe-0.77%C alloy) results in coarse pearlite micro-structure (see Chapter 7, Figure 7.8). It is evident in Figure 15.2 that full annealing

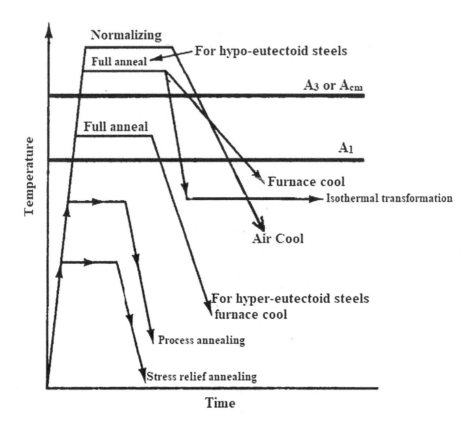

FIGURE 15.3   Thermal cycles for various heat-treatment processes in carbon steels.

of hypo-eutectoid steel would result in ferrite+pearlite microstructure (see Example 15.2). The purpose of full annealing is to improve machinability and to reduce hardness and brittleness.

15.2.5   NORMALIZING OF CARBON STEEL

Normalizing of steel involves the following steps:

1.   Slowly raising the temperature of steel about 30°C above the line A_3 (in case of hypo- eutectoid steels) or 30°C above line A_{cm} (in case of hypereutectoid steels) (see Figure 15.2);
2.   Holding steel at this temperature for sufficient time to allow complete phase transformation into austenite (see Figure 15.2);
3.   The steel is then cooled in room temperature in air with natural convection (see Figure 15.3).

Normalizing of eutectoid steel (Fe-0.77%C alloy) results in fine pearlitic microstructure. The purpose of normalizing is to improve strength and toughness.

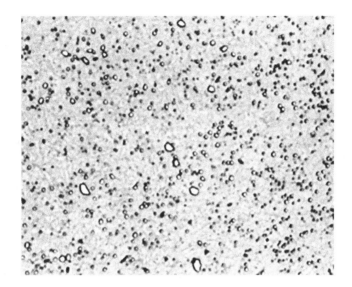

FIGURE 15.4 Microstructure of spheroidizing annealed steel.

15.2.6 Spheroidizing Annealing of Carbon Steel

Spheroidizing of steel involves heating a high-carbon steel to a temperature just below the line A_1, hold the temperature for a prolonged time followed by fairly slow cooling (see Figure 15.2). This heat treatment results in a microstructure in which all the cementite is in the form of small globules (spheroids) dispersed throughout the ferrite matrix (Figure 15.4). This microstructure improves machining as well as the resistance to abrasion.

15.2.7 Hardening of Steel

15.2.7.1 The Process of Hardening Steel

The purpose of hardening is to increase the hardness. Hardness is important not only in cutting tools, but also in many engineering components and structures.

Hardening of carbon steel involves the following steps:

1. Slowly raising the temperature of steel about 30°C above the line A_3 (in case of hypo- eutectoid steels) or 30°C above line A_1 (in case of hypereutectoid steels) (see Figure 15.2);
2. Holding steel at this temperature for sufficient time to allow complete phase transformation into austenite or austenite-cementite as the case may be;
3. Quenching in either cold water or oil.

The microstructure of hardened steel is either martensite (if cold-water cooled) or bainite (if oil cooled) (see Chapter 7, Figures 7.11–7.12). Unusual combination of properties (strength combined with ductility and toughness) can be achieved in

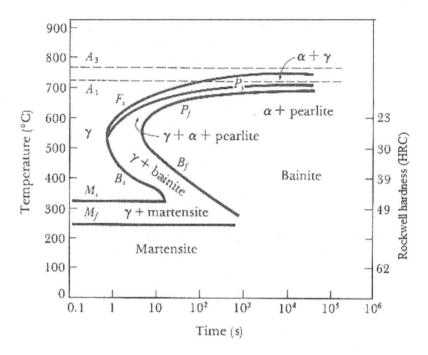

FIGURE 15.5   TTT diagram for a 1050 steel.

carbon steels by obtaining a dual-phase 50% ferrite-50% martensite microstructure; which requires special hardening heat treatment, as follows. A hypo-eutectoid steel is heated into the $\alpha + \gamma$ region of the phase diagram (Figure 15.2) and held for sufficient time to allow phase transformation; the steel is then quenched, thereby permitting the γ portion of the structure to transform to martensite (see Example 15.3).

It is important to mention that the design of a hardening heat treatment of a carbon steel component is mainly governed by TTT diagram. For example, the heat treatment of a 1050 steel motor-car axle hub requires reference to the TTT diagram, as shown in Figure 15.5 (see Example 15.4). This author has reported a successful heat treatment design for hardening a forged 1050 steel axle hub (Huda, 2012).

15.2.7.2   Hardenability and Jominy Test

The hardenability of a steel is the depth up to which it is hardened after the hardening process. It can be experimentally determined by the *Jominy test for hardenability*. In the *Jominy test*, a round steel bar of standard size is transformed to 100% austenite through heating for 30 minutes, and is then quenched on one end with room-temperature water. The water spray is extended vertically 2.5" (6.35 cm) without the sample in place, and the bottom of the sample should be 0.5" (1.3 cm) above the water opening (see Figure 15.6). Water is applied for at least 10 minutes. The cooling rate will be the highest at the end being quenched, and will decrease as the distance from the end increases. Then after cooling, a flat surface is ground on the test piece and the hardenability is determined by measuring the hardness along the bar.

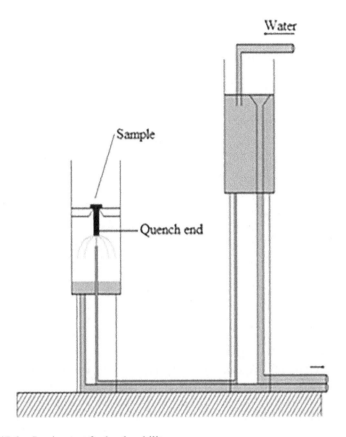

FIGURE 15.6 Jominy test for hardenability.

Hardenability curves can be obtained by plotting the Rockwell hardness number versus the distance from the quench end. The farther away from the quenched end that the hardness extends, the higher the hardenability. Hardenability measurement experiments have indicated that a value of $50HR_C$ can be obtained for various carbon and alloy steels samples (each containing 0.4 wt% C) as shown in Table 15.1 (Callister, 2007).

15.2.7.3 Calculating Hardenability of Structural Steels

Hardenability is expressed in terms of *ideal critical diameter* (D_i). It is the diameter at which 50% martensite can be obtained in the core in the case of ideal cooling. The ideal critical diameter $(D_i$, mm) can be calculated using the factors f_i, every one of which depends on the absolute content of the alloying elements in the steel, by (Nosov and Yurasov, 1995):

$$D_i = 25.4 \, f_C f_{Mn} f_{Si} f_{Mo} f_{Ni} f_V f_{Cu} \tag{15.3}$$

where C, Mn, Si, Mo, Ni, V, and Cu refer to the alloying elements contents corresponding to the factor f_i. The f_i values are computed as illustrated in Table 15.2 (see Examples 15.5–15.6).

TABLE 15.1

Hardenability Data for 50HR$_C$ for Various Steels

Steel AISI No.	1040	5140	8640	4140	4340
Distance from quenched end	1 mm	8 mm	12 mm	18 mm	50 mm

TABLE 15.2

Formulas for Calculating the Multipliers f_i for Ideal Critical Diameter

Element	Wt% element	Formula	Equation no.
Carbon	< 0.40	$f_C = 0.54 \ (\% \ C)$	(15.4)
Carbon	0.40–0.55	$f_C = 0.171 + 0.001 \ (\% \ C) + 0.265 \ (\% \ C)$	(15.5)
Manganese	< 1.2	$f_{Mn} = 3.3333 \ (\% \ Mn) + 1.00$	(15.6)
Silicon	< 2.4	$f_{Si} = 1.00 + 0.07 \ (\% \ Si)$	(15.7)
Nickel	< 2.0	$f_{Ni} = 1.00 + 0.363 \ (\% \ Ni)$	(15.8)
Chromium	< 1.75	$f_{Cr} = 1.00 + 2.16 \ (\% \ Cr)$	(15.9)
Molybdenum	< 0.55	$f_{Mo} = 1.00 + 3.00 \ (\% \ Mo)$	(15.10)
Copper	< 0.55	$f_{Cu} = 1.00 + 0.365 \ (\% \ Cu)$	(15.11)
Vanadium	< 0.2	$f_V = 1.00 + 1.73 \ (\% \ V)$	(15.12)

The computations from the hardenability curves (see subsection 15.2.7.2) by the J-406 SAE method involve calculating the hardness *IH* corresponding to the full hardening structure (100% martensite), and the hardness H_{50} corresponding to the structure with 50% martensite. The Rockwell hardness *IH* can be determined by (Nosov and Yurasov, 1995):

$$IH = 35.395 + 6.990x + 312.330x^2 \ 821.744x^3$$
$$+ \ 1015.479x^4 \ 538.346x^5 \quad (15.13)$$

where x is the carbon content (wt% C) in the steel (see Example 15.7).

The Rockwell hardness corresponding to 50% martensite in the structure can be estimated by:

$$H_{50} = 22.974 + 6.214x + 356.364x^2 \ 1091.488x^3$$
$$+ \ 1464.880x^4 \ 750.441x^5 \quad (15.14)$$

where x is the carbon content in the steel (see Example 15.8).

15.2.8 TEMPERING OF STEEL

Tempering is a heat-treatment process for slightly decreasing the hardness; and for increasing the toughness of hardened steels. Tempering involves heating the steel at a temperature in the range of 150–550°C for a while followed by slow cooling. During tempering, the carbon atoms diffuse out between the iron atoms in the martensite to form the iron carbide particles, thereby relieving the strain within the martensite. The resulting microstructure is *tempered martensite* (Figure 15.7). This author has

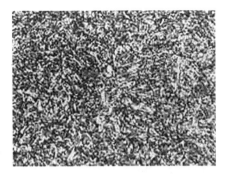

FIGURE 15.7 Microstructure of tempered martensite.

ensured production of tempered martensitic microstructure in forged 1050 steel for manufacturing the axle-hub of a motor-car (Huda, 2012).

Besides conventional tempering process, the following two modified tempering techniques for steels are in practice: (a) martempering, and (b) austempering. In *martempering*, the austenized steel is quenched in salt bath or lead bath; which also results in *tempered martensite* but with a lower and more uniform stress distribution. In *austempering*, the austenized steel is also quenched by lead bath or salt bath; but the resulting microstructure is *bainite*. The *martempering* and *austempering* procedures have considerable advantages over the conventional quenched and tempered procedure.

15.3 HEAT TREATMENT OF ALUMINUM ALLOYS

The selection of heat treatment approaches for *Al* alloys play an important role in strengthening *Al* alloys for aerospace and other stressed engineering applications (see Chapter 13). Notable examples of these alloys include: 2024-T3, 2024-T81, 7075-T6, and the like (see Table 13.1). The T6 temper is an artificial aging or precipitation hardening heat-treatment. For the T6 temper, the 7075 Al alloy (*Al*-5.6%Zn-2.5%Mg-1.6%Cu) is given a normal heat treatment at about 465°C and quenched in cold water. After it is precipitation heat treated at about 120°C for 24 hours, it becomes designated as 7075-T6. The tensile strength of the 7075-T6 alloy is around 750 MPa with fairly good ductility (11% elongation). Another, aerospace *Al* alloy is the 2024-T4 alloy; the heat-treatment process for this alloy is explained in the following paragraph.

The heat treatment for age-hardenable 2024-T4 alloy involves solution treatment (heating at temperature in the range of 450–550°C and holding for about 1 h followed by quenching). The quenched alloy may be cold-formed, if necessary. The solution treated alloy is then aged. Now we discuss the phase transformations during the heat treatment with reference to the partial aluminum-copper phase diagram (Figure 15.8). The first step in the heat-treatment is to heat the alloy at a temperatures in the range of 450–550°C; this temperature is chosen to obtain the single phase (α), provided the *Al* alloy contains up to 5.6% Cu (see Figure 15.8). The microstructure at this temperature will consist entirely of single phase: α, a solid-solution of Cu in FCC *Al* (see Example 15.8). The second step

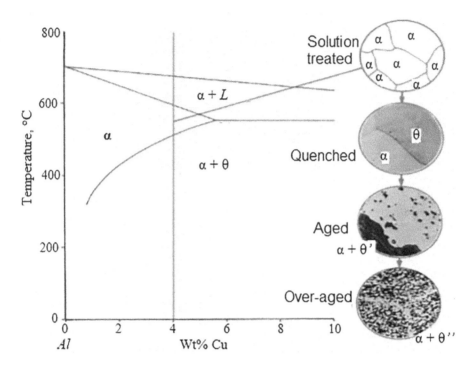

FIGURE 15.8   The steps in age-hardening of *Al-Cu* alloys.

is quenching (rapid cooling). On quenching the alloy, the Cu in the solid solution is retained thereby producing a super-saturated solid-solution (SSSS) of Cu in *Al* at room temperature. Finally the quenched alloy is aged; here, the quenched alloy is allowed to remain at ambient temperature (20°C) for a long-time duration, so the strength gradually increases and reaches maximum in about five days. This strengthening mechanism is called natural age-hardening of the Al-Cu alloys; which is a diffusion-controlled process (Huda, 2009). During the natural aging, Cu atoms in the SSSS diffuse out of the FCC *Al* lattice and form an intermetallic compound $CuAl_2$ (θ'); which is coherent with the lattice thereby imparting precipitation strengthening in the alloy (see Figure 15.8). Alternatively, the quenched alloy may be strengthened by artificial age-hardening; which involves tempering the quenched alloy at about 165°C for about 10 hours (see Chapter 9, Sect. 9.5). This thermally-activated precipitation-strengthening process increases the amount of θ' precipitates by accelerating the rate of diffusion (in accordance with the Arrhenius law) (see Chapter 4, Sect. 4.7), so the strength of the alloy rises further (see Examples 15.9–15.10). However, over-aging (heating at a temperature above 165°C) must be avoided since this practice would result in the precipitation of brittle θ'' phase thereby weakening the alloy (see Figure 15.8).

   The artificial aging (heating) of a quenched *Al-Cu* alloy accelerates the rate of diffusion, whereas its cooling to a very low temperature will retard the rate of diffusion of copper atoms from the aluminum lattice thereby preventing precipitation of θ' particles. A useful mathematical model for the kinetics of age hardening can be developed as follows.

The Arrhenius equation for the rate of a reaction is expressed by the following kinetic equation:

$$k = A \exp\left(\frac{Q_a}{RT}\right) \tag{15.15}$$

where k is the rate constant, A is the frequency factor, and Q_a is the activation energy.

By using the approach explained in section 4.7, we may derive the following formula:

$$t = A \exp\left(\frac{B}{T}\right) \tag{15.16}$$

where t is the time; T is the temperature (K); A is the frequency factor (a constant); and B is another constant (see Example 15.10).

15.4   HEAT TREATMENT OF COPPER ALLOYS

Copper and its alloys' products may be heat treated by processes, such as homogenizing, annealing, and precipitation hardening. These heat-treatment processes are explained as follows.

Homogenizing. It is applied to dissolve segregation and remove the coring effects found in some cast and hot-worked materials, chiefly those containing tin and nickel. In particular, tin bronzes, silicon bronzes, and copper nickels are subjected to prolonged homogenizing treatments at about 760°C before hot- or cold-working operations.

Recrystallization Annealing. Cold-worked copper and its alloys may be softened by recrystallization annealing. For copper and brass mill alloys, grain size is the standard means of evaluating a recrystallizing anneal. In copper-iron alloys (e.g. C19200, C19400, and C19500) and aluminum-containing brasses and bronzes (e.g. C61500, C63800, C68800, and C69000), the grain size is stabilized by the presence of a finely distributed second phase (see Table 15.2). These alloys maintain an extremely fine grain size at temperatures well above their recrystallization temperature, up to the temperature where the second phase finally dissolves or coarsens.

Precipitation Hardening. All precipitation-hardening copper alloys can be age hardened by solution treatment to a soft condition by quenching from a high temperature, followed by aging at a moderate temperature for a time-duration usually up to 3 h. In particular, the C19000 Cu alloy (containing about 1% nickel and about 0.25% phosphorus) is used for a wide variety of small parts requiring high strength, such as springs, clips, electrical connectors, and fasteners. This alloy is solution treated at 700 to 800°C (in case the metal must be softened between cold-working steps prior to aging, it may be satisfactorily annealed at about 620°C). Quenching is not necessary. For aging, the material is held at about 450°C for a time duration in the range of 1–3 h.

TABLE 15.3

Specifications, Compositions, Conditions, and Applications of Some Copper Alloys

UNS no.	Composition (wt%)	Condition	Applications
C11000	99.9%Cu	Annealed	Rivets, electrical wires, gaskets, roofing
C17200	Cu-1.9%Be-0.2%Co	Precipitation hardened	Springs, bellows, valves
C26000	Cu-30%Zn	Annealed cold-worked	Automobile radiator cores, lamp fixtures
C51000	Cu-5%Sn-0.2%P	Annealed cold-worked	Clutch disc, bellows, welding rods, springs
C71500	Cu-30%Ni	Annealed cold-worked	Condensers, heat exchangers, saltwater piping
C85400	Cu-29%Zn-3%Pb-1%Sn	As cast	Radiator fittings, light fixtures
C90500	Cu-10%Sn-2%Zn	As cast	Bearings, bushings, gears, piston rings
C95400	Cu-4%Fe-11%Al	As cast	Valve seats and guards, bearings

Table 15.3 presents the UNS specifications, composition, heat-treatment conditions, and applications of some copper alloys.

15.5 CALCULATIONS—WORKED EXAMPLES ON HEAT TREATMENT OF METALS

EXAMPLE 15.1 CALCULATING CARBON CONTENT IN STEEL FOR DESIRED MICROSTRUCTURE

It is desired to obtain ferrite-pearlite microstructure containing 8.9 wt% eutectoid ferrite in a hypo-eutectoid carbon steel. Calculate the appropriate carbon content in the steel so as to select the steel for obtaining the microstructure.

SOLUTION

Wt% eutectoid ferrite = 8.9 wt% = 0.089
Let wt% C in the hypo-eutectoid steel to be selected = x
In order to find total ferrite, we must consider the composition x to be between the ferrite (0.02%C) and cementite (6.67%C) compositions in the tie-line at the A_1 temperature in Figure 15.1–15.2. By using the *inverse lever rule* (Equation 6.15) (see Chapter 6), we can write:

$$W_{total\ ferrite} = \frac{C_{cem} - C_o}{C_{cem} - C_\alpha}$$

$$\text{Wt \% total ferrite} = \frac{6.67 - x}{6.67 - 0.02} = \frac{6.67 - x}{6.65}$$

In order to find proeutectoid ferrite, we must consider the composition x to be between the ferrite (0.02%C) and eutectoid (0.77%C) compositions in the tie-line at A_1 temperature in Figure 15.1. Again, by using the inverse lever rule, we obtain:

$$W_{proeutectoid\ ferrite} = \frac{C_{pearlite} - C_o}{C_{pearlite} - C_\alpha}$$

$$\text{Wt \% proeutectoid ferrite} = \frac{0.77 - x}{0.77 - 0.02} = \frac{0.77 = -x}{0.75}$$

By using Equation 15.1,
   Wt% total ferrite = wt% eutectoid ferrite + wt% proeutectoid ferrite

$$\frac{6.67 - x}{6.65} = 0.089 + \frac{0.77 - x}{0.75}$$

By solving the above equation for x, we get $x = 0.095$.
   Hence, we should select 0.095% carbon steel to obtain the desired microstructure.

EXAMPLE 15.2 DESIGN OF HEAT TREATMENT TO OBTAIN FERRITE-PEARLITIC MICROSTRUCTURE

By using the data in Example 15.1, design the heat treatment for the carbon steel to obtain the desired microstructure.

SOLUTION

Since, pearlite-ferrite is an equilibrium microstructure, the 0.095% carbon steel should be heat treated by full annealing. The design of the heat-treatment process is as follows. Heat the steel to austenitic range, hold for sufficient time to obtain 100% austenite, then slowly cool at the rate of about 20°C/h in a furnace to about 50°C into the ferrite-cementite range (see Figure 15.3). At this point, it may be cooled at room temperature in air with natural convection.

EXAMPLE 15.3 DESIGN OF HEAT TREATMENT FOR DUPLEX FERRITE-MARTENSITE MICROSTRUCTURE

Design a heat treatment to obtain a dual-phase steel with 50% ferrite and 50% martensite in which the martensite contains 0.65% C.

SOLUTION

By reference to the subsection 15.2.7.1, the heat-treatment temperature is fixed by the requirement that the composition of martensite is 0.65% C. By reference to Figure 15.2, we find that 0.65% C is obtained in austenite when the temperature is about 740°C. If the carbon content of steel is x, the use of the inverse lever rule (Equation 6.15) yields:

$$\% \, \gamma = \frac{x - 0.02}{0.65 - 0.02} \times 100 \; = \; 50$$

$$x = \; 0.34 \; \%C$$

Thus the heat-treatment design is as follows:

1. Select a carbon steel containing 0.34 wt% C;
2. Heat the steel to 740°C and hold for sufficient time duration (~ 1 h) to obtain 50% ferrite-50% austenite, with 0.65% C in the austenite;
3. Quench the steel to room temperature so as to transform the austenite to martensite, also containing 0.65% C.

EXAMPLE 15.4 DESIGN OF HEAT TREATMENT FOR 1050 STEEL AXLE

Design a heat-treatment process to produce a uniform microstructure and hardness of HRC 35 in a 1050 steel axle hub.

Solution

1050 steel contains 0.5% carbon.

By reference to Figure 15.5 we find that a hardness of 35 HRC is obtained by transforming austenite to bainite at about 450°C. From Figure 15.2, we find that for Fe-0.5%C alloy, A_3 temperature is 770°C. Hence, the design of hardening heat treatment for the 1050 axle hub is:

1. Austenize the steel at about 810°C (770 + 40 = 810°C), holding for 1 h so as to obtain 100% γ.
2. Quench the steel in oil at about 450°C and hold for a minimum of 100 s or 2 min. Austenite will transform to bainite.
3. Cool in air to room temperature.

EXAMPLE 15.5 CALCULATING THE MULTIPLYING FACTOR F_I FOR DETERMINING D_I FOR HARDENING

A steel contains 0.45%C, 0.9%Mn, 2.0%Si, 0.5%Cr, 0.2%Ni, 0.3%Mo, and 0.1%Cu. Calculate the multiplying factors for each of the elements for determining the critical ideal diameter.

Solution

By using Equation 15.5,

$$f_c = 0.171 + 0.001 \, (\% \, C) + 0.265 \, (\%C) = 0.171 + (0.001 \times 0.45)$$
$$+ (0.265 \times 0.45) = 0.291$$

By using Equation 15.6,

$$f_{Mn} = 3.3333 \ (\% \ Mn) + 1.00 = (3.3333 \times 0.9) + 1 = 4$$

By using Equation 15.7,

$$f_{Si} = 1.00 + 0.07 \ (\% \ Si) = 1.00 + (0.07 \times 2.0) = 1.14$$

By using Equation 15.8,

$$f_{Ni} = 1.00 + 0.363 \ (\% \ Ni) = 1.00 + (0.363 \times 0.2) = 1.07$$

By using Equation 15.9,

$$f_{Cr} = 1.00 + 2.16 \ (\% \ Cr) = 1.00 + (2.16 \times 0.5) = 2.08$$

By using Equation 15.10,

$$f_{Mo} = 1.00 + 3.00 \ (\% \ Mo) = 1.00 + (3 \times 0.3) = 1.9$$

By using Equation 15.11,

$$f_{Cu} = 1.00 + 0.365 \ (\% \ Cu) = 1.00 + (0.365 \times 0.1) = 1.036$$

EXAMPLE 15.6 CALCULATING THE IDEAL CRITICAL DIAMETER FOR HARDENABILITY

By using the data in Example 15.5, calculate the ideal critical diameter for the steel.

SOLUTION

$$f_C = 0.291, f_{Mn} = 4, f_{Si} = 1.14, f_{Ni} = 1.07, f_{Cr} = 2.08, f_{Mo} = 1.9, f_{Cu} = 1.036$$

By using Equation 15.3,

$$D_i = 25.4 \ f_C f_{Mn} f_{Si} f_{Mo} f_{Ni} f_V f_{Cu} = 25.4 \ (0.291 \times 4 \times 1.14 \times 1.9 \times 1.07 \times 1.036)$$
$$= 70.988 \ mm$$

The ideal critical diameter = 70.988 mm.

EXAMPLE 15.7 CALCULATING THE HARDNESS *IH* OF A CARBON STEEL

Calculate the hardness *IH* corresponding to the full hardening structure (100% martensite) for a carbon steel containing 0.5 wt%C.

SOLUTION

By using Equation 15.13,

$$IH = 35.395 + 6.990x + 312.330x^2 - 821.744x^3 + 1015.479x^4$$
$$- 538.346x^5$$
$$IH = 35.395 + (6.990 \times 0.5) + (312.330 \times 0.5^2)$$
$$- (821.744 \times 0.5^3) + (1015.479 \times 0.5^4)(538.346 \times 0.5^5)$$
$$IH = 60.52$$

EXAMPLE 15.8 CALCULATING THE HARDNESS H_{50} OF A CARBON STEEL

Calculate the hardness H_{50} of a carbon steel containing 0.5 wt%C.

SOLUTION

By using Equation 15.14,

$$H_{50} = 22.974 + 6.214x + 356.364x^2 - 1091.488x^3$$
$$+ 1464.880x^4 - 750.441x^5$$
$$H_{50} = 22.974 + (6.214 \times 0.5) + (356.364 \times 0.5^2)$$
$$(1091.488 \times 0.5^3) + (1464.880 \times 0.5^4)(750.441 \times 0.5^5)$$
$$= 22.974 + 3.107 + (356.364 \times 0.25) - (1091.488 \times 0.125)$$
$$+ (1464.880 \times 0.062) - (750.441 \times 0.031)$$
$$H_{50} = 46.3$$

EXAMPLE 15.9 DESIGNING A HEAT-TREATMENT PROCESS FOR AN AEROSPACE ALUMINUM ALLOY

An *Al*-5%Cu alloy is to be solution treated and age hardened. (a) Design a heat-treatment process for the alloy. (b) Can we successfully age harden this alloy by solution treatment at 430°C?

SOLUTION

The solution treatment of the alloy requires attaining α solid solution with reference to Figure 15.7.

For the *Al*-5%Cu alloy, the solution treatment temperature must be in the range of 530–550°C. (a) The design of the heat-treatment process may be as follows: (I) heating at 540 °C and holding for sufficient time to ensure 100% α, (II) quenching to room temperature, and (III) artificial aging at 120°C for about 10 hours.

(b) We may check the suitability of the 430°C solution-treatment temperature with reference to Figure 15.7. At this temperature, the alloy is in a two-phase region: α+θ. Since a single-phase homogeneous solid solution is not attained at 430°C, the solution treatment at 430°C for *Al*-5%Cu alloy would not result in successful age hardening.

EXAMPLE 15.10 CALCULATING THE TEMPERATURE FOR PREVENTING THE θ' PRECIPITATION

An *Al-Cu* alloy is solution treated and tempered for age hardening. During the natural aging treatment, the first signs of θ' precipitation from the super-saturated solid solution (SSSS) of *Cu* in *Al* are detected after four hours at 20°C. Another sample of the quenched alloy is artificially aged at 110°C for which the θ' precipitates are detected after two minutes. If it is required to prevent the precipitation of θ' particles in the alloy for two days, calculate the temperature at which the alloy must be held to achieve this effect.

SOLUTION

At $T = 20°C = 293K$, $t = 4\,h = 240$ min; At $T = 110°C = 383K$, $t = 2$ min.

By using Equation 15.16 for the two cases, we may write:

$$240 = A\exp\left(\frac{B}{293}\right) \qquad\qquad\qquad \text{(E-15.9a)}$$

$$2 = A\exp\left(\frac{B}{383}\right) \qquad\qquad\qquad \text{(E-15.9b)}$$

By taking the natural logarithms of both sides of these equations, we get:

$$ln\ 240 = ln\ A + \frac{B}{293}$$

and $ln\ 2 = ln\ A + \dfrac{B}{383}$

By simultaneously solving these two equations, we obtain:

$B = 5910$, and $A = 4.17 \times 10^{-7}$

Let T be the temperature for the initiation of θ' precipitation in 2 days ($= 48\,h = 2{,}880$ min), Equation 15.8 takes the form:

$$t = A\exp\left(\frac{B}{T}\right)$$

$$2880 = 4.17 \times 10^{-7} exp\left(\frac{5910}{T}\right)$$

$$T = 261K = -12°C$$

Hence, the quenched alloy must be maintained at the sub-zero temperature of –12°C.

QUESTIONS AND PROBLEMS

15.1. MCQs. Underline the most appropriate answers for the following statements:

(a) Which heat treatment of a eutectoid steel results in fine pearlite?
(i) full annealing, (ii) hardening, (iii) normalizing, (iv) tempering.

(b) Which heat treatment of a eutectoid steel results in coarse pearlite?
(i) full annealing, (ii) hardening, (iii) normalizing, (iv) tempering.

(c) Which heat treatment of a eutectoid steel results in martensite?
(i) full annealing, (ii) hardening, (iii) normalizing, (iv) tempering.

(d) Which heat treatment of carbon steel results in hardness reduction?
(i) full annealing, (ii) hardening, (iii) normalizing, (iv) tempering.

(e) Which property is mainly improved by spheroidizing annealing of steel? (i) machinability, (ii) strength, (iii) hardness, (iv) impact energy.

(f) What is the heating temperature for spheroidizing annealing of steel? (i) 700°C, (ii) 800°C, (iii) 900°C, (iv) 1,000°C.

(g) Which heat treatment results in removing the coring effect in copper alloys? (i) recrystallization annealing, (ii) homogenization, (iii) precipitation strengthening.

(h) What is the effect of solution treatment of Al-Cu alloy? (i) hardening, (ii) strengthening, (iii) softening, (iv) toughening.

15.2. Draw the partial Fe-Fe$_3$C phase diagram showing temperature ranges for various heat-treatment processes in carbon steels.

15.3. Draw a diagram showing thermal cycles for heat-treatment processes in carbon steels.

15.4. Why does the strength of a solution-treated *Al-Cu* alloy increase by aging treatment?

P15.5. Design a heat treatment to obtain a dual-phase steel with 40% ferrite and 60% martensite in which the martensite contains 0.65% C.

P15.6. Design a heat-treatment process to obtain ferrite-pearlite microstructure containing 8.8 wt% eutectoid ferrite in a hypo-eutectoid carbon steel.

P15.7. Design a heat-treatment process to produce a uniform microstructure and hardness of HRC 45 in a 1060 steel component.

P15.8. A steel component contains 0.47% C, 0.85% Mn, 2.1% Si, 0.6% Cr, 0.3% Ni, 0.31% Mo, and 0.11% Cu. Calculate the critical ideal diameter for hardening the steel.

P15.9. Calculate the hardness *IH* corresponding to 100% martensite for a carbon steel containing 0.6 wt% C.

P15.10. An *Al-Cu* alloy is solution treated and tempered for age hardening. During the natural aging treatment, the first signs of θ' precipitation from the super-saturated solid solution (SSSS) of *Cu* in *Al* are detected after five hours at 15°C. Another sample of the quenched alloy is artificially aged at 100°C for which the θ' precipitates are detected after three minutes. If it is required to prevent the precipitation of θ' particles in the alloy for three days, calculate the temperature at which the alloy must be held.

REFERENCES

Callister, W.D. (2007) *Materials Science and Engineering: An Introduction.* John Wiley & Sons Inc, Hoboken, NJ.

Huda, Z. (2007) Development of heat treatment process for a powder metallurgy (P/M) superalloy. *Materials and Design*, 28(5), 1664–1667.

Huda, Z. (2009) Precipitation strengthening and age hardening in 2017 aluminum alloy for aerospace application. *European Journal of Scientific Research*, 26(4), 558–564.

Huda, Z. (2012) Reengineering of manufacturing process design for quality assurance in axle-hubs of a modern motor-car—A case study. *International Journal of Automotive Technology*, 13(7), 1113–1118.

Nosov, V.B. & Yurasov, S.A. (1995) Calculation of hardenability of structural steels from their chemical composition. *Metal Science and Heat Treatment*, 37(1), 16–21.

Totten, G.E. & Howes, M.A.H. (1997) *Steel Heat Treatment Handbook.* Marcel Dekker, New York.

Zakharov, B. (1962) *Heat Treatment of Metals.* Peace Publishers, Moscow.

16 Surface Engineering of Metals

16.1 SURFACE-HARDENING PROCESSES

16.1.1 Basic Principle of Surface Hardening

Surface hardening or *case hardening* is generally applied to steels; it involves hardening the steel surface to a required depth without hardening the core of the component. A distinct advantage of surface hardening of steel is that the outer skin (case) is hard whereas the inner core is left untouched so that the material still possesses the toughness/ductility properties (see Figure 16.1).

The surface hardening of steel has wide industrial applications; these include: case-hardened gears, case-hardened grinding balls, and the like (Davis, 2003; Schneider and Chatterjee, 2013). There are two groups of surface-hardening techniques: (a) surface hardening keeping composition unchanged, and (b) surface hardening involving composition change; these techniques are discussed as follows.

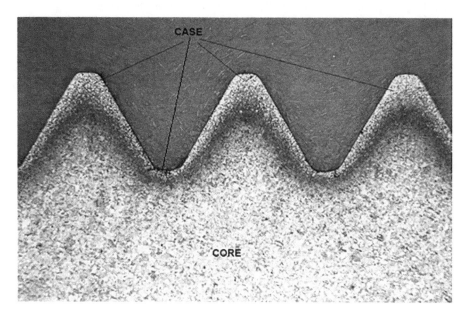

FIGURE 16.1 Surface hardened component (gear).

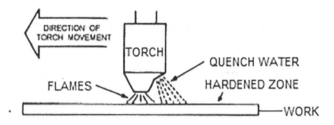

FIGURE 16.2 Schematic of flame hardening process.

16.1.2 SURFACE HARDENING KEEPING COMPOSITION THE SAME

There are two surface-hardening processes that allows us to keep the composition unchanged: (a) flame hardening and (b) induction hardening.

Flame hardening. *Flame hardening* is a localized surface-hardening heat-treatment process to harden the surface of steel containing moderate carbon contents. In *flame hardening*, a high intensity oxy-acetylene flame is applied to the selective region so as to raise the surface temperature high enough in the region of austenite transformation (see Figure 16.2). The overall heat transfer is restricted by the torch and thus the core (interior) never reaches the high temperature. The heated region (case) is quenched to achieve the desired hardness. Tempering can be done to eliminate brittleness, if desired.

Induction Hardening. *Induction hardening* involves the use of electric inductor coils to locally heat the steel component followed by fast cooling (quenching). The high frequency electric fields rapidly heat the surface of the component; which is then quenched using water. This surface treatment results in a localized hardened layer at the surface.

16.1.3 SURFACE HARDENING INVOLVING CHANGE IN COMPOSITION

Principle. Steels may be surface hardened involving a change in composition by any of the following methods: (a) *carburizing*, (b) *nitriding*, and (c) *carbonitriding*. Surface hardening involving a change in composition is a diffusion-controlled process in which atoms of carbon, nitrogen, or both carbon and nitrogen are added to the surface by diffusion. For example, in carburizing, carbon atoms are added to the surface by a diffusion-controlled mechanism that is governed by the Fick's law of diffusion.

Carburizing and Its Mathematical Modeling. There are two methods that are generally practiced for carburizing steels: (i) pack carburizing, and (ii) gas carburizing. In the pack carburizing process, components to be surface hardened are packed in an environment with high carbon content (such as charcoal or barium carbonate). The components are heated to a temperature range of 850 to 950°C so that the carbon is diffused into the outer surface of the component (see Figure 16.3). The diffusion rate of carbon depends upon the composition of the part, temperature, and time. The outer surface of the material is hardened whereas the core remains ductile. The hardened case depth varies from 0.1 to 2 mm.

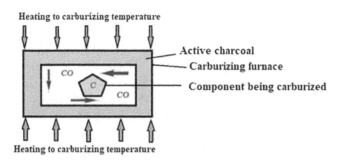

FIGURE 16.3 The process of *pack carburizing of steel* (CO = carbon monoxide gas; C = component being carburized).

The time required for carburizing a steel component can be calculated by:

$$t_{\text{carb.}} = \frac{d_c}{0.2} + 3 \qquad (16.1)$$

where $t_{carb.}$ is the carburizing time, h; and d_c is the case depth, mm (see Example 16.1).

The diffusion of carbon during carburizing is based on the differential of concentration principle; which is expressed by Fick's Second Law of diffusion (see Chapter 4). The useful mathematical expressions (resulting from the integration of Fick's Second Law) have been discussed in detail in Chapter 4 (see Equations 4.4–4.13). In particular, Examples 4.2–4.6 illustrate the problems and their solution related to carburizing of steel; which is of great industrial importance to metallurgists.

Nitriding. It is the process of diffusion enrichment of the surface layer of a component with nitrogen. In *gas nitriding*, an alloy steel part is heated at a temperature in range of 500–600°C for a duration in the range of 40–100 hours in the atmosphere of ammonia (NH_3), which dissociates to hydrogen and nitrogen. As a result, the nitrogen diffuses into the steel forming nitrides of iron, aluminum, chromium, and vanadium thereby forming a thin, hard surface.

Carbonitriding. It involves diffusion enrichment of the surface layer of a part with carbon and nitrogen. *Carbonitriding* is widely applied to the surface treatment of gears and shafts; and is capable of producing a case depth in the range of 0.07–0.5 mm having hardness of 55–62 HR_C.

16.2 SURFACE COATING—AN OVERVIEW

Surface coating process involves cleaning/surface preparation of a part (called substrate) followed by the application of a coating material; finally the coated part is dried either in air or oven. Surface coatings provide protection, durability, and/or decoration to part surfaces. Metals that are commonly used as coating material (for coating on steels) include: nickel, copper, chromium, gold, and silver. A good corrosion-resistant surface is produced by chrome plating; however their durability is just fair. The nickel electroplated parts have excellent surface finish and brightness; they are durable coating, and have fair corrosion resistance.

There are many different surface coating processes; these include electroplating (nickel plating, chrome plating, etc.); hot dip coating, electroless plating, conversion coating, physical vapor deposition (PVD), chemical vapor deposition (CVD), powder coating, painting, anodizing, and the like. Among these processes, electroplating is the most commonly practiced surface coating process, and is discussed and mathematically modeled in the following section; the other coating processed are described elsewhere (Huda, 2017; Hughes *et al.*, 2016; Tracton, 2006).

16.3   ELECTROPLATING AND ITS MATHEMATICAL MODELING

16.3.1   ELECTROPLATING PRINCIPLES

Electroplating process involves the cathodic deposition (plating) of a thin layer of metal on the base metal or other electrically conductive material. The electroplating process requires the use of direct current (*d.c*) as well as an electrolytic solution consisting of certain chemical compounds that make the solution highly conductive. The positively charged plating metal (anode) ions in the electrolytic solution are drawn out of the solution to coat the negatively charged conductive part surface (cathode). Under an electromotive force (*e.m.f*), the positively charged metal ions in the solution gain electrons at the part surface and transform into a metal coating (see Figure 11.12).

In modern surface engineering practice, electrolytic solutions usually contain additives to brighten or enhance the uniformity of the plating metal. The amount of plating material deposited strongly depends on the plating time and current levels to deposit a coating of a given thickness. The electroplating process is governed by Faraday's laws of electrolysis.

16.3.2   MATHEMATICAL MODELING OF ELECTROPLATING PROCESS

Mass of Metal Deposited. The mass of plated metal can be computed by the formula:

$$W = \frac{I\,t\,A}{n\,F} \tag{16.2}$$

where W is the mass of the plated metal, g; I is the current, amperes; t is time, s; A is the atomic weight of the metal, g/mol; n is the valence of the dissolved metal in electrolytic solution in equivalents per mole; and F is the Faraday's constant (= 96,485 Coul/equiv.).

Thickness of Deposit. The *thickness* of the plated metal, T, can be determined as follows:

$$T = \frac{W}{\rho\,S} \tag{16.3}$$

where T is the thickness of the metal plated, cm; W is the mass of the plated metal, g; ρ is the density of the plating metal, g/cm³; and S is the surface area of the plated work-piece, cm².

Current Density. In electroplating industrial practice, *current density* is more important than current. A good quality coating requires *smaller current densities but it takes more electroplating time.* The *current density* is related to *current* by:

$$J = \frac{I}{S} \tag{16.4}$$

where J is current density, A/cm^2; I is current, A; and S is surface area of plated work-piece, cm^2.

Electroplating Time. The determination of *electroplating time* is also of great industrial importance since it has a direct impact on direct labor cost. A useful mathematical modeling can be developed by rearranging the terms in Equation 16.1, as follows:

$$t = \frac{W n F}{I A} \tag{16.5}$$

where t is the electroplating time, s; W is the mass of the plated metal, g; I is the current, amperes; A is the atomic weight of the metal, g/mol; n is the valence of the dissolved metal in electrolytic solution in equivalents per mole; and $F = 96,500$ Coul/equiv.

Cathode Current Efficiency. It is defined as "the ratio of the weight of metal actually deposited to that, which would have resulted if all the current had been used for deposition". Mathematical models for *cathode current efficiency* are presented as Equations 11.18–11.19.

Electroplating Rate. It is the weight of the metal deposited per unit time. The electroplating rate can be determined by rearranging the terms in Equation 11.8 as follows:

$$R = \frac{W}{t} = \frac{IA}{nF} \tag{16.6}$$

where R is the rate of electroplating, g/s; and other symbols have their usual meanings.

16.4 CALCULATIONS—EXAMPLES IN FINISHING/ SURFACE ENGINEERING

EXAMPLE 16.1 CALCULATING THE CARBURIZING TIME FOR STEEL

Calculate the carburizing time required to obtain a case depth of 1.7 mm in a steel component.

SOLUTION

By using Equation 16.1,

$$t_{carb.} = \frac{d_c}{0.2} + 3 = \frac{1.7}{0.2} + 3 = 11.5 \text{ h}$$

The carburizing time = 11.5 hours.

EXAMPLE 16.2 COMPUTING THE MASS OF PLATING METAL IN ELECTROPLATING

A current of 1.5 ampere is passed for 45 minutes to a steel work-piece during a silver plating electrolytic process. Compute the mass of the silver metal that is plated on the steel work-piece.

SOLUTION

$I = 1.5$ A; $t = 45$ min = 2700 s; $A = 107.8$ g/mol; $F = 96,485$ Coulombs/equiv.; $n = 1$; $W = ?$ g

By using Equation 16.2,

$$W = \frac{ItA}{nF} = \frac{1.5 \times 2700 \times 07.8}{1 \times 96485} = 4.5g$$

The mass of silver plated = 4.5 grams.

EXAMPLE 16.3 CALCULATING THE THICKNESS OF THE PLATING METAL IN ELECTROPLATING

The 25 cm² surface area of a steel work-piece was electroplated with 1.2 gram of copper. Calculate the thickness of the plated copper. The density of copper = 8.9 g/cm³.

SOLUTION

$S = 25$ cm², $W = 1.2$ g, $\rho = 8.9$ g/cm³, $T = ?$

By using Equation 16.3,

$$T = \frac{W}{\rho S} = \frac{1.2}{8.9 \times 25} = 0.00539 \text{cm}$$

The thickness of the plated copper = 0.00539 cm = 53.9 μm.

EXAMPLE 16.4 COMPUTING CURRENT WHEN CURRENT DENSITY AND WORK DIMENSIONS ARE GIVEN

The electroplating of copper on a cylindrical work-piece (length = 15 cm, r = 5 cm) was found to require a current density of 2 mA/cm². Calculate current for the electroplating process?

SOLUTION

$r = 5$ cm, $h = 15$ cm, $J = 2$ mA/cm^2, $I = ?$

First we find the surface area of cylinder, S, as follows:

$S = (2\pi\, r\, h) + (2\pi\, r^2) = (2\pi \times 5 \times 15) + (2\pi \times 5^2) = 471.2 + 157 = 628.3$ cm^2

By using Equation 16.4,

$$I = JS = 2 \times 628.3 = 1256.5 \cdot mA$$
$$\text{Current} = 1256.5 \times 10^{-3} A = 1.25A$$

EXAMPLE 16.5 CALCULATING THE ELECTROPLATING TIME

It is required to electroplate 27 grams of zinc on a steel work-piece by using a 15 amp current flow through a solution of $ZnSO_4$. Calculate the electroplating time.

SOLUTION

$W = 27$ g, $I = 15$ A, Atomic mass of Zn $= A = 65.4$ g/mol, $F = 96{,}500$ Coul/ equiv., $t = ?$

The $ZnSO_4$ chemical formula indicates that zinc is divalent (Zn^{2+}) i.e. $n = 2$. By using Equation 16.5,

$$t = \frac{W\,n\,F}{I\,A} = \frac{27 \times 2 \times 96500}{15 \times 65.4} = 5312s$$

The electroplating time $= 5312$ s $= 1.47$ h.

EXAMPLE 16.6 CALCULATING THE ELECTROPLATING RATE

A 10-amp current flows for electroplating nickel on a steel work-piece. Calculate the electroplating rate for the process.

SOLUTION

$I = 10$ A, $A = 58.7$ g/mol; $n = 2$, $F = 96{,}500$ Coul/equiv., $R = ?$

By using Equation 16.6,

$$R = \frac{I\,A}{n\,F} = \frac{10 \times 58.7}{2 \times 96500} = 0.003 g/s$$

Electroplating rate $= 0.003 \times 3600$ g/h $= 11$ g/h.

EXAMPLE 16.7 CALCULATING THEORETICAL WEIGHT OF METAL DEPOSIT

A 10-amp current flows for 1.5 h for electroplating nickel on a steel work-piece. What is the *theoretical weight of the deposit?*

SOLUTION

$I = 10$ A, $t = 1.5$ h $= 5400$ s, $M = 58.7$ g/mol; $n = 2$, $F = 96,500$ Coul/equiv.

Electric Charge $= Q = I t = 10$ x $5,400 = 54,000$ C
By using Equation 11.19,

$$W_{theo.} = \frac{QM}{Fn} = \frac{54000 \times 58.7}{96500 \times 2} = 16.4g$$

The theoretical weight of the nickel deposit $= 16.4$ g.

QUESTIONS AND PROBLEMS

16.1.   What are the objectives of surface engineering of metals?
16.2.   (a) Define the term "surface hardening" with the aid of a diagram.
        (b) What is the distinct advantage of surface hardening?
        (c) What are the two groups of surface-hardening processes?
16.3.   Briefly explain *flame hardening process* with the aid of a diagram.
16.4.   (a) List down the surface-hardening processes that involve change in composition.
        (b) Explain gas carburizing process with the aid of a diagram.
16.5.   (a) Define *"surface-coating process."*
        (b) What are the objectives of surface coating?
        (c) Compare the merits and demerits of *chrome plating* and *nickel plating.*
        (d) List down the various techniques of surface-coating processes.
16.6.   Explain *electroplating* with the aid of a diagram.
P16.7. A carburizing heat treatment of a steel alloy for a duration of 12 hours raises the carbon concentration to 0.55% at a depth of 2.3 mm from the surface. Estimate the time required to achieve the same concentration at a depth of 4.5 mm from the surface for an identical steel at the same carburizing temperature.
P16.8. A component made of 1010 steel is to be case-carburized at 927°C. The component design requires a carbon content of 0.6% at the surface and a carbon content of 0.25% at a depth of 0.4 mm from the surface. The diffusivity of carbon in steel at 927°C is 1.28 x 10^{-11} m²/s. Compute the time required to carburize the component to achieve the design requirements.
P16.9. A current of 2.5 A is passed for 35 min to a steel work-piece during a nickel-plating electrolytic process. Compute the mass of the nickel that is plated on the steel work-piece.

P16.10. The 15 cm^2 surface area of a steel work-piece was electroplated with 0.9 grams of gold. Calculate the thickness of the plated copper. The density of gold = 19.3 g/cm^3.

P16.11. Calculate the carburizing time required to obtain a case depth of 2 mm in a steel component.

REFERENCES

Davis, J.R. (2003) *Surface Hardening of Steels: Understanding the Basics.* ASM International, Materials Park, OH.

Huda, Z. (2017) *Materials Processing for Engineering Manufacture.* Trans Tech Publications, Switzerland.

Hughes, A.E., Mol, J.M.C., Zheludkevich, M.L. & Buchheit, R.G. (2016) *Active Protective Coatings.* Springer Materials Science Series, Dordrecht, Netherlands.

Tracton, A.A. (2006) *Coating Materials and Surface Coatings.* CRC Press Inc., Boca Raton, FL.

Schneider, M.J. & Chatterjee, M.S. (2013) *Introduction to Surface Hardening of Steels.* ASM International, Materials Park, OH.

Answers to MCQs and Selected Problems

CHAPTER 2

P2.9. $a = 0.286$ nm, $\rho = 7.89$ g/cm^3
P2.11. [001] = directionally solidified gas-turbine blade
  [100] = grain oriented electrical steel
P2.13. 8.89 g/cm^3
P2.15. (a) 1, (b) 6, (c) 0.52
P2.17. 27.2 nm

CHAPTER 3

P3.9. 1.76 kPa
P3.11. 67.9 x 10^{-7}. There are 68 vacancies in each 10^7 atomic sites.
P3.13. 2.5 MPa
P3.15. 62.5

CHAPTER 4

P4.9. 2.43 g
P4.11. 1.0 x 10^{18} atoms/m^2-s
P4.13. 1.69 mm
P4.15. 27.5 hours

CHAPTER 5

P5.2. (MCQs): (a) iii, (b) ii, (c) iv, (d) v, (e) v, (f) i, (g) iii, (h) ii, (i) v, (j) iii
P5.9. 28.8 μm
P5.11. 37.7 μm
P5.13. 1.86 x 10^8 m/s
P5.15. 0.01 nm or 10 pm
P5.17. 7.93 g/cm^3

CHAPTER 6

P6.9. (a) 1280°C, (b) 1230°C, (c) 50%Cu-50%Ni, (d) 70%Cu-30%Ni
P6.11. Just above the eutectic, composition of liq.: Fe-4.27 wt%C
  Just below the eutectic: composition of γ: Fe-2%C, and composition of
    Fe$_3$C: Fe-6.67%C
  Just above the eutectic: weight fraction of liquid = 1
  Just below the eutectic: weight fraction of γ = 0.514, and weight fraction of
    Fe$_3$C = 0.486
P6.13. (a) Wα = 0.674, Wβ = 0.326 (b) Vα = 0.56, Vβ = 0.44

CHAPTER 7

P7.9. (a) The heterogeneous nucleation is not extensive, (b) extensive, (c) not extensive.

P7.11. 20 x 10^{-5} cm; *coarse pearlite*

P7.13. Austenize at 770°C; then quench and hold at 260°C for 3 h. Cool to room temperature.

CHAPTER 8

P8.1. (MCQs): (a) ii, (b) i, (c) iii, (d) iii, (e) ii, (f) iv, (g) iv, (h) i.

P8.7. σ_{eng} = 708.6 MPa, ε_{eng} = 0.129, σ_{true} = 800 MPa, ε_{true} = 0.121

P8.9. S_{ut} = 580 MPa, Breaking strength = 580 MPa

P8.11. 0.345

P8.13. 0.195

P8.15. 205 VHN

CHAPTER 9

P9.9. 328 MPa

P9.11. 2 x 10^8 cm^{-2}

P9.13. 51.05 MPa

P9.15. Yield strength = 800 MPa, Tensile strength = 1,200 MPa, % Elongation = 23

CHAPTER 10

P10.13. (a) 2,448 MPa, (b) 16.3

P10.15. 130.5 MPa$\sqrt{m}$.

P10.17. a_c = 4.3 cm; Since the crack is embedded, the NDT technique may be either X-ray radiography or ultrasonic testing.

P10.19. 130.38 MPa

P10.21. (a) 17 MPa, (b) 8 x 10^{-5} h^{-1}

P10.23. 612 kJ

CHAPTER 11

P11.1. (T/F): (a) T, (b) F, (c) F, (d) T, (e) F, (f) F, (g) F, (h) T, (*i*) T, (j) T, (k) T

P11.11. 0.88 V; since the EMF value is positive, the reaction would occur spontaneously.

P11.13. 2.4633 V

P11.15. 6.7 mg/s (The electrolyte is silver nitrate).

P11.17. 7 min

P11.19. 1.46 x 10^{-7} kg/m^2×s

P11.21. 0.6 mm

CHAPTER 12

P12.2. (MCQs): (a) iv, (b) iii, (c) i, (d) i, (e) i, (f) iv, (g) iv, (h) iv, (*i*) iii, (j) iv, (k) ii, (*l*) ii, (m) iv, (n) ii, (*o*) ii, (p) i, (q) iii, (r) iv, (s) i.

P12.5. $W_{Pearlite} = 89.3\%$, $W_{Fe3C} = 10.7\%$

P12.7. 1,475°C

P12.9. 2,465 kg

CHAPTER 13

P13.1. (MCQs): (a) ii, (b) iv, (c) i, (d) iii, (e) ii, (f) i, (g) iv, (h) i, (*i*) ii, (j) ii, (k) i, (*l*) iii, (m) ii, (n) iv, (*o*) ii, (p) i, (q) iii, (r) iv.

P13.13. α-$Al = 88.2\%$; Si phase = 11.8%

P13.15. The 7075-T6 *Al* alloy weighs one-third as compared to steel with a higher yield strength.

P13.17. (a) 31.78 wt%, (b) α phase.

P13.19. 11 grams

CHAPTER 14

P14.7. (a) –34.6°C; (b) 268°C

P14.9. 0.44

CHAPTER 15

P15.1. MCQs (a) iii, (b) i, (c) ii, (d) iv, (e) i, (f) i, (g) ii, (h) iii

P15.5. 1. Select a carbon steel containing 0.398 wt% C;
2. Heat the steel to 740°C and hold for sufficient time duration (~ 1 h);
3. Quench the steel to room temperature.

P15.7. 1. Austenize the steel at 780°C (740 + 40 = 780°C), holding for 1 h so as to obtain 100% γ.
2. Quench the steel in oil at about 350°C and hold for about 1,000 s or 20 min;
3. Cool in air to room temperature.

P15.9. 164.27

CHAPTER 16

P16.7. 13 h

P16.9. 1.6 g

P16.11. 3.36 mA/cm²

Index

A

abrasive paper, 79–80
activation energy
    for creep, 214, 223
    for diffusion, 65–66, 72, 136
    for growth, 137
    for thermal process, 329
additive manufacturing (AM), 8
aerospace alloys, 286, 288, 291–294
age hardening, *see* precipitation strengthening
air hardening LAHS steels, 266
albite-anorthite ceramic system, 114–115
alloyed cast iron, 274
alloys
    formation rules, 41–42
    solid solution alloys, 41
alloy steels
    classification of, 263, 265
    designations and applications of, 264
    effects of elements on, 264
    Hadfield manganese steels, 266
    high silicon electrical steels, 268
    LAHS alloy steels, 1, 265–266
    maraging steels, 266
    stainless steels, 266–267
    tool and die steels, 266
    TRIP steels, 268
aluminum, 285
aluminum alloys, 285–288
amorphous material, 11–12
anisotropic, 164
annealing
    full annealing, 321–322
    grain growth, 310–312
    recovery, 309
    recrystallization, 309–310
    spheroidizing, 323
    stages of, 309
anode, 230–231, 245
anodizing, 342
Arrhenius Law, 65, 214, 329
ASTM, 90
atomic packing factor (APF), 15–16, 24–26
austempered ductile iron, 271, 274
austempering, 326–327
austenite, 112, 138–142
austenitic stainless steel, 267
Avogadro's number, 17
Avrami relationship, 309–310

axle hub heat treatment, 327
axle material, 266

B

bainite, 142–143, 274
bimetallics, 291
bio-compatible Ti alloys, 292
body centered cubic (BBC) crystal, 14–17, 20, 24–25
Bragg angle, 21
Bragg's law, 21–22
brazing, 7
breaking strength, 163, 205
Brinell hardness test, 165–166
brittle fracture, 199–202
Burgers vector, 43–44

C

carbon equivalent, 270
carbonitriding, 341
carbon steels
    microstructures, 261
    types and properties, 259–260
carburizing
    defined, 64, 340
    Fick's 2nd law, 63
    surface hardening, 340–341
case depth, 324, 341
casting (metal casting), 6–7
cast irons
    defined, 270
    ductile or nodular cast iron, 272
    gray cast iron, 271
    malleable cast iron, 273
    properties, 271
    white cast iron, 271
cathode, 230–231
cathode current efficiency, 245
cathodic protection (CP), 243–244
cementite, 138–139, 261–262
ceramics, 4
chemical metallurgy, 6
coatings, 244–246
coherent precipitates, 187–188
cold working, 48–49
composite materials, 4
composition determination, 87–89
computerized image analysis, 91–92

concentration cell, 243
concentration gradient, 57–58
continuous cooling transformation (CCT), 143–144
coordination number (CN), 14–16, 288
copper alloys, 81, 289–290
coring effect, 107, 329
corrosion
    defined, 199, 229
    fatigue, 240
    forms of, 236–241
    inhibitors, 246
    potential, 232
    prevention of, 242–245
crack propagation, 208–209, 242
creep
    behavior, 213
    defined, 212
    rate of, 213–214
    strain rate, 238
    test, 212–213
crevice corrosion, 238
critical cooling rate (CCR), 144, 264
critical nucleus radius, 134–135
critical resolved shear stress, 47–48
crystal defects
    definition and types, 39
    line defects, 42–44
    point defects, 39–41
crystallites, 24
crystallographic direction, 18–19
crystallographic plane, 19–20
crystal structure
    analysis, 15–17
    data, 17
    defined, 12
    types, 14, 45
crystal systems, 12
cubic (crystal system), 13
current density (J), 234–235

D

de-alloying corrosion, 239
deformation processing, 48
dendritic structure, 77
density computation, 17, 20–21, 89
designation systems
    for aluminum alloys, 285–286
    steel AISI system, 259–260
    for titanium alloys, 294
design of metal-assembly, 242–243
design philosophy (of fracture mechanics), 208
diffraction, see X-ray diffraction
diffusion
    defined, 57
    mechanisms, 40, 59
    types, 60

diffusion coefficient, 61
directional solidification (DS), 19
dislocation
    defined, 39, 42
    density, 43, 185–186, 192
    movement of, 43
    types, 42–44
dispersion strengthening, 188–189
dopants, 57
draft, 49, 54
ductile cast iron, 272, 281
ductile fracture, 200–201
ductile-to-brittle transition temperature (DBTT), 202
ductility, 157, 161, 163
duplex stainless steel (DSS), 267

E

edge dislocation, 42
elasticity, 157
elastic limit, 162
elastic properties, 164
elastic strain, 162
electrical steels (high-silicon steels), 268
electrochemical cell, 230
electrochemical corrosion, 230
electrode potential, 231
electrolyte, 230
electromotive force (EMF), 231
electron probe micro-analysis (EPMA), 87
endurance limit, 210
engineer, 1
engineering stress and strain, 159
environmentally assisted corrosion (EAC), 240
equiaxed grains, 309
erosion corrosion, 240
eutectic
    alloys, 109
    composition and temperature, 108
    defined, 104, 108
    phase diagram, 108–109
    reaction, 104
eutectoid reaction, 104
extensometer, 161
extractive metallurgy, see metallurgy

F

face centered cubic (FCC) structure, 15–16
factor of safety, 205
failure
    classification of, 199–202
    defined, 199
Faraday's constant, 246

fatigue
    defined, 208–209
    life, 210–211
    strength, 211
    test, 210
ferrite, 261
ferritic stainless steel, 267
Fick's 1st law of diffusion, 61
Fick's 2nd law of diffusion, 61–62
fracture
    defined, 199
    mechanics, 206–208
    types, 200–201
fracture toughness (K_{IC}), 208
Frenkel defect, 40
fretting corrosion, 240

G

galvanic corrosion, 232, 237
galvanic series, 232
gas carburizing, *see* carburizing
Gibb's phase rule, 103
glass, 4, 11
Goodman's law, *see* modified Goodman's law
grain, 77
grain boundaries, 77
grain-boundary strengthening, 183
grained microstructure, 77
grain growth
    defined, 310
    kinetic behavior, 310
grain-growth kinetics
    effect of second phase, 311
    effect of temperature, 311
    effect of time, 311
grain-oriented electrical steels (GOES),
        268
grain size, 77, 184–185
graphitization, 271
gray cast iron, 271
Griffith crack theory, 203
growth and kinetics, 136

H

Hadfield manganese steels, *see* alloy steels
Hall-Petch equation, 184
hardenability, 266, 324
hardening heat treatment process, 323
hardness, 165
    conversion, 170
    test, 165–169
heat treatment
    applications, 319
    defined, 319
    types, 319, 321–327

hexagonal close packed (HCP) crystal structure,
        16–17
high cycle fatigue (HCF), 211
homogenization annealing, 246
Hook's law, 162
hot shortness, 263
hot working, 309
Hume-Rothery rules, 41
hydrogen embrittlement (HE), 241

I

image analysis, 90–92
impact behavior, 170, 199–202
impact energy, 170
impact testing, 170
impact toughness, 170
impressed voltage, 243
impurities in steel, 262
indexing in X-ray diffraction, 23–24
industrial applications of
    crystallographic directions, 19
    crystallographic planes, 20
    diffusion, 57, 64
inoculants in cast irons, 272
insulators for corrosion protection, 243
intergranular (IG) corrosion, 239
inter-lamellar spacing, 139–140
intermetallic compound, 41
inter-planar spacing, 21
interstitial solid solution, 41
isomorphous alloy, 106
isomorphous binary phase diagram, 106–107
isothermal transformation (IT) diagram,
        140–141
isotropic behavior, 164

J

joining processes, 7
Jominy (hardenability) test, 324

K

kinetics, 137
    of solid-state phase transformation,
        137–138
Knoop hardness test, 169

L

lamellar pearlite, 138
Larson-Miller (LM) parameter, 214
lattice (crystal lattice), 12
leaded copper alloys, 290–291
lever rule (of phase diagram), 107
light microscopy, 83

limited solid solubility, 109
linear atomic density, 20–21
linear kinetic constant, 235
line defect, *see* dislocation
line-intercept method of grain size measurement, 90–91
liquid metal embrittlement (LME), 11.5.6.3
low cycle fatigue, 211

M

magnesium alloys, 295
malleable cast iron, *see* cast irons
manufacturing processes in metallurgy, 6–7
marageing steels, *see* alloy steels
martempering, 327
martensite, 142–143
martensitic stainless steels (MSS), 267
material(s)
    aerospace applications, 285, 291–292
    analysis/characterization, 87
    classification of, 2
    selection, 208, 246
    technology, 1
matrix, 4
mean stress, 209
mechanical properties, 157–170
metallography, 78–81
metallurgical failures, 199–200
metallurgically influenced corrosion, 239
metallurgist, 1
metallurgy
    classification, 5
    extractive metallurgy, 5
    mechanical metallurgy, 6
    physical metallurgy, 5
metals/alloys, 2
microconstituents of steel, 260
microhardness test, 169
microscopy, 82–86
microstructural transformation, *see* phase transformation
microstructure, 77–78, 139
Miller indices, 18–20
mixed dislocation, 42
modified Goodman's law, 211
modulus of elasticity, 164
modulus of resilience, *see* resilience

N

necking, 158, 163
Nernst equation, 232
nickel alloys, 291–292
nitriding, 341
nodular cast iron, 272
non-equilibrium cooling, 142–143

nonferrous alloys, 285
non-steady-state diffusion, 61
normalized activation energy for creep, 291
normalizing of steel, 322
notch, 170, 212
notch sensitivity factor (q_n), 212
nucleation, 131–136

O

optical microscope, 83
orthorhombic (crystal system), 12
oxidation, 229
oxidation kinetics, 235

P

pearlite, 139, 260
percent cold work, 48
percent elongation, 162
percent reduction in area, 162
peritectic reaction, 103
peritectoid reaction, 103
phase, 103
    diagram, 103
    transformation, 103
planar atomic density, 20
plastic deformation, 48
plastics, 4
point defects (crystal defects)
    defined, 39
    interstitial, 41
    substitutional, 41
    vacancy, 40
Poisson's ratio, 164
polycrystalline material, 77
polymers, 4
powder metallurgy, 8
precipitation strengthening, 187
pressure-temperature phase diagram, 105
process annealing, *see* annealing, recrystallization
proportional limit, 162–163

Q

quartz, 11
quenching, 187, 323

R

rapid solidification, 77
recovery, 309
recrystallization
    defined, 309
    fraction, 309

rate of, 309
temperature, 309
reduction reaction, 229
refractories, 4
residual stresses, 212
resilience, 164
resolution (of a microscope), 84
Rockwell hardness test, 167
rolling, 48
rubber, 4

S

sacrificial anode, 243
safety in design, 205
sand casting, 6–7
scanning electron microscopy (SEM), 84
Scherrer's formula, 24
Schmid's law, 47
screw dislocation, see dislocation
sensitized steel, 247
shear modulus, 160
single crystal gas-turbine blades, 19
slip
    deformation by, 43
    direction, 19
    plane, 20
    system, 44
S-N curve, 210–211
soldering, 7
solid solubility, 109
solid solution, 41, 181
    strengthening, 181
solid-state phase transformation, see phase
        transformation
solution treatment, 327–328
solvus line/curve, 109
specific weight, 288
spheroidizing annealing, see annealing
S-shape curves for eutectoid reaction, 138
stainless steels, 266–267
standard electrode potential (E°), 231
steady-state diffusion, 60
steel
    alloy (see alloy steels)
    carbon (see carbon steels)
    classification, 259
    defined, 259
stiffness, 164
strain, 157–158
strain hardening, 185
strain-hardening exponent, 185
stress, 157–158
    amplitude, 209
    range, 209
    ratio, 209
stress corrosion cracking (SCC), 241

stress-rupture life, 214
stress-strain diagram, 160–162
stress concentration factor, 204
structural materials, 259, 288
substitutional solid solution, 41
superalloy, 291
super-saturated solid solution (SSSS), 328
surface engineering, 339
surface hardening, 339

T

temperature
    ageing, effect on, 327–328
    fatigue, effect on, 212
    grain size, effect on, 309
    mechanical properties, effect on, 309
    nucleation and growth, effect on, 309
temper designation of aluminum alloys, 285
tempered martensite, 326
tempering, 285, 326
tensile strength, 162
tensile test, 160–161
tensile mechanical properties, 162
tetrahedral structure, 12
theoretical weight of metal deposit, 342
thermal barrier coating (TBC), 4
thermally activated diffusion, 65
thermoplastics, 4
thermosetting polymers, 4
time-temperature transformation (TTT) diagram,
        140
titanium alloys, 292
tool steels, 266
toughness (tensile toughness), 170
transformed induced plasticity (TRIP)
        steels, 268
transgranular fracture, 200–201
transmission electron microscopy (TEM), 86
triple point (in a phase diagram), 105
true strain, 157–158
true stress, 157–158

U

unary phase diagram, 105
uniform corrosion, 236
unit cell, 12
unit cell volume, 15
unlimited solid solubility, 106

V

vacancy(ies), 40
    fraction, 41
    number of, 41
Vickers hardness test, 168

W

water quenching, 323–324
wear, 199
weight fraction, 117
welding, 7
white cast iron, 271
work hardening, *see* strain hardening
wrought iron, 259

X

X-ray diffraction, 21–23

Y

yielding, 162
yield point, 162
yield strength, 162
yield stress, 162
Young's modulus, *see* modulus
    of elasticity

Z

zinc alloys, 295